Studies in Computational Intelligence

Volume 876

Series Editor

Janusz Kacprzyk, Polish Academy of Sciences, Warsaw, Poland

The series "Studies in Computational Intelligence" (SCI) publishes new developments and advances in the various areas of computational intelligence—quickly and with a high quality. The intent is to cover the theory, applications, and design methods of computational intelligence, as embedded in the fields of engineering, computer science, physics and life sciences, as well as the methodologies behind them. The series contains monographs, lecture notes and edited volumes in computational intelligence spanning the areas of neural networks, connectionist systems, genetic algorithms, evolutionary computation, artificial intelligence, cellular automata, self-organizing systems, soft computing, fuzzy systems, and hybrid intelligent systems. Of particular value to both the contributors and the readership are the short publication timeframe and the world-wide distribution, which enable both wide and rapid dissemination of research output.

The books of this series are submitted to indexing to Web of Science, EI-Compendex, DBLP, SCOPUS, Google Scholar and Springerlink.

More information about this series at http://www.springer.com/series/7092

Vladimir Mashtalir · Igor Ruban ·
Vitaly Levashenko
Editors

Advances in Spatio-Temporal Segmentation of Visual Data

Springer

Editors
Vladimir Mashtalir
Kharkiv National University
of Radio Electronics
Kharkiv, Ukraine

Igor Ruban
Kharkiv National University
of Radio Electronics
Kharkiv, Ukraine

Vitaly Levashenko
Faculty of Management Science
and Informatics
University of Žilina
Žilina, Slovakia

ISSN 1860-949X ISSN 1860-9503 (electronic)
Studies in Computational Intelligence
ISBN 978-3-030-35482-4 ISBN 978-3-030-35480-0 (eBook)
https://doi.org/10.1007/978-3-030-35480-0

This Springer imprint is published by the registered company Springer Nature Switzerland AG
The registered company address is: Gewerbestrasse 11, 6330 Cham, Switzerland

Preface

The subject of the book is to present systematized research, development and implementation of visual data spatio-temporal segmentation techniques. Video streams as unstructured or poorly structured data issue a challenge to create a unified, desirably online, framework capable to deal with visual semantic. The video-structure level is usually organized to capture temporal properties of a video in terms of such continuities as scenarios, events, episodes, etc., what is based on homogeneous consecutive sequences of frames. Image and video contents in terms of middle-level features like spatial and time segmentations, of course along with other attributes, provide promising tools for sufficiently understandable and measurable visual content analysis. The goal is to equip the reader with the necessary knowledge about trends in crucial improvement of spatial and temporal partitions and coverings exploration.

Chapter 1 proposes a new range of adaptive models and algorithms that are aimed at a coordinated solution of filtering problems and segmentation of object boundaries and lines by contrast, the key elements of which are the ring mask of the boundary detector and statistical methods for dropping extreme observations. Such results can be used to improve the machine vision systems, including improvements obtained due to segmentation of the boundaries of objects and lines by contrast. Such methods are considered the most stable since they are invariant to fairly significant variations in the brightness of images.

Chapter 2 deals with the mathematical problem of image segmentation of onboard systems of optical and electronic surveillance. It is established that the segmentation of such images results in the separation of the image into artificial objects (objects of interest) and natural objects (background). To segment the images of onboard systems of optical and electronic surveillance, the use of the ant colony method and the method of the artificial bee colony is proposed. The entity of the simplest ant colony optimization method of image segmentation is presented. The process of segmentation of the image is established as a set of sections of agents' movement taking into account the attractiveness of the route and the concentration of pheromones on it. The entity of the improved ant colony optimization method of image segmentation is presented, where only the best agents increase the level of pheromone on their routes. The entity of the method of the artificial bee colony is described, which provides for determining the positions of agents, their migration and the conditions for stopping the iterative process according to the criterion of the minimum of the objective function and determining the optimal value of the threshold level. The results of the experimental study of the application of methods to segment an optical and electronic image are presented; they confirmed the efficiency of segmentation methods. It is established that the use of segmentation methods will improve the quality of processing of optical and electronic images.

Chapter 3 describes an approach to interpreting heterogeneous data from sensors based on a new perception model, which implements such cognitive functions as generalization and abstraction of data, context tracking and switching attention. The model of representation of the space–time segment of heterogeneous data by abstract concepts of different levels of generalization is introduced. The model takes into account the dynamic characteristics of the data, their completeness, ageing rate and fuzziness. On the basis of fuzzy L-R numbers, a new mechanism of the knowledge-based inference is built by abstracting and generalizing. The universality of the perception model is shown in the context of the intelligent machine interpreting heterogeneous spatio-temporal data from sensors on the situation at the intersection. The functions of tracking the context and switching attention are shown in the context of the interpretation of data on the geometric shape of the motion trajectory of a mobile intelligent machine.

Chapter 4 considers approaches to video skimming into semantically consistent segments of video streams, which are highly redundant and weakly structured data. In such a way, one of the promising ways is spatio-temporal segmentation as frame partitions represent certain spatial image content. Also, properties were formulated and proved which ultimately determine the characteristics of permissible segmentation transformations when searching for a compromise between over and undo segmentation. Temporal segmentation of multidimensional time series has been examined, which enables structuring video streams and significantly reduces the amount of data that will require online processing. For this, multidimensional time series analysis theory was used, since a completely natural video representation is a sequence of frames, followed by their combination into groups of frames (shots) with the same content. It was shown that various approaches could be used to detect shots with homogeneous characteristics, which are based on VAR models, exponential smoothing and predictive models.

Chapter 5 considers new approaches based on computational intelligence methods for solving the tasks of fuzzy clustering–segmentation of data streams sequentially fed for processing. The known methods of probabilistic and potential clustering under conditions of overlapping classes are analysed, and their adaptive recurrent modifications designed for solving problems arising in the framework of data stream mining are introduced. The reasonability of a parallel probabilistic-potential approach is shown, where centroid-prototype clusters are specified using the optimization gradient procedure. For situations where the processed data is "polluted" by abnormal outliers, adaptive fuzzy clustering methods based on robust target functions that suppress outliers are proposed. The close connection of the proposed procedures with self-learning algorithms of T. Kohonen self-organizing cards is shown.

Chapter 6 considers and theoretically justifies the original method for implementing fuzzy models using relational systems. Unlike the known approaches, the proposed method enables solving the problems of data mining of integrated systems—databases and fuzzy systems. A method and technology are developed for organizing, storing and searching for fuzzy data by means of relational systems. An original information technology to search for fuzzy data by means of relational systems is proposed, which, unlike the well-known approaches, enables using a Structured Query Language (SQL) to solve a complex data mining problem. Since standard self-organizing maps solve the problems of clear clustering, neuro-fuzzy self-organizing maps are introduced, and it is shown that the neighbourhood function is actually a membership function in fuzzy inference systems. A robust modification of a neuro-fuzzy self-organizing card and an algorithm for its self-learning is introduced.

Kharkiv, Ukraine — Vladimir Mashtalir
Kharkiv, Ukraine — Igor Ruban
Žilina, Slovakia — Vitaly Levashenko

Contents

Chapter 1
Adaptive Edge Detection Models and Algorithms

Kirill Smelyakov, Sergiy Smelyakov and Anastasiya Chupryna

Abstract The effective development of machine vision systems currently requires the development of adaptive models and image processing algorithms, starting with solving preprocessing tasks and ending with the tasks of recognizing objects in images. The basis of their successful application is the possibility of effective segmentation of objects of interest to us. And first of all segmentation of the boundaries of objects and lines by contrast. Such methods are considered the most stable, since they are invariant to fairly significant variations in the brightness of images. The chapter proposes a new range of adaptive models and algorithms is proposed, which are aimed at a coordinated solution of filtering problems and segmentation of object boundaries and lines by contrast, the key elements of which are the annular mask of the boundary detector and statistical methods for eliminating extreme observations.

Keywords Digital image · Edge detector · Mask · Pixel neighborhood · Contrast estimation · Filtering · Smoothing · Sharpening · Boundary segmentation

1.1 The Current State of the Issue of Detecting the Object Boundaries in the Image

At present, it is observed a distinct tend to expand the scope of artificial intelligence systems to solving of a wide range of relevant application problems [1–4]. At this, the functioning of an ever increasing number of such systems [5–7] is based on using the results of segmentation, recognition and tracking of images which are obtained using the specialized computer vision systems [8, 9].

K. Smelyakov (✉) · S. Smelyakov · A. Chupryna
Kharkiv National University of Radio Electronics, 14 Nauky Ave, Kharkiv 61166, Ukraine
e-mail: kyrylo.smelyakov@nure.ua

S. Smelyakov
e-mail: astrotheos@yahoo.com

A. Chupryna
e-mail: anastasiya.chupryna@nure.ua

V. Mashtalir et al. (eds.), *Advances in Spatio-Temporal Segmentation of Visual Data*, Studies in Computational Intelligence 876,
https://doi.org/10.1007/978-3-030-35480-0_1

It is well known that for several decades the systems of computer vision are being used to identify the bank cards and cars by their license plates, to control the quality of products regarding the defect detection, tolerance monitoring, component measurements etc. [10, 11]; as well, they are used in text interpretation systems [12].

Besides, the computer vision systems are now increasingly used for identification of individuals by biometric data [13–16], and in the augmented reality systems [17]. More and more extensive studies are being carried out with the aim to make it possible to use the computer vision systems for medical purposes; e.g. to diagnose diseases by the magnetic resonance imaging [18–20]. And this is only a small part of essential problems being solved by the modern artificial intelligence systems which are grounded on the use of machine vision systems [10, 21, 22].

The basis for the successful application of modern computer vision systems lies in the possibility of efficient segmentation of objects, presenting interest to us, in the considered digital images [10, 23, 24].

Modern methods of segmentation of digital images [10, 25], which are used in such systems, focus on detection of either the interiority, or boundary of object of interest in the image [26–28]. By most specialists the methods of the latter type are considered to be more robust, since they are focused on detection of object boundaries and lines by contrast and, therefore, are invariant to sufficiently significant brightness variations of images [10, 28–30]. The development of namely such segmentation methods by many experts is considered a priority also because they allow us to ensure the robust use of modern robotic systems (for example, drones [31] used in modern systems for monitoring an alfresco situation or systems for reliability evaluation [32]), as well as autonomous vision systems intended for registration, identification and monitoring [10, 29].

It is also important to note that at present the digital image processing algorithms are increasingly being used as auxiliary ones in many "non-traditional" areas; e.g. for increasing the efficiency of transfer of data (images) [33]. In such applications, the filtering and smoothing algorithms are most often used, allowing to increase the level of uniformity of brightness distribution of the objects and background and, by that, to increase the compression ratio of images while maintaining the quality level.

Note also that for convenience of practical implementation of image processing algorithms, the respective computer vision libraries have been developed [34–36].

1.1.1 Classical Edge Detection Approach

Object boundary segmentation by contrast is based on the use of mask, which, on the one hand, determines the geometric shape and dimension of neighborhood of the considered pixel, and on the other—specifies the distribution of weight coefficients of pixels' brightness of this neighborhood for obtaining a convolution by the mask, the value of which is taken as an estimate of the pixel's contrast. Today, the anthology of the issue has been studied quite well [10, 29].

For better understanding of the further material, it suffices to consider a typical example of application of edge detector, for example—the well-known Sobel operator [10, 11, 29].

Before using a detector, the shape and the radius of the pixel neighborhood are determined: on the basis of its brightness distribution B the contrast is evaluated; for the default one, the "square" neighborhood of 3×3 pixel is considered.

Let the coordinates (ξ, η) of the image pixel under consideration be fixed and the brightness distribution B in its neighborhood be as follows

$$B = \begin{bmatrix} B_{1,1} & B_{1,2} & B_{1,3} \\ B_{2,1} & B_{2,2} & B_{2,3} \\ B_{3,1} & B_{3,2} & B_{3,3} \end{bmatrix}, \tag{1.1}$$

where the pixel (ξ, η) is corresponded by the central brightness $B_{2,2}$.

In such conditions, the Sobel edge detector works in this way.

Step 1. By using masks of the Sobel operator

$$G_{\xi} = \begin{bmatrix} -1 & 0 & 1 \\ -2 & 0 & 2 \\ -1 & 0 & 1 \end{bmatrix}, \quad G_{\eta} = \begin{bmatrix} 1 & 2 & 1 \\ 0 & 0 & 0 \\ -1 & -2 & -1 \end{bmatrix}, \tag{1.2}$$

the partial derivatives are calculated

$$s_{\xi} = \sum_{i=1}^{3} \sum_{j=1}^{3} B_{(i,j)} \cdot G_{\xi(i,j)}, \quad s_{\eta} = \sum_{i=1}^{3} \sum_{j=1}^{3} B_{(i,j)} \cdot G_{\eta(i,j)}. \tag{1.3}$$

Step 2. The gradient module is calculated

$$M = \sqrt{(s_{\xi})^2 + (s_{\eta})^2}. \tag{1.4}$$

Step 3. At last, it is applied the threshold criterion

$$Cr = \begin{cases} 1, & if\ M > T, \\ 0, & else. \end{cases} \tag{1.5}$$

which relates the pixel under consideration to the edge if its contrast estimate being given by the brightness gradient module M is greater than the predetermined threshold value T, viz. $M > T$.

Other edge detectors are applied in a similar typical pattern. Their main distinctions are defined by the principles of setting the weight coefficients of the detector mask. Also, the rule for evaluating contrast by mask may differ. So, for turn-invariant detectors, e.g.—for detectors based on the use of Laplace operator, the gradient is not calculated. The features of application of the majority of widely known detectors of object boundary pixels are quite fully described in [10, 11, 29].

1.1.2 Main Problems of Modern Edge Detectors Application

Apart from the object edge detectors, the similar detectors, which are focused on detection of object corners (e.g. Canny Detector) and line fragments, are currently being widely used. In addition, a truly great number of algorithms for constructing connected components of an object's boundary (line), including models and algorithms for tracing, thinning, as well as eliminating of boundary gaps [10, 11, 29], have been developed by now.

Under conditions of sufficiently high contrast, the existing detectors are able to efficiently select the pixels of boundary or line, thereby ensuring the efficacy of subsequent construction of connected components of the objects' boundaries in the image. However, under conditions of low contrast, the boundary pixels are not always distinguished effectively [29], which is often expressed by the appearance of discontinuities in the object boundary. On the other hand, an attempt to reduce the threshold with the aim to adequately segment the boundary in low-contrast conditions leads to a significant thickening of the fragments of the object boundary due to the detection of pixels inside the object, its shadow and background.

In such a situation, it is required either to eliminate the boundary breaks, or to thin the boundary, or to do both. Usually, if these negative effects manifest themselves to a small extent, then their elimination is not a problem, although it may take a considerable amount of time. With the growth of the number and length of the boundary gaps, as well as the level of manifestation of the boundary thickening effect, the adequate elimination of these effects for objects of regular form (the shape of which is a priori known) becomes more and more problematic. The computational complexity increases nonlinearly. For objects of irregular form (the form of which is not known a priori), the adequate elimination of these negative effects becomes almost impossible.

The results of mathematical modeling and the experiment performed allow us to draw a number of important conclusions about the reasons for ineffectiveness of segmentation of boundaries of objects in low-contrast images. In the main this is due to:

(a) lack of an adequate consideration of the metric properties of objects, since the form of detector masks, as a rule, is the square, not the circle;
(b) lack of allowing for the effect the object shadow exert on the estimation of boundary pixel contrast, when building the mask and detector models;
(c) use of the averaging rule for evaluating a pixel contrast when performing the convolution by mask (1.3), (1.4);
(d) use of a weak criterial foundation oriented to segmentation of boundary pixel just by its contrast level (1.5);

(e) ungrounded simplification of computational models; for example, when instead of estimating the contrast by gradient (1.4), the following approximation is used

$$M = \max\{|s_\xi|, |s_\eta|\}, \tag{1.6}$$

which is argued just by the expediency of reducing the computational complexity.

In such a situation, the main task of the work is the construction of an adaptive model and an effective algorithm for detecting pixels of object boundary, which are maximally free from these drawbacks. At this, it should be noted that the ineffectiveness of boundary pixel detection by contrast can often be caused by the presence of significant brightness variations due to the noise and texture effects. In this regard, in order to ensure the overall computational efficiency of segmentation, we will complex the preprocessing stages (mainly the filtering/smoothing stages) and detecting the boundary of the object.

Begin with a brief description of the process of obtaining a digital image in order to introduce the most important terminology, key models and features of obtaining a digital image in relation to formation of the shadow of an object, the branches of the border, etc.

1.1.3 Basic Terminology, Definitions and Models

The process of obtaining a digital image is to subject the given analog image of the scene in question to operations of spatial discretization and quantization by brightness levels. By using the standard digitizing devices, one can get a binary (sometimes called the "black and white"), greyscale, or colour image [10, 11, 29]. Unless otherwise specified, the further consideration presumes dealing with the greyscale input images.

Note that with a relatively small number of brightness levels, false contours appear on the quantized image. They arise as a result of a jump-like change in the brightness of a quantized image and are especially noticeable on flat sections of brightness variations. Therefore, with a uniform quantization of images, at least 64 levels are required. In this regard, if not specified otherwise, we head for the standard 8-bit greyscale, or 256 gradation scale of brightness $[0, \ldots, 255]$ being widely used now.

As a result of digitization, we obtain an image that is defined by a matrix of brightness values $F = \{f_{ij}\}_{ij}$ on the input image area D, which is represented by a set of nodes $D = \{d_{ij} = (i, j)\}$, $i = 0, \ldots, n$, $j = 0, \ldots, m$, formed by the intersection points of a regular rectangular grid L on the plane R^2, $L \subset R^2$, the generatives of which are parallel to the coordinate axes xOy. In the theory of digital image processing these nodes d_{ij} are called pixels (Fig. 1.1).

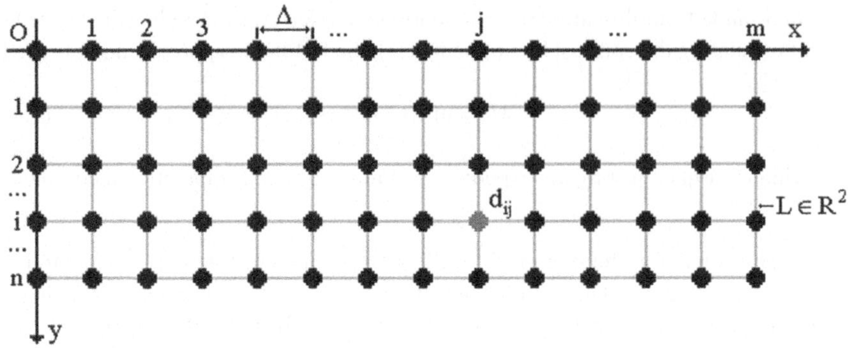

Fig. 1.1 A discrete field of digital image

At this, the magnitude of the discretization step, Δ, is usually chosen from the condition of distinguishing the smallest objects of interest in the image, taking into account the scale of the image and the angle of view [10, 29].

To estimate the distances and sizes of objects in the domain D, it is natural to consider it as a discrete set of point-pixels with coordinates defined by their indices, which is immersed in Euclidean space R^2 with the metric ρ_E. In this case, the distance between the adjacent pixels along each axis is 1, and the distance between arbitrary pixels d_{ij} and $d_{\xi\eta}$ is determined by the expression

$$\rho_E(d_{ij}, d_{\xi\eta}) = \sqrt{(i-\xi)^2 + (j-\eta)^2}. \tag{1.7}$$

At this, other metrics in the domain D may be used as well, but as auxiliary one's; e.g. for integer estimating the length of local gap of object boundary the following metrics are often used

$$\rho_M(d_{ij}, d_{\xi\eta}) = |i-\xi| + |j-\eta|, \tag{1.8}$$

$$\rho_O(d_{ij}, d_{\xi\eta}) = \max\{|i-\xi|, |j-\eta|\}. \tag{1.9}$$

However, unless otherwise specified, in the below consideration the Euclidean metric (1.7) is presumed.

Since the topology and geometry of discrete space D differ from that of the plane R^2, for dealing with the boundaries, lines and other types of objects in a discrete set D it is necessary to determine the corresponding topological analogs—a neighborhood, a boundary et al.

In mathematics, a neighborhood of a point is understood as an open domain without a boundary, and it is allowed to consider the neighborhoods of infinitely small radius. However, in view of the practical unrealizability of the latter concept on a finite set of points and a relatively small number of pixels correlated with the segmented object, for image processing purposes it is more convenient to consider a neighborhood as a domain with its boundary.

So, let $\varepsilon > 0$ be the given radius. Then in the below consideration the ε-neighborhood of pixel $d_{\xi\eta}$ is understood as the set of pixels $\{d_{ij}\}_{ij}$ satisfying the condition

$$O_\varepsilon(d_{\xi\eta}) = \{d_{ij} | \rho(d_{ij}, d_{\xi\eta}) \leq \varepsilon\}. \tag{1.10}$$

It can also be denoted $O_\varepsilon(\xi, \eta)$ and called simply neighborhood, if the value ε is assumed to be known or insignificant.

It is clear that for any pixel in the domain D the number of its various neighborhoods is finite, and for different values of ε the neighborhoods of the pixel may coincide.

With the aim to eliminate such ambiguity when working with pixels of a digital image, for the values of ε only typical between-pixel distances ρ are used, which are calculated as follows

$$\rho(i, j) = \sqrt{(i - \xi)^2 + (j - \eta)^2}, \quad i, j = 0, 1, 2, \ldots, \tag{1.11}$$

where (i, j)—are two coordinates of pixels (Fig. 1.2), at least one of which differs from the central pixel coordinates (ξ, η). In these conditions, the admissible neighborhood radii belong to the following set of typical distances pertaining to the discrete field of the image

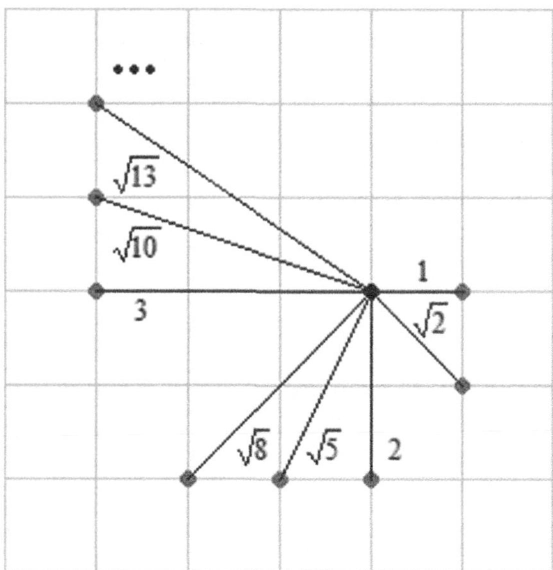

Fig. 1.2 The typical between-pixel distances

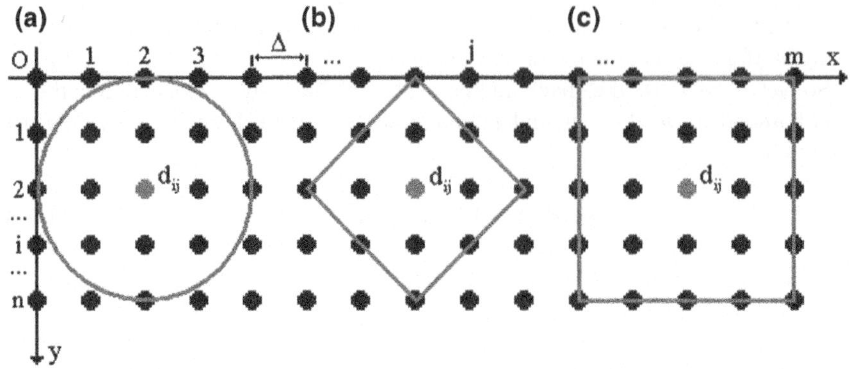

Fig. 1.3 Geometric objects which describe the neighborhoods defined by metrics ρ_E (**a**), ρ_M (**b**), and ρ_0 (**c**) for radius $\varepsilon = 2$

$$\{\rho_\varsigma\}_{\varsigma=1,2,\ldots} = \{1, \sqrt{2}, 2, \sqrt{5}, \sqrt{8}, 3, \sqrt{10}, \sqrt{13}, \ldots\}. \tag{1.12}$$

Notice that when the metrics (1.8) and (1.9) are used the distances will always be integer, and therefore both the value ε and the typical distances ρ_ς will always be represented by natural numbers

$$\{\rho_\varsigma\}_{\varsigma=1,2,\ldots} = \{1, 2, 3, \ldots\}. \tag{1.13}$$

Notice that at a fixed radius ε the neighborhoods, defined by the specified metrics, differ in their shapes: for metric ρ_E the (closed) neighborhood is bounded by a circle (Fig. 1.3a), for metric ρ_M—by a rhombus (Fig. 1.3b), and for metric ρ_0—by a square (Fig. 1.3c).

Although at first glance the metrics ρ_M and ρ_0 may seem unnatural, they are very useful in estimating the length of a boundary gap by the number of pixels, as well as in solving of other problems of this kind.

In addition, of all the typical distances $\{\rho_\varsigma\}_{\varsigma=1,2,\ldots}$ the first two are of particular importance, as are the neighborhoods they define. So, if for two pixels d_{ij} and $d_{\xi\eta}$ the following condition is fulfilled

$$|i - \xi| + |j - \eta| = 1, \tag{1.14}$$

they are called 4-connected (Fig. 1.4); accordingly, the pixels are called 8-connected, if they satisfy the condition

$$\max\{|i - \xi|, |j - \eta|\} = 1. \tag{1.15}$$

Note that the set of pixels, comprising the central one and all 4-connected with it, forms a 1-neighborhood, i.e. a neighborhood with radius $\rho_1 = 1$, which, due to its fundamental value for subsequent analysis, will also be called elementary.

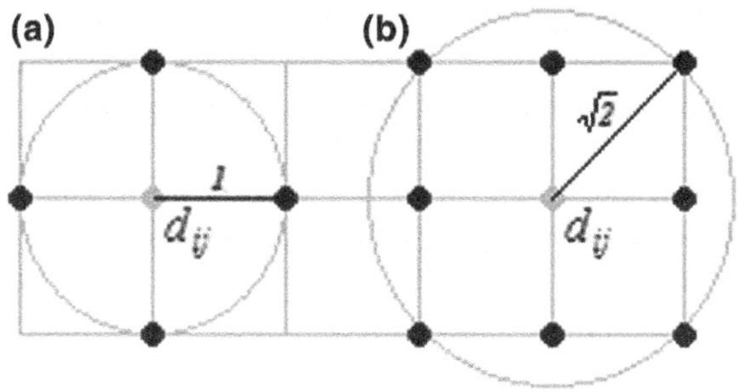

Fig. 1.4 Pixels marked with black dots are 4- (**a**) and 8-connected (**b**) with the pixel d_{ij} (marked in gray)

Accordingly, in the case of an 8-connection, we obtain a neighborhood with radius $\rho_2 = \sqrt{2}$. The circles, which bound the closures of these neighborhoods, in Fig. 1.4 are marked in gray.

Further on, the concept of connectivity is generalized to pixels lying in an arbitrary neighborhood. In this case, the pixels d_{ij} and $d_{\xi\eta}$ are called ε-connected if the following condition is fulfilled

$$1 \leq \rho(d_{ij}, d_{\xi\eta}) \leq \varepsilon. \tag{1.16}$$

This type of connectivity is essential to ensure adequate handling of sets that are not 4- or 8-connected.

Now, make use of topological notions relative to connectivity and neighborhood for defining the discrete analogs of concepts, which are required for adequate segmentation of the interior and the boundaries of objects, as well as lines in digital photos. In the ideal case, as a result of segmentation, it is desirable to obtain such sets of pixels that represent a discrete, more precisely—a point-like analog of a two-dimensional or one-dimensional sets of canonical types.

In the first case, we are talking about a closed set S with interior $I = Int(S)$, whose closure $Cl(S)$ gives the original closed set S, whereas the difference between them determines the boundary $\Gamma = S \backslash I$ of the set S.

In the second case we are talking about the line as a connected set of points without a two-dimensional interior.

However, with respect to the set Z obtained as a result of the "initial" segmentation, there is no possibility to talk about the exact fulfillment of the specified conditions, and therefore conducting an adequate identification of segmented objects becomes extremely complicated, unless the set Z is represented as a combination of discrete sets of canonical type.

That is why for solving this problem in the discrete domain D, it is required to introduce a topology that allows to adequately define the principles of identification of discrete sets of various types. But it should be remembered that the objects with corresponding topological properties does not arise by themselves, and therefore at various stages of image processing, the discrete sets being detected from the image may be subject to correction.

Meanwhile, the use of such a topology should not cause problems in terms of an excessive increase in the complexity of algorithms intended for analyzing the arising situations. A reasonable compromise in this regard can be considered the following system of concepts that have shown their effectiveness in terms of building models and methods of segmentation, as well as analyzing the results of segmentation for subsequent identification of the detected objects. To bring terminology closer together, the elements of a discrete domain are called a point rather than a pixel.

A point $d_{\xi\eta}$ of a set $G \subset D$ is called internal if in its ε-neighborhood it contains only the points of the set G. The set of the interior points of the set $M \subset G$ forms its interior $Int(M)$.

As mentioned above, the set G correlated with the identified object should be considered closed. Therefore, if $G_{Int} = Int(G)$, then we should take the set G to be the closure of its interior, $G = Cl(G_{Int})$. In this case the difference $Fr(G) = G\backslash G_{Int}$ between these sets gives the boundary $\Gamma = Fr(G)$ of the set G, identified with the boundary of the segmented object.

Accordingly, the exterior of the set G may be defined as $\overline{G} = Ext(G) = D\backslash G$, while the points of $\overline{G}$ may be called the external ones.

However, since during the segmentation, the fulfillment of all these rules can only be carried out with some accuracy, the following definitions are introduced.

The point $d_{\xi\eta} \in G$ is called the boundary one, if its ε-neighborhood contains at least one point from each of the sets: $Int(G)$, $\overline{G}$, and Γ. The combination of these points and forms the boundary of the object defined by the set G.

The point $d_{\xi\eta} \in G$ is called the adjoining one, if its ε-neighborhood contains at least one point from each of the sets $\overline{G}$ and Γ, but does not contain any point of $Int(G)$. The combination of these points forms the adjoining set for the set G. In general, it can have several connected components, which (with more or less qualitative segmentation) appear to be discrete analogs of lines.

Thus, if a pixel of a set G is not internal, then it is a boundary pixel or belongs to an adjoining set.

As a one-dimensional set, lines are usually considered. Their analogue in the discrete space D is 4- or 8-connected pixel chains without interiors, i.e. a sequence of pixels $\{p_\zeta\}_{\zeta=1,2,\ldots,m}$, each pair of which is separated from each other by a distance not exceeding the typical one: $\rho_1 = 1$, or $\rho_2 = \sqrt{2}$. According to the distance ρ_k, $k \in \{1, 2\}$, call such chain k-line.

At this, the topological and metric properties of k-lines can be refined additionally, according to the conditions of the problem under consideration.

For example, the k-lines can be used for determining the linear connectivity of components and sets. This is due, for example, to the fact that the interior and the boundary of a set G can be connected, or not connected, that is, they can decompose into connected components. Since in the framework of this work, i.e. in terms of segmentation of object boundary pixels, this aspect is not essential, it is not specifically considered. But in this case it is important that k-lines can be used to describe the boundaries of objects.

At the same time, when considering a two-dimensional set (Fig. 1.5a), it should be

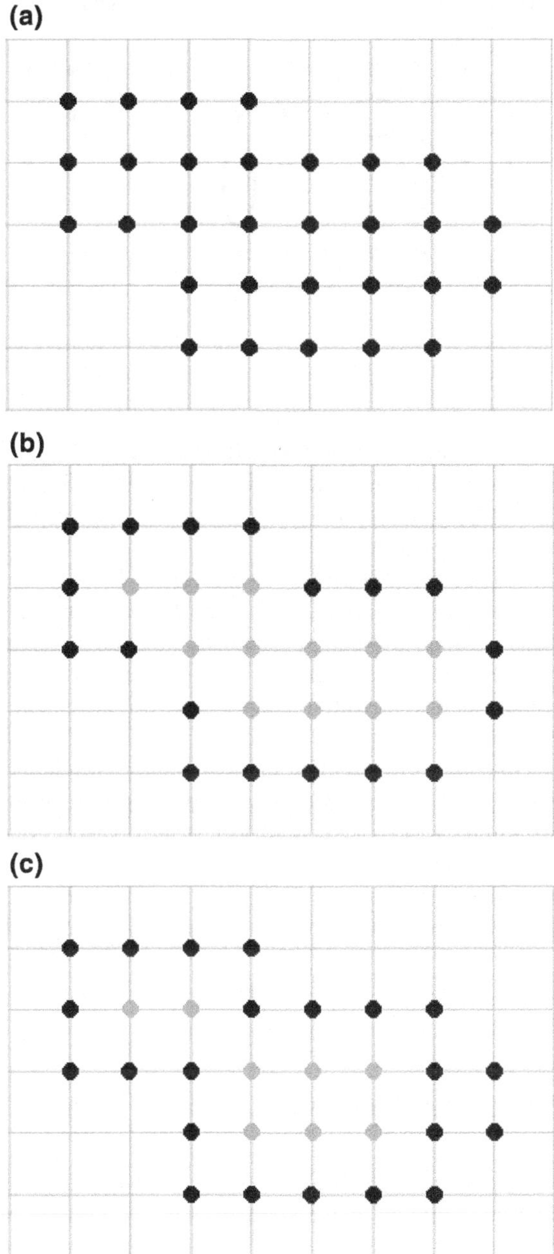

Fig. 1.5 The segmented set (**a**) and its 8-connected (**b**) and 4-connected (**c**) boundaries marked with black dots

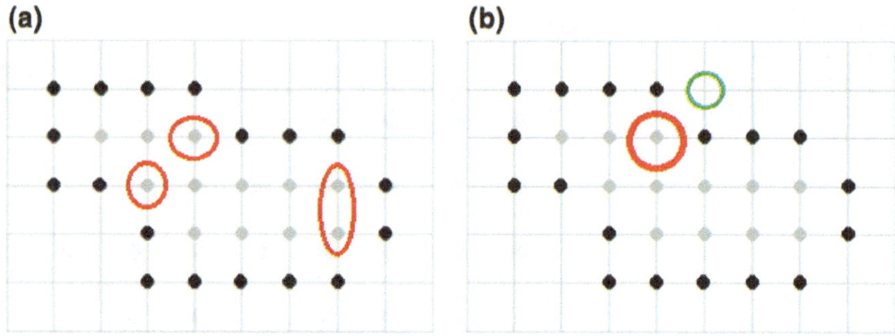

Fig. 1.6 The pixels of the object's interior which, when using the overwhelming number of standard masks with dimension 3 × 3, will contain a background pixel, are marked in red

borne in mind that the 8-connected boundary is engendered by a 4-connected interior with the neighborhood radius of $\rho_k = 1$ (Fig. 1.5b), and vice versa: the 4-connected boundary is engendered by 8-connected interior with the neighborhood radius of $\rho_k = \sqrt{2}$ (Fig. 1.5c).

To ensure uniformity of models of the boundaries of objects and lines, in some cases they are described by 1-lines. However, in this case, when the center of the mask of radius 2 is superposed with the pixel belonging to the interior of the object (marked with a red circle in Fig. 1.6), the background pixel (marked with a green circle) can also fall into the mask area. But in principle, this situation is not critical, since the 8-connected boundary can be obtained by tracing [29].

Note in this respect that in the literature, the choice of a neighborhood radius for determining the inner pixel of a set is being made from various considerations. With the aim to preserve the commonness of the definitions of the boundaries of objects and lines, and also, taking into account the positive features of existing models, we take $\varepsilon = 1$ (i.e. $\rho_k = 1$) in order to provide the most accurate description of the interior as prerequisite for improving the adequacy of the selection of boundaries and more accurate modeling of individual lines. Recall that in this case the interior points are determined by the 4-connectedness principle, and the boundary is 8-connected.

1.1.4 The Features of Obtaining a Digital Image

Above, we considered a basic model of a digital image, in which it is assumed that after digitizing an image is represented by a matrix of integer values of brightness $F = \{f_{ij}\}_{ij}$. Meanwhile, the recognition of objects represented in the image requires an object-oriented analysis of the distribution of brightness $F = \{f_{ij}\}_{ij}$. Consider the features of this issue in more detail, and begin with the effect of the object's shadow—the most important for the purpose of segmentation of the boundaries of objects by contrast.

In many modern studies and models, it is assumed that each pixel of a digital image can be uniquely assigned either to an object or to a background. However, this assumption is erroneous, since the formation of the image is accompanied by the action of a number of factors causing its distortion.

The key factor limiting the possibility of unambiguous distinction between the pixels of an object and the background is the presence of an object shadow which is understood as the set of pixels in a digital image, the brightness of which, when digitized, was formed simultaneously by the radiation of both the object, and the background. The appearance of the shadow pixels is due to the fact that the radiation of the object and the background may simultaneously hit the same element of the analog-to-digital converter (ADC) matrix.

Suppose that in the area of the border, the brightness of the object pixels is at the level f, and the brightness of the background pixels is at the level b; for definiteness, suppose that $f < b$. Let the fraction of the radiation of an object on an ADC matrix element be S_f, and the fraction of the background radiation on the same element be $S_b = 1 - S_f$. In this situation the conditions $0 < S_f < 1, 0 < S_b < 1, S_f + S_b = 1$ fulfill, and the brightness h of a pixel of object shadow takes the value

$$\mathrm{h} = \lfloor f \cdot S_f + b \cdot S_b \rfloor, f \le \mathrm{h} \le b, \tag{1.17}$$

where $\lfloor \cdot \rfloor$ is the operation of rounding to the integer.

On a digital image, a chain of object's shadow pixels usually disposes between the pixels of object boundary and background. Under normal data acquisition conditions, the width of the shadow is one pixel only, measured in metrics ρ_M (1.8), or in metrics ρ_0 (1.9). At this, due to the rounding in (1.17), the brightness of some shadow pixels may be identical to that of the boundary pixels of the object or background (Fig. 1.7).

After digitization, the small-sized objects can be greatly smoothed or even disappear completely if their radiation, when digitized, does not provide obtaining of neither interior, nor boundary, but only a shadow. As a rule, for those classes of objects that need to be recognized, the requirements for resolution and the scale of their presentation are set up. However, for other objects of the field of view such a requirement is not made. Therefore, quite often the field of view may contain quite a few small objects and objects on the verge of resolution, which are represented by a shadow only.

This situation is especially characteristic when working with aerial photographs of objects on the ground. The main negative effect of the presence of a large number of such objects in the image manifests itself in an effect similar to a finely dispersed texture, or noisiness of an image. That is why for increasing the effectiveness of object segmentation and recognition it is often required to smooth such variations of brightness in the image.

Due to occurrence of digitization errors, as well as due to a presence of shadow, the boundary of an object is characterized by the presence of branches, which in some cases may take the form of loops. As a result, the topology of the object boundary acquires the structure represented by its true component and branches of

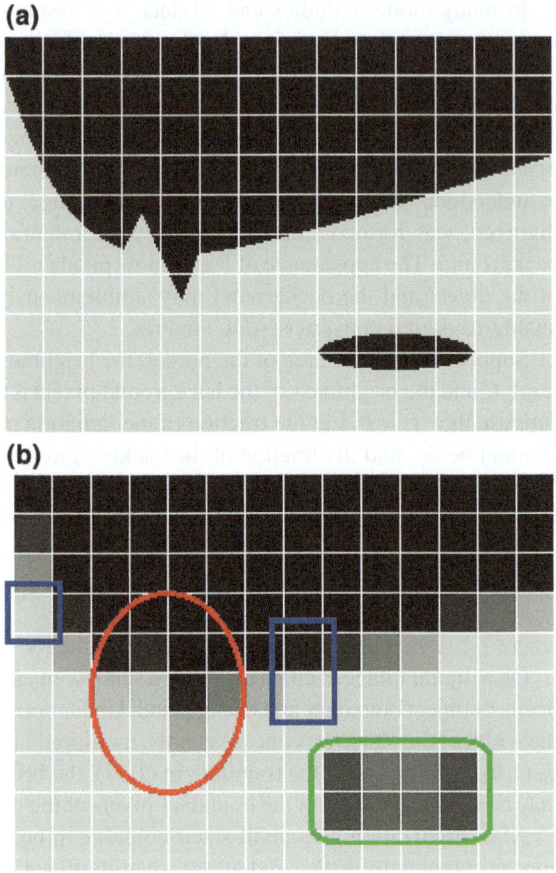

Fig. 1.7 Optical (**a**) and digital (**b**) images of a black object with a shadow on a gray background. Here, the blue rectangle highlights those pixels of the shadow, the brightness of which is almost identical to that of the adjacent pixels of the object's boundary or background; the red ellipse highlights the pronounced branches of the object's boundary; and the green rounded rectangle highlights the object on the verge of resolution, which after digitization is represented by the set of shadow pixels only

the boundary, which causes the visual effect of so-called irregularity of boundary (Fig. 1.8).

Given the presence of an object shadow, the pixels of a digital image are divided into four main categories:

1. the pixel of an object interior is the pixel which in its 1-neighborhood has the object pixels only;
2. the pixel of an object boundary is a pixel which in its 1-neighborhood has (a) at least one pixel belonging to the object interior, and (b) at least one pixel not belonging to the object;
3. the pixel of a line (including the linear component of the object boundary)—a pixel that has no pixels of the object interior in its 1-neighborhood, but has in its 1-neighborhood a) pixel(s) of the object not belonging to its interior, and b) pixel(s) not belonging to the object;
4. the pixels of the object shadow are pixels that are not uniquely related to with objects and lines.

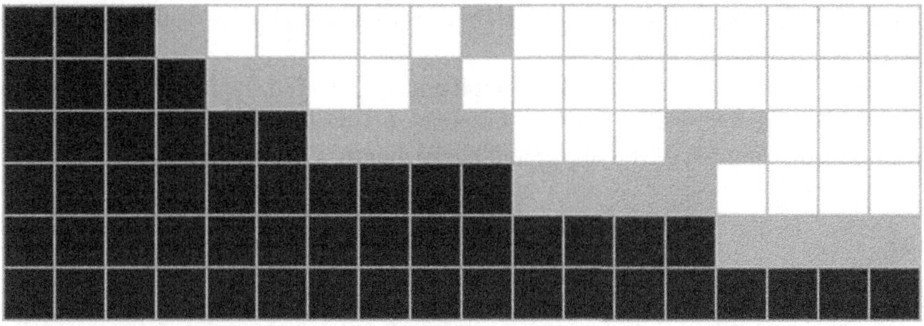

Fig. 1.8 The emergence of border branches after digitizing an image, where black (■) marks the objects pixels, white (□)—the background pixels, and gray (▪)—the objects shadow pixels

The background is understood as a set of pixels, which belong to the objects that are not of interest to us and their shadows.

Due to a number of features of formation of the shadow, it cannot be excluded that the pixels of the shadow may be indistinguishable from the pixels of the object boundary or background in brightness. From this point of view, the fact that the shadow pixels are detected during a boundary segmentation does not put a problem, but increases the complexity of computations; therefore, in order not to thin the border [29] later, it is advisable to minimize the number of segmented pixels of the object shadow.

Meanwhile, it is the presence of the shadow that is one of the most important factors which reduce the contrast of objects in the image and the stability of their segmentation, since the shadow pixels are located between the object and the background thus contributing to smoothing the brightness transitions between them. In this regard, the stability of boundary segmentation by contrast can only be achieved by constructing such models of masks and other elements of the boundary detector, which allow us to estimate the contrast of the object (line) by the drop of brightness between its boundary pixels and the boundary pixels of the background, without taking into account the shadow between them.

In addition to appearance of an object shadow, it should also be taken into account that all objects of real scenes have a certain surface texture, and their brightness distribution may be noised. Besides, various distortions of object brightness distribution are possible, for example, due to JPEG compression [10, 29]. All these factors of influence can produce greater or smaller variations of brightness in the pixels of the objects and background.

The presence of brightness variations in adjacent pixels of an object may not reduce its level of contrast with the background.

However, in conditions where the level of such variations is comparable to the magnitude of brightness drop between the object boundary and the background, it becomes simply impossible to unambiguously classify the object pixels by contrast, viz. to separate them into those belonging to the interior and to the boundary. In such a situation, in order to obtain an adequate result of segmentation of object boundary pixels, it is necessary to pre-filter and/ or smooth the brightness distribution of the image. But first things first.

Figure 1.9 shows an airfield photograph and an enlarged fragment of an airplane wing with a clearly visible shadow, branches of the boundary, and variations in brightness distribution of the object and background.

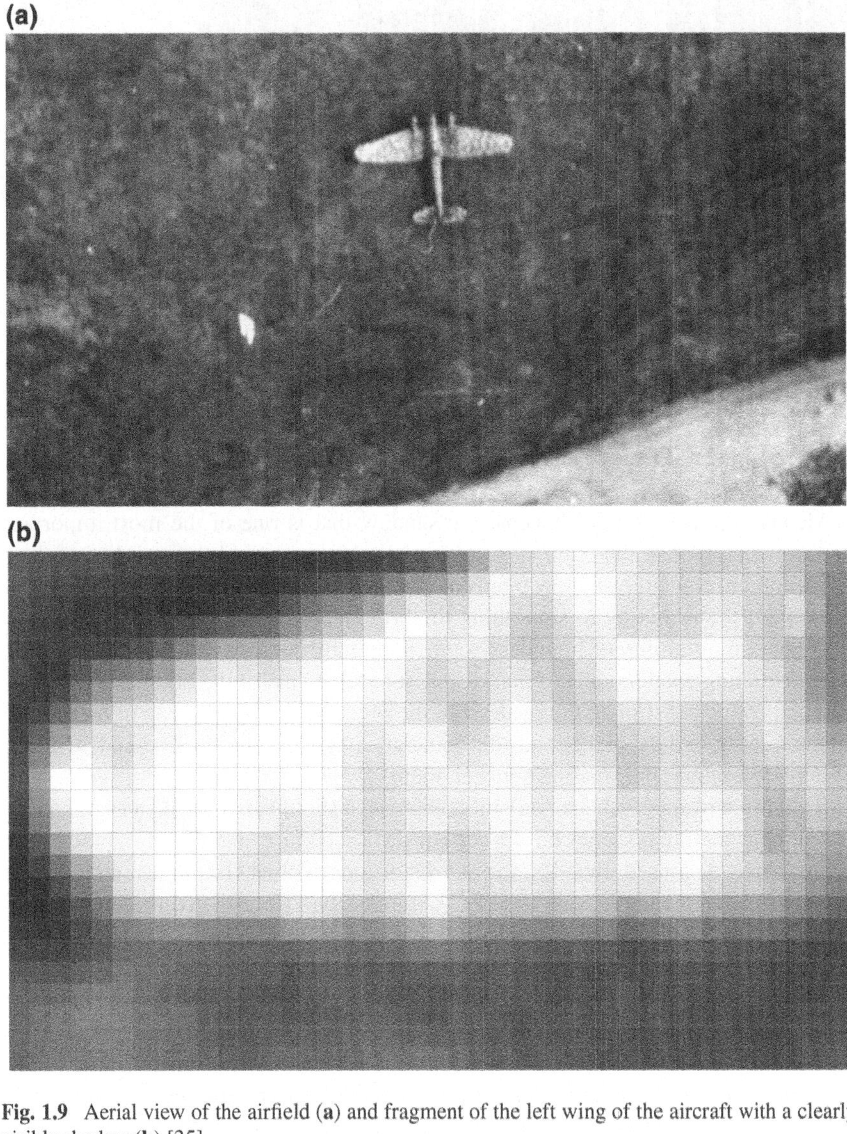

Fig. 1.9 Aerial view of the airfield (**a**) and fragment of the left wing of the aircraft with a clearly visible shadow (**b**) [35]

1.1.5 *Model of the Boundary Profile of an Object in the Image*

Classically, the construction of the edge detector begins with the choice of the model of the object's edge profile (of brightness) and adjusting the detector to the form of the latter. At present, as a rule, such profiles are considered [29] that do not explicitly take into account the presence of an object shadow (Fig. 1.10). Sometimes the profiles are specified in a "noised" form [29], but even in this case they do not show the presence of object's shadow in an explicit form.

In such a situation, the mask of an edge detector, however, like the convolution rule, is adjusted to the boundary of the object without taking shadow into account, which naturally does not contribute to the effectiveness of boundary segmentation. So, what an object boundary profile is to be considered? To answer this question, draw a line perpendicular to the object boundary with the background and consider the brightness distribution along this line (Fig. 1.11).

From consideration of this brightness distribution it can be seen that a shadow three pixels wide is present between the object and the background. In this situation, the profile of the boundary of the object, taking into account the presence of this shadow, is shown in Fig. 1.12. It is namely that profile to which the mask of the edge detector should be configured with the aim to enable effective segmentation of the aircraft wing boundary in the presence of a shadow width of 3 pixels. At this, to take into account in principle the presence of a shadow, an "averaged" profile shown in

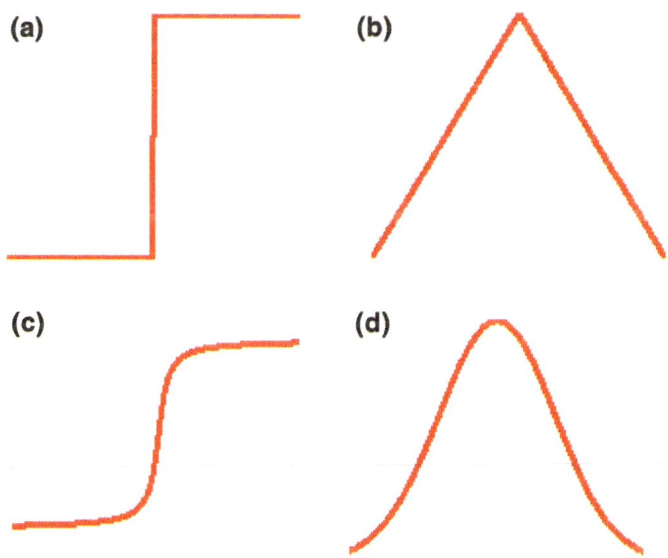

Fig. 1.10 Artificial profiles of an object border of type Step (**a**, **c**), and line of type Roof (**b**, **d**) in discrete (**a**, **b**) and continuous (**c**, **d**) form [29]

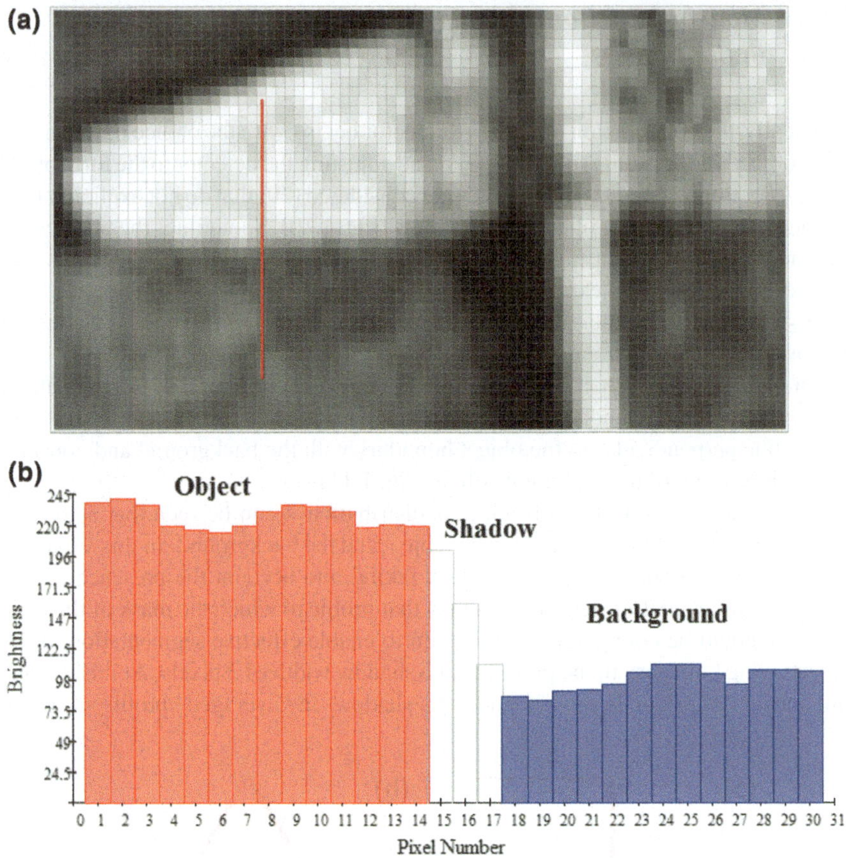

Fig. 1.11 The line perpendicular to the boundary of the left wing of the aircraft with the background (**a**) [35] and brightness distribution along this line (**b**)

Fig. 1.12b should be considered. And only then to analyze the profile details and to more fully take into account the parameters of the shadow of the object (Fig. 1.12a).

Under normal data acquisition conditions, the width of an object shadow (measured along the perpendicular to the boundary) is only one pixel.

However, in unfavorable data acquisition conditions, the shadow width may increase. In this case, the number of steps corresponding to the shadow on the profile of the object's boundary increases proportionally.

The use of corresponding profile of the boundary in an explicit form (Fig. 1.12) indicates the presence of the object shadow, the parameters of which should be taken into account for construction an adequate mask model and edge detector.

At the same time, replenishing such a profile with variations of brightness inside the object and background is not advisable at the initial stage of modeling, since it would mask the effect of having an object shadow.

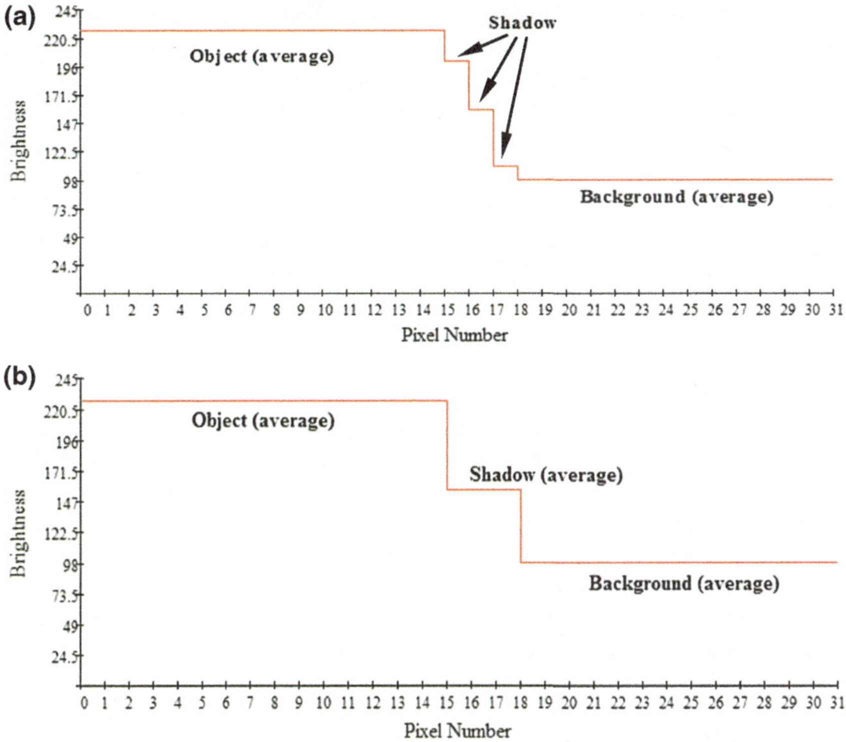

Fig. 1.12 Model of an object border profile with a background, taking into account the presence of a shadow three pixels wide (**a**), and on average (**b**)

To consider the brightness variations (Fig. 1.11) is advisable in the next step—for building an effective procedure for image pre-processing and setting the parameters of the edge detector.

1.2 Adaptive Image Preprocessing Model

Variations in brightness of an object due to co-manifestation of effects of quantization, texture and noise pollution can make it difficult to adequately segment the border pixels. If such variations are comparable in order of magnitude with the drop in brightness at the boundary of the object with the background, this becomes a real problem. The thing is that due to the similarity of the contrast estimates of the interior and boundary pixels, the results of segmentation in such conditions are usually characterized by the appearance of significant gaps and/or thickening of the object boundary, that is, of a large number of false pixels and boundary components.

In such a situation, to eliminate these negative effects, after segmentation of the border pixels it is necessary to apply methods for eliminating gaps and/or boundary thinning techniques; for example, nonmaximum suppression [29]. For objects of a regular type, the shape of which is known in advance, the use of such methods can produce adequate results. For objects of an irregular type, the shape of which is not known in advance, all these methods are partly heuristic in nature and it is difficult to predict the consequences of their application. In both cases, the use of such methods to compensate for the distortion is quite time-consuming.

Even if the average level of brightness variations in a picture is relatively low, the presence of such variations will complicate the segmentation of the borders of low contrast objects, the detection of the boundaries of which is often a very important task.

In such a situation, before the segmentation begins, the image is pre-processed. To ensure the stability of the segmentation in the preprocessing stage, it is important to filter the noise and smooth the brightness of the pixels. In this relation the distinction between the filtering and smoothing is understood as follows. The filtering is meant a two-step procedure, which consists in finding and then—smoothing the extremal brightness values of the image. The smoothing is understood as a simple application of a smoothing filter to all pixels of a snapshot without parsing. In this regard, when considering filtering algorithms, we assume that in the respective neighborhood the brightness of the pixel in question is extremal [10].

If there is such an opportunity, the filtering and smoothing operations are advisable to perform as part of the procedures of accumulation [10] and/or zooming out. A standard in both situations is the use of averaging filter [10] which finds the arithmetic, geometric, harmonic or other mean; to eliminate the effect of noise, a filter of a truncated average is often used, which works according to the following algorithm.

For the considered series of brightness values $\{f_i\}_{i=1,\dots,n}$, the first k minimal values are put in the beginning of the series, whereas the k maximal ones—to the end of this series; after that, the arithmetic mean filter is applied on the set of intermediate brightness values $\{f_i\}_{i=1+k,n-k}$. Thus, after sorting the considered pixel brightness values, with the use of the truncated average filter the new smoothed brightness Z of the filtered pixel Z is calculated as follows

$$h = \left\lfloor \frac{1}{n - 2 \cdot k} \sum_{i=1+k}^{n-k} f_i \right\rfloor, 2k < n. \tag{1.18}$$

By controlling the parameter k, the truncated average filter (1.18) provides: (1) adaptation to the level of noise by eliminating the extremal observations, and (2) reduction of the degree of smoothing of image boundaries and lines, compared to the averaging filters, due to the high-quality smoothing of the interior.

At the limit values of parameter k, the filter of the truncated average becomes the arithmetic averaging (at $k = 0$), or median one (at $k = (n - 1)/2$) [10].

The use of the accumulation procedure requires the fulfillment of specific conditions for obtaining a series of snapshots of the same field of view [10], which is simply impossible when solving the overwhelming majority of practical tasks. On the other hand, although in many cases the scaling is technically allowable, it is not always admissible. In situations where the objects of interest are presented in the image at the required scale, whereas the image also contains linear or small-sized objects, the scaling is simply inadmissible. Thus, the accumulation and scaling can not be perceived as a kind of standard procedure for pre-processing of the image, and in this situation one have to look for other techniques. So, if the objects of interest are quite large (that is, they have both a boundary and an interior), a smoothing is currently used without accumulating and scaling the image. Where it leads? An example of arithmetic mean filtering is shown in Fig. 1.13.

The estimates of mean and dispersion for the object and background brightness distributions before and after filtering are presented in the listing shown in Fig. 1.14.

Smoothing leads to that the brightness values of the pixels of the object rather quickly approach the average value. However, the boundary of the object is significantly smoothed, and the shadow of the object expands in proportion to the doubled radius of the smoothing filter mask (Fig. 1.13). The main advantage of such smoothing is that it brings the object and background brightness distributions closer to the average, which allows later on to obtain high contrast estimates between these average levels.

In the considered example, before smoothing the brightness difference between the pixels of the object (calculated at a distance of 2 pixels) makes $29 - 21 = 8$ units, $18 - 14 = 4$—between the background pixels, and $21 - 19 = 2$ units—between the background and object boundary pixels (Fig. 1.13). In such a situation, it is impossible to segment the boundary of the object adequately.

After a single smoothing, the situation has changed dramatically. The brightness variations have significantly decreased, although the shadow has expanded to 3 units. In the practice of smoothing, most often the 3×3 mask dimension is used. Therefore, in most practically significant cases, the width of the shadow will expand from 1 to 3 pixels.

When using a mask of radius 4, the maximum difference in brightness between the pixels of the object is $26 - 23 = 3$ units, between the pixels of the background—$17 - 14 = 3$ units, and between the pixels of the background and object boundary—$23 - 17 = 6$ units (Fig. 1.13). In such a situation, to adequately segment the boundary of the object by contrast already becomes possible.

But it is not precisely the true boundary of the object, but the boundary obtained after expanding the shadow. Therefore, after segmentation, it is required still to expand the boundary to the outside. To do this, in the course of tracing the border [29], it will be necessary to construct its external neighborhood, and so it will contain the pixels of the boundary.

This is the minus of image smoothing. Another significant disadvantage of such smoothing is that the linear and small-sized objects will be entirely represented by a shadow.

The use of a mask of radius 4 leads to a significant increase in the computational complexity of segmentation. Not to mention that the enlarging of the mask radius

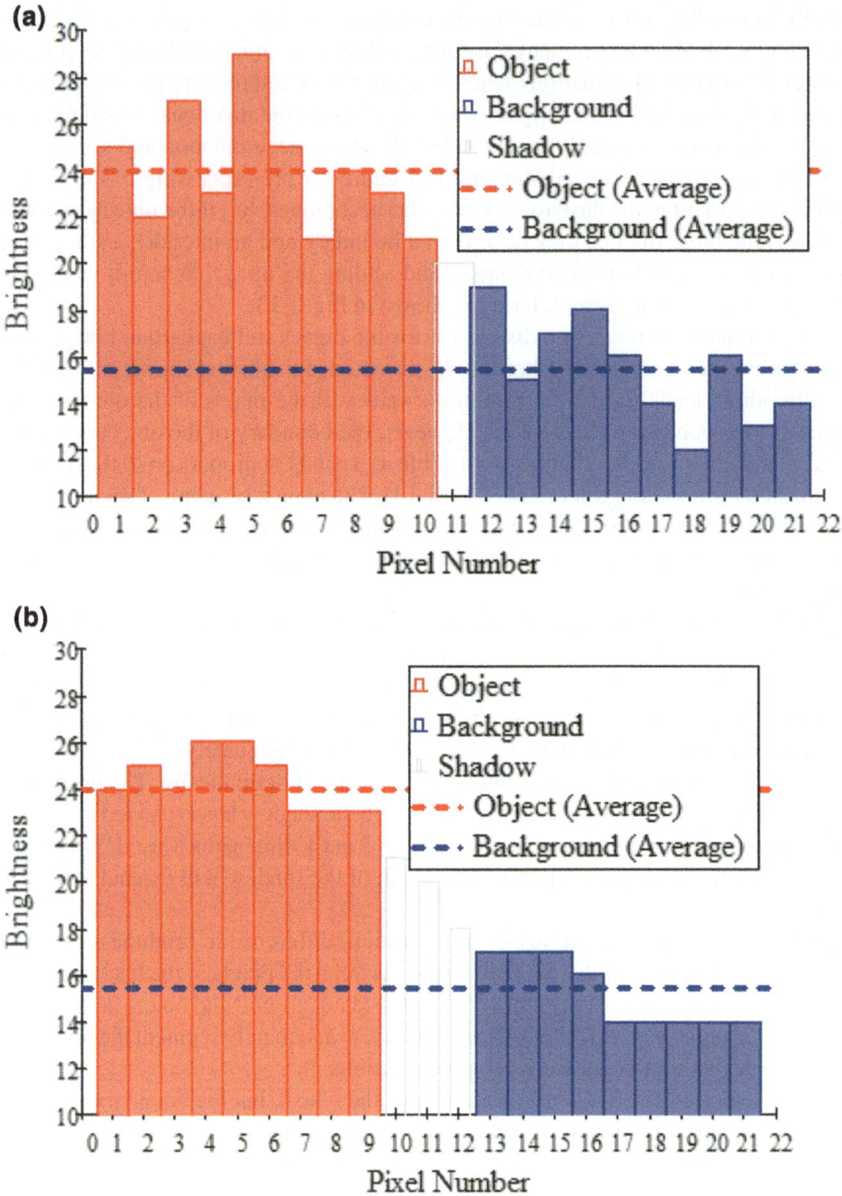

Fig. 1.13 Fragment of the brightness distribution of the object in the slice (**a**) and the same fragment after smoothing with an arithmetic mean filter (**b**)

(a)

$$y1 := (25\ 22\ 27\ 23\ 29\ 25\ 21\ 24\ 23\ 21)^T \qquad a1 := (24\ 25\ 24\ 26\ 26\ 25\ 23\ 23\ 23)^T$$

$$My := \frac{1}{\text{length}(y1)} \cdot \sum_{i=0}^{\text{length}(y1)-1} y1_i \quad My = 24 \qquad Dy := \frac{1}{\text{length}(y1)} \cdot \sum_{i=0}^{\text{length}(y1)-1} \left(y1_i - My\right)^2 \quad Dy = 6$$

$$Ma := \frac{1}{\text{length}(a1)} \cdot \sum_{i=0}^{\text{length}(a1)-1} a1_i \quad Ma = 24.333 \qquad Da := \frac{1}{\text{length}(a1)} \cdot \sum_{i=0}^{\text{length}(a1)-1} \left(a1_i - Ma\right)^2 \quad Da = 1.333$$

(b)

$$y2 := (19\ 15\ 17\ 18\ 16\ 14\ 12\ 16\ 13\ 14)^T \qquad a2 := (17\ 17\ 17\ 16\ 14\ 14\ 14\ 14\ 14)^T$$

$$My := \frac{1}{\text{length}(y2)} \cdot \sum_{i=0}^{\text{length}(y2)-1} y2_i \quad My = 15.4 \qquad Dy := \frac{1}{\text{length}(y2)} \cdot \sum_{i=0}^{\text{length}(y2)-1} \left(y2_i - My\right)^2 \quad Dy = 4.44$$

$$Ma := \frac{1}{\text{length}(a2)} \cdot \sum_{i=0}^{\text{length}(a2)-1} a2_i \quad Ma = 15.222 \qquad Da := \frac{1}{\text{length}(a2)} \cdot \sum_{i=0}^{\text{length}(a2)-1} \left(a2_i - Ma\right)^2 \quad Da = 1.951$$

Fig. 1.14 Listings of estimates of mean M and dispersion D of object brightness distributions before $y1$ and after $a1$ smoothing by an arithmetic mean filter (**a**), and a similar listing for the brightness distributions of the background $y2$ and $a2$, respectively (**b**)

significantly increases the probability of identifying the object shadow pixels as the object's ones.

It is known that inefficient execution of low-level operations (e.g. filtering or smoothing) leads to the need to compensate for distortion by performing higher-level operations (e.g. expanding the boundary, eliminating the boundary gaps, etc.). At the same time, whereas the low-level operations, as a rule, are performed on individual pixels, the adequate performing of the high-level operations, most often, requires consideration of pixel groups belonging to neighborhood or boundary, which is at least much more time consuming.

In this regard, to ensure the overall computational efficiency of detecting the boundary of an object, it is advisable first of all to increase the efficiency of smoothing in the sense of the following three main conceptual demands:

1. bringing the object and background brightness values closer to the average levels; in other words—elimination of significant brightness differences between them;
2. not expanding the shadow of the object;
3. not smoothing the lines and small objects.

To make this 100% is unlikely. However, quite high results can be achieved using an adaptive filter.

Thus, to improve the quality of smoothing and improve the efficiency of post-processing, it is advisable to develop an adaptive smoothing filter, the use of which

(Fig. 1.15a) satisfies the formulated three demands. The estimates of mean and dispersion for the object and background brightness distributions after the adaptive filtering are presented in the listing shown in Fig. 1.15b.

After a single adaptive smoothing the situation has changed dramatically. Now, when using a mask of the same radius, the maximal difference in brightness between the pixels of the object (calculated at a distance of 2 pixels) is $26 - 23 = 3$ units, $17 - 14 = 3$ units—between the pixels of the background, and $22 - 17 = 5$ units—between the pixels of the object boundary and the background. In such a situation it is now possible to adequately segment the boundary of the object by contrast. In addition, there is no need to engage in the expansion of the object's boundary.

(a)

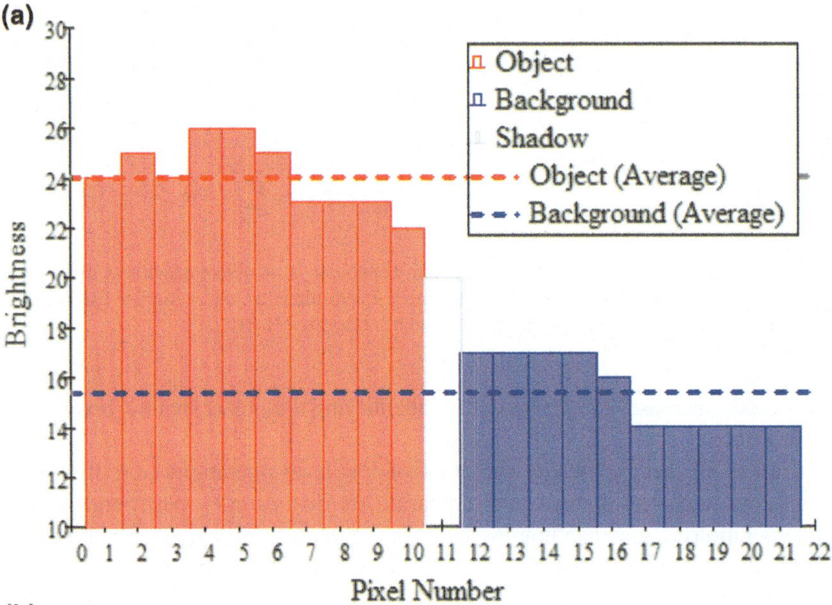

(b)

$y11 := (24\ 25\ 24\ 26\ 26\ 25\ 23\ 23\ 23\ 22)^T \qquad y22 := (17\ 17\ 17\ 17\ 16\ 14\ 14\ 14\ 14\ 14)^T$

$$My := \frac{1}{\text{length}(y11)} \cdot \sum_{i=0}^{\text{length}(y11)-1} y11_i \quad My = 24.1 \qquad Dy := \frac{1}{\text{length}(y11)} \cdot \sum_{i=0}^{\text{length}(y11)-1} \left(y11_i - My\right)^2 \quad Dy = 1.69$$

$$Ma := \frac{1}{\text{length}(y22)} \cdot \sum_{i=0}^{\text{length}(y22)-1} y22_i \quad Ma = 15.4 \qquad Da := \frac{1}{\text{length}(y22)} \cdot \sum_{i=0}^{\text{length}(y22)-1} \left(y22_i - Ma\right)^2 \quad Da = 2.04$$

Fig. 1.15 Adaptive smoothing: fragment of the brightness distribution of the object in the slice after smoothing with an adaptive filter: the initial distribution is shown in Fig. 1.13a (**a**) and listings of estimates of mean M and dispersion D of brightness distributions of object y11 and background y22 after adaptive smoothing with an arithmetic mean filter; the initial brightness distributions and their estimates are presented in Figs. 1.13 and 1.14, respectively (**b**)

1.2.1 Adaptive Filtering of Extremal Observations

The fundamental difference between adaptive and non-adaptive filtering is that the adaptive filtering algorithm first selects a set of homogeneous pixels in the considered neighborhood and only after that performs filtering on this set. Such a procedure can be several times, or even an order of magnitude more laborious, than simply finding the average in a neighborhood. However, on the other hand, an adequate choice of parameters of an adaptive filter allows us to not smooth the object boundaries, lines and small objects presented in the image.

The simplest approach to building an adaptive filter is truncation being used in the truncated average filter (1.18). However, this approach is too rigid because it does not take into account the local form of an object in a neighborhood. How it can be improved? Perhaps the simplest way—is to use the rule of averaging over the closest observations.

So, assume that $f_{\xi\eta}$ is the extremal brightness that in some area is reached in the pixel $d_{\xi\eta}$. Further on, in the given neighborhood $O_\varepsilon(\xi,\eta)$ of pixel $d_{\xi\eta}$ find n brightness values being closest to the brightness $f_{\xi\eta}$ and then replace the latter one by the mean h obtained by averaging these n closest values pertaining to the pixels in the neighborhood $O_\varepsilon(\xi, \eta)$.

The value n is adjusted with the size of the neighborhood and the topology of the object, and then—optimized experimentally. When using a neighborhood of radius $\sqrt{2}$, for the boundary of the object it is initially assumed that, as a rule, $n = 2$, and then this value may be increased as a result of training. At this, the main thing is to control this value to be not overestimated for the formed set to remain homogeneous (viz. to contain only the pixels of the object without shadow and, especially, without the background).

In this case, the averaging rule (that is, the filter) for obtaining the mean h is also optimized at the learning stage. If the parameter n is found adequately, then further it is sufficient to use an arithmetic mean filter.

After filtering, the pixels that were not extremal may become so. In such situation, it may take several iterations to eliminate the more or less significant extremal observations in the image. At this, to minimize the computational complexity it is advisable to act as follows.

Prior to filtering, all pixels of the image are labeled "one": by default, they are all considered potentially extremal.

Iteratively, the image pixels with the label "one" are considered. If the pixel is not extremal, it gets a label "zero". Otherwise, the pixel brightness filtering is performed; at this, all pixels of its $\sqrt{2}$-neighborhood are marked "one". This means that at the next iteration each of them should be analyzed for giving an extremal value.

This way the image filtering iterations are carried out, which can be several. How many such iterations are required to filter an image with a high quality? There are several criteria for completing the filtering process.

Firstly, at the training stage, a fixed number of image filtering iterations can be set up.

Secondly, when replacing the old brightness with a new one, you can evaluate the module of brightness difference $|h - f_{\xi\eta}|$, so that when a certain threshold is reached, the pixel $d_{\xi\eta}$ be not considered extremal any more; in other words, on condition $|h - f_{\xi\eta}| < T$ the pixel should be neither considered extremal, nor be filtered, and the pixel is assigned the label "zero" for it to be not considered at the next iteration.

The use of these simple rules will allow us, with a reasonable computational complexity, to eliminate all significant brightness spikes in an image in just a few iterations.

1.2.2 Adaptive Smoothing of Brightness Variations in the Image

During filtering, we first find atypical extremal observations, and then replace them with the mean brightness of a homogeneous group of pixels in the specified neighborhood according to a given rule.

But when smoothing, it is assumed the absence of significant noise in the image. All pixels of the image are being smoothed indiscriminately. In such a situation, it is technologically expedient to firstly perform filtering in order to find and eliminate significant noise, and only then to perform smoothing of the image. Suppose that the filtering is done. And it's time to smooth. How best to do this?

Firstly, we can apply averaging over the nearest observations. When filtering, in obtaining a new brightness estimate the extremal value was not taken into account. Now, it is possible either to include the brightness of the considered pixel into a homogeneous group, or not. In the second case, the larger variations in brightness will be smoothed faster. But in both cases the brightness of the object pixels will, iteration by iteration, approach the average, though at different speeds. But what happens to the shadow pixels? They may be absorbed by the object/background, or form brightness lines of almost the same level.

An important feature of filtering and smoothing is that to achieve the desired result, they can (and should) be performed iteratively, step by step, by firstly reducing the noise level, and then—the level of brightness variations of objects and background. In modern conditions, this approach not only pays for itself in terms of heightening the quality of preparation of image for segmentation, but also allows for efficient parallelization. In this regard, the filtering and smoothing differ in that that after the first iteration the filtering of residual brightness spikes will be performed an order of magnitude (and more) faster, because on smoothing the image is processed in whole.

For improving the computational efficiency and smoothing quality, it may be advisable to increase the size of the neighborhood (and the parameter n) which reduces the number of algorithm iterations.

As an alternative to averaging over the n closest observations, it is possible to perform smoothing in a homogeneous neighborhood by constructing it with the use of the classical (known) statistical algorithms.

The essence of such an algorithm is firstly to form in the vicinity $O(\xi, \eta)$ a set of pixels being homogeneous relative to the considered value $f_{\xi\eta}$, and then to apply the specified smoothing filter to this homogeneous set. The algorithm that provides construction of such homogeneous set of pixels in the given vicinity is as follows.

First, by a given criterion (by n closest observations, and/or by the threshold criterion) a homogeneous reference set H is formed which contains pixels in the neighborhood $O(\xi, \eta)$. Then the set H is iteratively extended at the expense of the unused pixels of the neighborhood, while it is possible (by default—by the $k\sigma$-criterion). At each iteration of extension, the brightness f' to be the closest to $f_{\xi\eta}$ is searched among the pixels which have not been yet used and lie within the neighborhood $O(\xi, \eta)$; after that, the degree of heterogeneity is assessed.

$$|f_{\xi\eta} - f'| < k \cdot \sigma, \sigma^2 = \frac{1}{m} \cdot \sum_{i=1}^{m} (f_{\xi\eta} - f_i)^2. \quad (1.19)$$

The extension is performed if condition (1.19) is satisfied. This extension procedure is performed whilst possible, i.e. while condition (1.19) is fulfilled and the considered neighborhood still contains unused pixels. In addition, the fulfillment of the constraint $m \leq n$ on the size of the set H can be checked.

In comparison with smoothing over n closest observations, this algorithm has both advantages and disadvantages. The main benefit is that by virtue of the control over the heterogeneity measure (1.19) permits to increase the volume H and achieve a high-quality smoothing in one pass. Meanwhile, only one threshold Z is used for the entire image. And this may not allow to adequately process objects with different levels of contrast and brightness variation at the boundary. In addition, with a large value of n and relative smoothness of object brightness distribution, a brightness not belonging to the object may be attributed to the set H. This is the effect of false brightness masking. To combat it, instead of condition (1.19), the fulfilling of the condition $(f_{\xi\eta} - f')^2 < k \cdot \sigma^2$ may be demanded, where the use of the squared deviation provides greater sensitivity to appearance of extremal observations.

1.3 Indicators and Quality Criteria of Boundary Segmentation

If we talk about the problem of filtering (smoothing), then the quality indicator is quite simple and unambiguous—it is a measure of the approximation of pixels brightness to the average for an object. As for the problem of segmentation of object boundary pixels being considered below, the situation is the opposite. Therefore, in order to understand the principles of construction and application of edge detectors, it is advisable to first introduce into consideration a unified system of formalized indicators and criteria for assessing the quality of boundary segmentation.

1.3.1 *The Basic Model of an Object Boundary Template for a Digital Image*

Any assessment of the quality of segmentation makes sense in the conditions when a template of an object boundary is specified. Only in such a situation can we compare the result obtained with the standard and draw the conclusion about the adequacy or inadequacy of segmentation.

At first glance, it may seem that the template of the boundary of a simply connected object can be given by an 8-connected closed line. One could expect this to be that outcome which we expect to obtain as the perfect result of segmentation of object boundary. However, this is not entirely true.

If we ask several experts to construct the boundary of the same object, the obtained boundaries may not coincide. And this is due to pixel brightness variations in the object shadow. In such a situation, the template of the boundary of a simply connected object should be considered as an 8-connected closed line that is not contractible to a point on the set of potentially segmented pixels; for example, in Fig. 1.16—it is represented by an annular band. In this case, the searched line can vary within the given area of adequate segmentation; in the above example—within the specified ring.

With the aim to specify the tolerances for the result a' relative to the boundary template line a, the expert must specify not only the line a itself, but also the values ε and δ, ($\varepsilon \geq \sqrt{2}$, $\varepsilon \in \{\rho_\varsigma\}_\varsigma$), where ρ_ς is defined in (1.12); these values determine the maximal limiting segmentation errors for the boundary pixels for them to fit the internal and external template neighborhoods (Fig. 1.16), viz. for the pixels of the object's interior and its shadow. Without loss of generality, we further assume that $\varepsilon = \delta$.

As a result, the boundary $Fr\,(E)$ is corresponded by a ring-shaped region G, formed by a set of pixels of two-sided ε-band representing the template line a, which

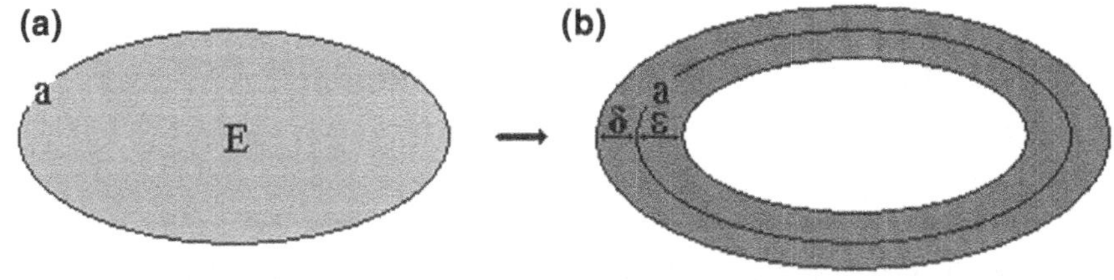

Fig. 1.16 The template a of the object boundary $Fr\,(E)$ (**a**), and the band G of an adequate segmentation of the object image boundary being represented by grey strip (**b**)

is understood as an area of adequate boundary segmentation, both in terms of pixels and in the sense that any 8-connected closed line in G being not contractible to a point, is considered as an adequate approximation of the boundary $Fr(E)$. The introduced concept of template is further used to assess the adequacy of segmentation results according to the specified criteria.

Most often (by default) the boundary is understood as an 8-connected line; subject, however, from a formal point of view, whether the boundary is a 4- or 8-connected line completely depends on the adopted system of definitions of the interior and the boundary (Fig. 1.5).

1.3.2 The Quality Criteria for Boundary Segmentation by Template

At a high level, the adequacy of the segmentation is determined by the following criteria.

(1) The criterion Cr_{ASC} of adequacy of segmentation of object boundary component. An object boundary component is considered to be segmented adequately if for a given limiting error $\Delta_L = \varepsilon$ this component of the boundary is represented by a not contractible 8-connected closed line in the region of adequate segmentation.
(2) The criterion Cr_{AS} of adequacy of segmentation of object boundary. The segmentation of the boundary of an object that has several components of connectedness is considered adequate if each component of the boundary is segmented adequately (viz. according to Cr_{ASC}), and the specified restrictions on the level of false segmentation of boundary are fulfilled.

However, this result is still required to be obtained. In this regard, it is necessary to introduce into consideration the indicators and local criteria of segmentation quality for individual pixels and the boundary as a whole so that it becomes possible:

- to formulate the requirements to models, detectors and algorithms,
- to analyze the results of their use, and
- to adjust the parameters according to the conditions of the problem.

1.3.3 Unified System of Quality Indicators for Boundary Segmentation

When using the boundary template for quality assessment, it is natural to base on using key indicators of partial and false boundary segmentation. In this work the false segmentation is understood as the one which leads to detection of false pixels (components) outside the area G (an error of the first kind); and the partial segmentation—is the one, in the result of which the boundary of the object image has been detected in the area G, but not to a full extent (an error of the second kind). The presence of partial segmentation leads to appearing of discontinuities of the boundary (viz. to its fragmentation), and consequently—to impossibility of constructing an 8-connected closed line that could be contracted into a point in the region G of adequate segmentation.

Let the object boundary be described by the model $G = \cup G_i$, $(i = 1, \ldots, n)$, and as a result of its segmentation a set of pixels S be obtained. Divide it into two subsets T and F, $(T \cup F = S, T \cap F = \emptyset)$, so that the first consists of pixels that belong to the area G, whereas the second contains the pixels not belonging to this area.

For each component of the object image boundary, select the set $T_i = T \cap G_i$ with an area $\mu(T_i)$ defined by the number of pixels; the number of non-empty sets of this type denote m, $m \leq n$. In its turn, select the connected sets F_j, $(j = 1, \ldots, k)$, with the areas $\mu(F_j)$ from the pixels of the set F. In these notations, the quality of the object boundary segmentation is characterized by the following two main integral indicators.

The partial boundary segmentation indicator

$$b_p = (n - m)/n, \tag{1.20}$$

characterizes the fraction $\mu(T_i)$ of the number of components of image boundary with the zero measure of segmentation.

The false boundary segmentation indicator

$$b_f = \mu(F)/[\mu(T) + \mu(F)], \tag{1.21}$$

characterizes the fraction of the falsely segmented pixels

Notice, that by construction these indicators can be interpreted as the probabilities of the corresponding events.

In the case when these indicators take zero values $b_p = 0$, $b_f = 0$, the criterion of adequacy Cr_{AS} is satisfied, provided that the criterion Cr_{ASC} is also satisfied for each component T_i.

A nonzero indicator (1.20) means that some of the image boundary components are not segmented, which is inadmissible; therefore, an adequate boundary segmentation method should be configured to provide a zero value for the indicator (1.20).

However, an appearance of false boundary components, although undesirable, is not a problem if they can be filtered at the subsequent stages of image processing.

In general, the false pixels of the boundary can be divided into two sets—$F^{(0)}$ and F^*. The former has no pixels 8-connected with the domain G, while the latter has them. The presence of false boundary components of the first type often does not create fundamental problems, since in most cases the sets $F_j \in F^{(0)}$ separable from the regions G_i can be relatively easily filtered at subsequent stages of image processing by their topological or geometric parameters.

Essentially more difficult is the truncation of false components of the second type, $F^* = \{F_j^*\}_j$, which have pixels 8-connected with the region G. Therefore, when evaluating the level of false segmentation according to (1.21), it is advisable to consider the estimates for the sets $F^{(0)}$ and F^* separately, by minimizing firstly the measure of the set F^*.

Consider now the quality indicators of the boundary component segmentation from the viewpoint of existence of a closed 8-connected line in the set of pixels T_i, $(i = 1, \ldots, n)$, which is homotopic to the boundary template a_i.

In general, the set T_i is represented by partitioning $\{T_{i(\xi)}\}_\xi$ into connected sets lying in the domain G_i. If at least one of them contains the line $a_{i(\xi)}$ being homotopic to the boundary template a_i, then it is considered an adequate approximation of the template a_i of the corresponding component of the image boundary. For brevity, such a simply connected closed line being not contracted into a point is called a contour. Otherwise, only fragments of an open-ended boundary component can be detected

Let m be the number of segmented boundary components $\{T_i\}_{i=1,\ldots,m}$ for which such lines exist. Then the indicator of partial segmentation of boundary components is as follows

$$c_p = (n - m)/n. \tag{1.22}$$

This indicator is equal to 1 if for none of the segmented boundary components it is impossible to construct a contour being homotopic to the respective component of the boundary, and 0 if in each of these components such line exists.

The indicator (1.22) determines the proportion of open-ended contours among the segmented components of the object boundary. To estimate the measure of partial segmentation of boundary components (re to such contours), the maximal gap is presently used as one of the main indicators of this type

$$r_p = \max_i\{\rho_i\max\}, \tag{1.23}$$

where $\rho_i\max$ is the estimate of the maximal length of the gap for the segmented component of the boundary T_i.

It is used to separate the boundary gaps arising during segmentation into local, with a length of not more than the given value τ, which contain several pixels, and regional, with a length of more than the given value τ, consisting of tens or even hundreds of pixels.

Elimination of local boundary gaps is not a problem for images of any type; on the other hand, an adequate elimination of regional boundary gaps for images of irregular objects is almost impossible. In such a situation, the use of the indicator (1.23) allows you to check for the presence of difficult-to-eliminate regional boundary gaps. Therefore, an adequate boundary segmentation detector must be tuned so that the inequality $r_p \le \tau$ holds. In the case when $r_p = 0$, the criterion of adequacy Cr_{AS} is satisfied. If $r_p \le \tau$, the criterion of adequacy Cr_{ASC} is satisfied relative to the absence of regional boundary breaks.

If the number of pixels of the segmented boundary component T_i exceeds the number of pixels approximating the boundary template a_i in the region G_i, then the segmentation redundancy takes place that is to be eliminated. Let m be the number of segmented components of the boundary $\{T_i\}_i$, in each of which, T_i, there is an adequate approximant a_i^* of the template a_i containing $\mu(a_i^*)$ pixels from T_i.

Then the indicator of segmentation redundancy takes the form

$$c_s = \frac{1}{m}\sum_{i=1}^{m} c_{s(i)}, \tag{1.24}$$

$$c_{s(i)} = \begin{cases} \mu(T_i)/\mu(a_i^*), & if \quad \mu(T_i) > \mu(a_i^*), \\ 0, & else. \end{cases} \tag{1.25}$$

Looking a little ahead, it should be noted that when using a mask with not a minimal radius in the presence of shadow and brightness variations, as a result of segmentation it is always allocated a band of pixels among which, apart from the boundary pixels, there may be the internal near-boundary pixels of the object and pixels of the object shadow. This situation is normal. The presence of such redundancy is not an inadequate result of segmentation for many reasons; in particular, since it can be eliminated at the post-processing stage, for example, using the non maximal suppression method, or by tracing the boundary [29]. However, the use of these methods often requires significant additional time. Therefore, it is advisable to adjust the model parameters and the boundary detector so as to minimize the value of redundancy indicator (1.24).

As the result, we obtain the system of basic indicators of segmentation quality which is presented in Table 1.1.

If all the quality indicators discussed above are zero, then the segmentation quality of the boundary is ideal. However, in the presence of shadow and the influence of other negative effects of digitization, segmentation with zero values of the above said quality indicators is simply not possible. In this situation, there are many pressing

Table 1.1 The table of indicators of boundary segmentation quality

Design	Name of quality indicator
Category 1. Segmentation error	
Δ_n	Limiting segmentation error
Category 2. Partial segmentation	
b_p	Partial boundary segmentation indicator (1.20)
c_p	Partial contour segmentation indicator (1.22)
r_p	Indicator of maximal gap size (1.23)
Category 3. False segmentation	
b_f	Indicator of false boundary segmentation (1.21)
Category 4. Segmentation redundancy	
c_s	Indicator of segmentation redundancy (1.24), (1.25)

issues: what quality indicators to minimize in the first place, why, and what it will result in.

If at a high level the quality of segmentation is determined by fulfillment of the above formulated criteria Cr_{ASC} and Cr_{AS}, then at a low level the adequacy of segmentation is determined by criteria imposing the restrictions on the levels of quality indicators which are listed in Table 1.1.

In this regard, it is important to understand that the low-level operations are performed with a view to the subsequent application of high-level operations; for example, smoothing is performed for segmentation of boundary pixels, while segmentation—for constructing a connected boundary.

In such a situation, at a low level the criterion of segmentation quality should be formed taking into account the properties of high-level methods. Therefore, if the greatest problem is the elimination of regional boundary gaps, then at a low level, first of all, it is necessary to minimize the length of the boundary gap to some given value.

In practice, such optimization is performed at the expense of reducing the threshold of edge detector, since there is a known regularity: a lowering of the threshold decreases the level of partial segmentation, but increases the levels of false and redundant segmentation. And vice versa. However, such a change in threshold does not automatically mean obtaining the acceptable levels of partial and/ or false segmentation.

In general, the low-level operations such as filtering the noise, smoothing the brightness variations and identification of boundary pixels are an order of magnitude (and sometimes two to three orders of magnitude) less labor-intensive than algorithms for constructing a connected boundary and even much less labor-intensive than the algorithms that provide restoring a distorted boundary, which include the redundancy elimination, filtering the false boundary components, and restoring the boundary gaps. Why? Because the high-level techniques are based on consideration of sets of pixel combinations, i.e. represent the combinatorial problems on the pixel sets.

In such a situation, it is natural to try to improve the models and algorithms for low-level image processing so that the pixels of the object boundary can be segmented more efficiently. And especially—of the low contrast objects and objects of irregular shape.

The use of such low-level algorithms pays off even in a situation when they work several times slower than their counterparts not providing the respective boundary segmentation quality, since, with a significantly higher quality of results, in the future it will not be necessary to apply time-consuming algorithms for restoring a distorted boundary.

Namely for achieving this goal, the adaptive noise filtering and smoothing algorithms have been proposed as the base stage, which serve to ensure a high quality of segmentation of boundary pixels.

The continuation of implementation of this approach lies in construction of an effective detector for adaptive segmentation of the object boundary pixels.

1.4 Adaptive Edge Detector

Consider now the model and algorithm of application of edge detector; by resting upon the results of filtering and/or smoothing, it allows us to segment the boundary pixels of various types of objects for different data acquisition conditions relative to a shadow width and contrast level.

We begin with a description of the detector mask model, which provides taking into account the topological and metric properties of an object and its shadow.

1.4.1 Constructing the Models of Trace and Image of a Circle

Consider a parametrically defined circle on the plane

$$\begin{cases} x = \xi + \rho \cdot \cos(2\pi\tau), \\ y = \eta + \rho \cdot \sin(2\pi\tau), \end{cases} \tau \in [0, \ldots, 1], \tag{1.26}$$

the center (ξ, η) of which is defined by integer coordinates (or zero), and the radius ρ belongs to the set of typical distances $\{\rho_\varsigma\}_\varsigma$.

The set of pixels of a digital image belonging to a circle of the form (1.26) call the trace of this circle (Fig. 1.17).

Take the set of intersection points of the circle (1.26) with the generatrixes of a discrete digital image field and round their coordinates to the integers. The resulting set of pixels is called the image of this circle (Fig. 1.18).

If necessary, any image of a circle can be represented by the union of a finite number of traces of circles of different radii. For example, the image of a circle presented on Fig. 1.18c can be represented by combining the traces of circles of radii 2 and $\sqrt{5}$ (Fig. 1.19).

That is why the trace of a circle is taken to be the basic structural element of image of a circle, neighborhood, and ring mask model which are the considered below.

The image of a circle is a 4- or 8-connected set of pixels by construction. At this, the image of a circle is 4-connected, only when the condition of 4-connectedness is

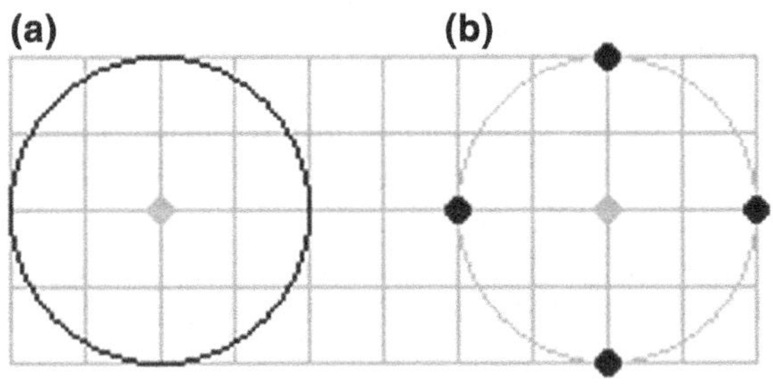

Fig. 1.17 The circumference of the radius 2 (**a**) and its trace represented by four pixels, which are highlighted by black dots (**b**)

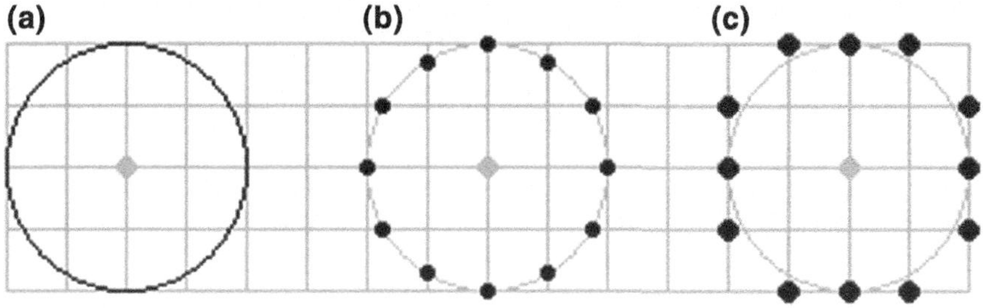

Fig. 1.18 The circle of radius 2 (**a**), its points of intersection with the generatrixes (**b**) and its image (**c**) are highlighted in black

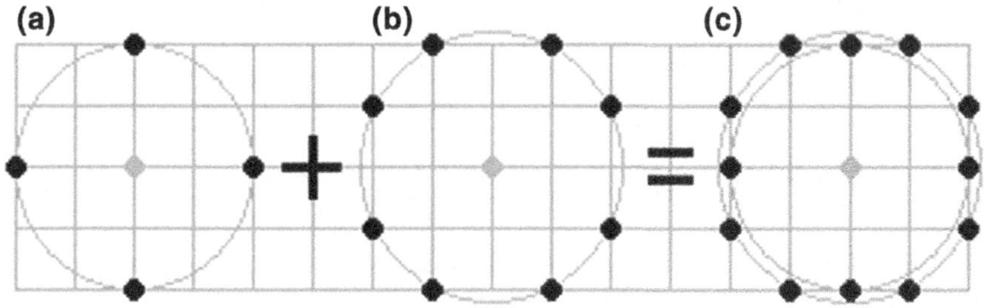

Fig. 1.19 Traces of circles of radii 2 (**a**) and $\sqrt{5}$ (**b**), and the image of a circle of radius 2 (**c**) which is formed by their combination are highlighted in black

fulfilled for all pairs of adjacent pixels. If at least for one pair of its adjacent pixels the condition of 4-connectedness is not satisfied, the image of the circle is 8-connected.

To calculate the geometric location of the points (pixels) that define the image of circle, these points are convenient to store as a matrix M of the displacements of coordinates relative to the center (ξ, η) of the circle of (1.26), on the basis of which this image is constructed. Below, these coordinates are represented according to the growth of the parameter $\tau \in [0, \ldots, 1]$ (Fig. 1.20).

Fig. 1.20 The image of a circle of radius 2 and the offset of its pixel coordinates

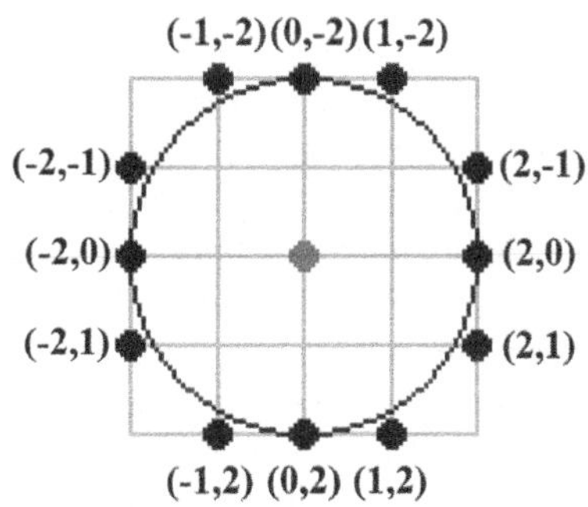

M =

#	1	2	3	4	5	6	7	8	9	10	11	12
1	2	2	1	0	-1	-2	-2	-2	-1	0	1	2
2	0	1	2	2	2	1	0	-1	-2	-2	-2	-1

Fig. 1.21 The matrix of coordinate displacements of circle image pixels, for the circle of radius 2 which is represented in Fig. 1.20; the first line corresponds to coordinate on the abscissa axis, and the second line—on the ordinate axis

If the coordinates of circle image pixels are stored in a form being invariant to displacement (Fig. 1.21), then, knowing the coordinates of the center, one can quickly construct the image of a circle anywhere in the digital image. And this is just one of many advantages of such matrix form of storing the coordinates of circle image pixels. In particular, this matrix method of storing the coordinates of pixels is used for the trace of a circle and for the considered below mask model.

1.4.2 A Model of the Ring Mask of Edge Detector

The mask of the edge detector is called the image of a circle of the given radius ρ belonging to the set of typical distances $\{\rho_\varsigma\}_\varsigma$.

The mask is required for determining the annular fragment of the neighborhood of the considered pixel (ξ, η), the brightness distribution of which will be analyzed to determine the type of pixel (ξ, η).

Reasoning from the definitions of pixels' belonging to boundary and interior, which are given in Para. 1.1.3, one may get the impression that it is enough to use an image of a circle of radius 1 or $\sqrt{2}$. However, this is not the case, since for adequate segmentation it is required to evaluate the contrast of the object boundary pixels relative to the background, by "overcoming" the shadow (Para. 1.1.4). But for this you need to use masks of radii larger than $\sqrt{2}$.

An object with an interior is a 4-connected set (not counting the branches of its boundary and linear components, if any). For a stable segmentation of the boundary of such object, it is sufficient to use an 8-connected mask, because when the center of such mask with an adequately chosen radius is imposed on a pixel of the object boundary, the mask will necessarily include both the object's, and the background pixels. This provides a possibility for adequate evaluating the contrast of an object boundary pixel relative to the background.

For the purposes of stable segmentation of lines (linear components of the object boundary), only a 4-connected mask should be used, which can easily be obtained based on replenishing of the 8-connected image of a circle by one or several circle traces (Fig. 1.22).

Note that most of the images of the circle can be replenished up to a 4-connected mask in several ways. Which one of them to choose? This question can be answered only through learning experimentally.

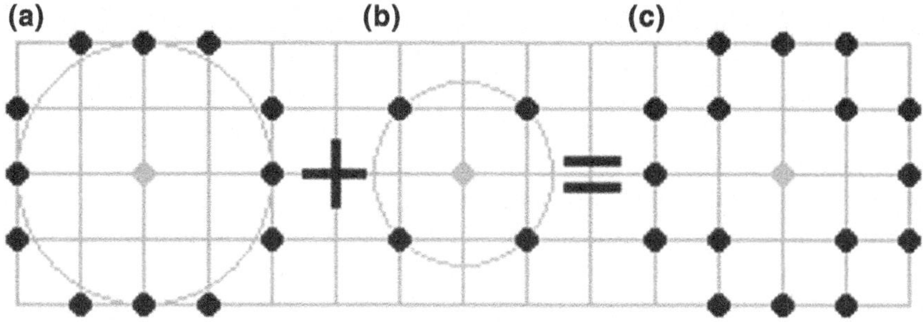

Fig. 1.22 Building a 4-connected mask based on the image of a circle of radius 2 and its replenishment represented by a trace of circle of radius $\sqrt{2}$

When constructing a replenished 4-connected mask, it is advisable to supplement the 8-connected image of the circle with a minimal number of traces until reaching 4-connectedness while using the traces of adjacent radii out of the set of typical distances $\{\rho_\varsigma\}_\varsigma$. This is so because an increase in the number of mask traces after reaching 4-connectedness will only lead to an increase in the computational complexity of segmentation. In addition, for an adequate segmentation by a mask based on the use of an excessively large number of traces of different radii, it becomes increasingly problematic to select the threshold for the edge detector.

Thus, by mask we mean either an image of a circle, or an 8-connected image of a circle being complemented by the minimal number of traces up to reaching of 4-connectedness. The mask of the first type is called simple, and of the second type—composite.

Guided by the results of numerous experiments, we can conclude that in most practically significant cases, the mask radius is advisable to be chosen equal to 2. Using a mask of a smaller radius leads to estimating the difference in brightness between the object and its shadow, and not with the background. On the other hand, increasing the mask radius leads to a growth in false and redundant segmentation.

So, the proposed ring mask model is built. What advantages does it have in comparison with the analogues?

The proposed mask model is based on the image of circle. Therefore, its form, approximately, will have the shape of a circle, which allows more adequately to take into account the metric properties of objects in a digital image. With increasing of mask radius this feature will appear more and more significant.

Since we deal with a radius rather than a diameter, the linear size of the mask will always be odd, which is required for an unambiguous determining the center and convenience of working with neighborhood pixels.

The mask replenishment operation allows us to effectively configure it onto analysis of objects of various topological types, including the lines.

If there is a "clearance"—an intermediate pixels between the center and the mask pixels, by adjusting its width and parameterizing the mask radius, the mask can be adjusted for analyzing the objects with shadows of different widths. Under normal conditions—with one pixel width of the shadow—for adequate boundary segmentation it is sufficient to use a mask with radius 2, because if the center of the mask is imposed onto a boundary pixel, at least one mask pixel belongs to the background; this makes it possible to estimate the difference in brightness between the pixels of object and background, and not with the object's shadow. Unless otherwise specified, hereinafter we consider the masks with radius 2.

Storing the mask in an invariant matrix form (Fig. 1.21) allows you to effectively segment the pixels of object boundaries and lines in any area of the image, and without the need to calculate the coordinates of pixels by using the formula of a circle (1.26). Besides, storing the coordinate displacements with respect to the growth of the parameter $\tau \in [0, \ldots, 1]$, makes it possible, in subsequent analysis, to effectively analyze the vector diagram of the contrast distribution in the neighborhood of the center coinciding with the pixel in question.

1.4.3 Pixel Contrast Estimating Model

Suppose that for evaluating the contrast of the image pixels the edge detector mask is specified, for which the coordinate displacement matrix is defined (Fig. 1.21). In this situation, the estimating of the image pixel contrast is carried out as follows.

First, we find the pixel contrast distribution $U = \{u_i\}_{i=1,\ldots,n}$ over the mask, where the contrast u_i of the pixel with coordinates (ξ, η) is defined as follows

$$u_i = f(\xi, \eta) - f(\xi + M[1, i], \eta + M[2, i]), i = 1, \ldots, n, \tag{1.27}$$

and, then, the minimal and maximal values of this distribution are determined.

$$p = \min_i \{u_i\}_i, q = \max_i \{u_i\}_i, i = 1, \ldots, n. \tag{1.28}$$

After that, the following values are calculated:

(1) pixel contrast level

$$c = \max\{|p|, |q|\}, \tag{1.29}$$

(2) pixel homogeneity level

$$g = \min\{|p|, |q|\}. \tag{1.30}$$

When using the classical approach (Paras. 1.1.1 and 1.1.2), an estimation of contrast over a mask can be performed for a different combinations of pixels belonging to object, shadow and background, the brightness of which can vary in a great range. Besides, this gives a huge amount of contrast values. And what is worse, for the entire image this approximation yields the poorly separable contrast distributions for the interior and object boundaries (Fig. 1.23a). Actually, this is where the main problem of choosing an adequate threshold for segmentation of boundary pixels by contrast

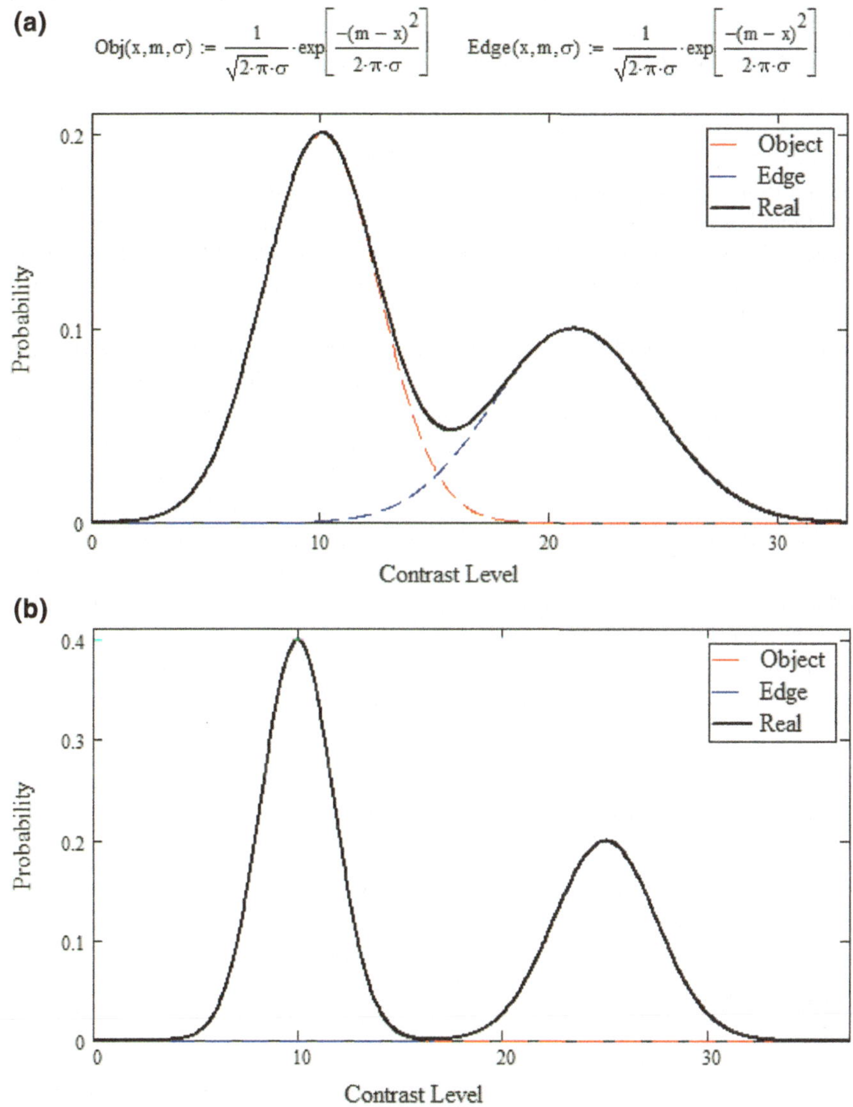

Fig. 1.23 The approximations of contrast density distributions in case of classical estimation (**a**) and when using the rule (1.29) (**b**)

is hidden. In most practically significant situations, whatever threshold we choose, partial and false (excessive) segmentation will still be present at a significant level.

In comparison with analogues, using the rule (1.29) with an adequately selected mask radius allows evaluating the contrast of an object explicitly—by the brightness difference between the pixels of its boundary and background, and not by a gradient or convolution calculated on the set of object, shadow, and background pixels (re to brightness), which are simultaneously covered by the detector mask. And in most practically significant cases (Fig. 1.23b) this leads to separability of the laws of contrast distribution for the object interiority and boundary, what is the most important factor in ensuring the effectiveness of segmentation.

In comparison with analogues, the use of the proposed approach to estimating the contrast allows one to adapt to the actual boundary profile in the neighborhood and not to use neither averaging nor ineffective geometric templates (for example, Prewitt and Sobel masks), which contributes to the growth of segmentation quality. In particular, the application of the proposed approach does not lead to the masking of low-contrast and uneven boundaries, containing in their neighborhood a small fraction of the object pixels.

Remind at this, that before estimating the contrast we have performed filtering and smoothing; therefore, the contrast is evaluated: (1) in the absence of significant noise and variations of brightness, (2) between the average levels of the object and the background. Therefore, such an assessment will be not only effective, but also resistant to variations in the brightness of the input image, which may be caused by noise, as well as by manifestation of texture effects and image digitization.

As we will see further, the use of the homogeneity measure (1.30) along with the contrast measure (1.29) makes it possible to significantly reduce the levels of false and excessive segmentation owing to the use of classical boundary pixel definition given in Para. 1.1.3.

The use of the differential rule for contrast estimating, having the form (1.29), makes it possible to propose the following classification of images by contrast.

1.4.4 Contrast and Non-contrast Images of an Object

Suppose that for a mask of a given radius ρ_ζ the value $a = \min|u_i|$ represents the minimal contrast between the object boundary and the background, and $b = \max\left|u'_j\right|$—is the maximal contrast between a pair of the object pixels.

Then, if the condition $b < a$ is satisfied, the object is considered to be contrast; otherwise the object is considered non-contrast (Fig. 1.24).

If the image is contrast the segmentation of the pixels of its boundary is trivial. For such an object, the threshold T of the edge detector is chosen from the condition $b < T \leq a$. For a non-contrast object, it is no longer possible to choose a threshold that would allow us to uniquely discriminate between the interior pixels and boundary ones [30].

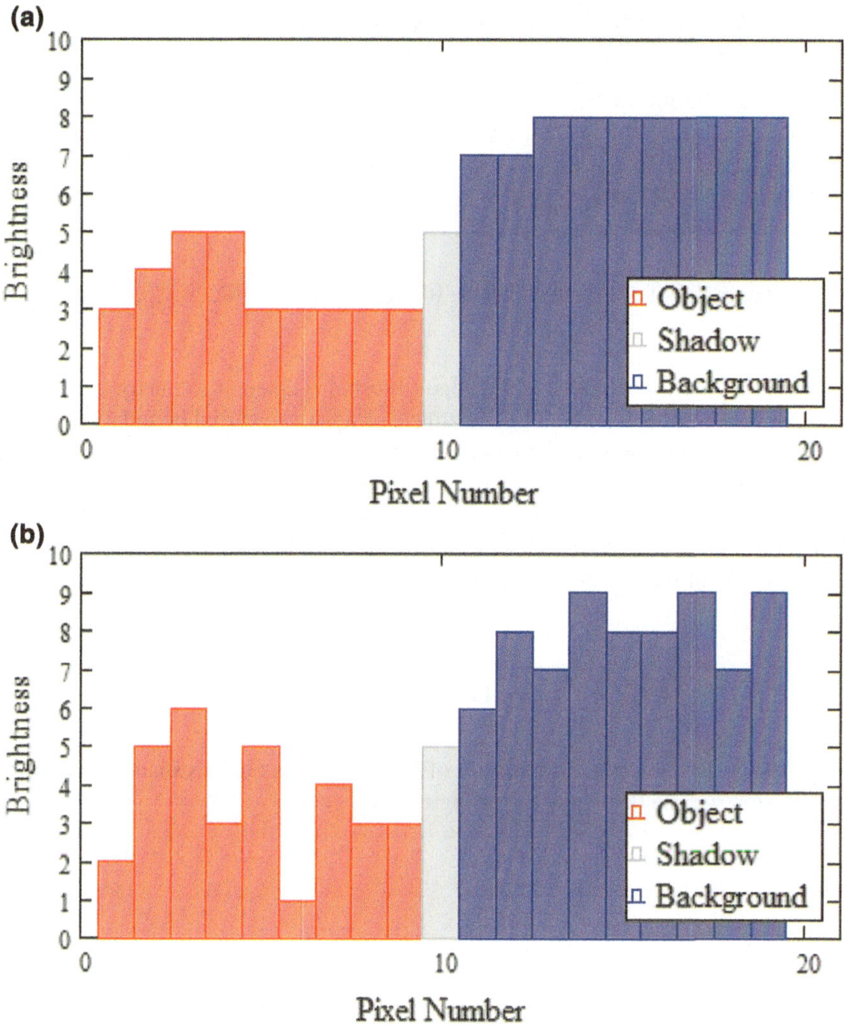

Fig. 1.24 The histogram of brightness in the slice perpendicular to the boundary of the object with the background: for a contrast object, where $a = 4$, $b = 2$ (**a**) and for a non-contrast object, where $a = 3$, $b = 4$ (**b**)

However, in practice we often encounter the situation with a non-contrast image. Nevertheless, in many practically significant cases this problem can be solved by applying the system approach proposed in this work. Thus, for example, Fig. 1.24b shows the case of a low-contrast image. After smoothing, the image will already be a contrast (Fig. 1.24a), but only using the rule (1.29). However, making use of the classical convolution rule does not allow us to achieve such a separability of the object's boundary from its interior.

1.4.5 Margin of Safety of an Edge Detector

When using the proposed image processing model, the densities of the contrast distributions can be not only simply separated (Fig. 1.23): even a gap may appear between the object interior's maximal contrast estimate $c_{\max}$ and the object boundary minimal contrast estimate $c'_{\min}$: $r = c'_{\min} - c_{\max}$, $r > 0$, $c'_{\min} > c_{\max}$. In such a situation, the choice of the threshold at the level

$$T = c_{\max} + 0.5 \cdot r = c_{\max} + 0.5 \cdot (c'_{\min} - c_{\max}) = 0.5 \cdot (c'_{\min} + c_{\max}), \tag{1.31}$$

will turn the margin of safety of the edge (threshold) detector to maximum. In this relation, with respect to excluding the effects of false and partial boundary segmentation, the maximization of safety margin, all other things being equal, is one of the most important factors in ensuring the stability of boundary pixel segmentation by contrast. Therefore, at first glance, an excessive smoothing of brightness distribution of an image at the preprocessing stage may actually be necessary to increase the safety margin of the threshold detector to the required level.

1.4.6 Basic Model of Edge Detector

Based on the adopted above definitions of line, as well as object's shadow, boundary and interior, we use below the following main detectors:

(1) the interior detector

$$F_i = \begin{cases} 1, \; if \; (c \leq T), \\ 0, \; otherwise; \end{cases} \tag{1.32}$$

In contrast to an edge detector, for detection of an object interior it is not required a mask which simultaneously covers both the object and the background; so if we are interested in detecting the pixels of the interior, the mask radius can be reduced to $\sqrt{2}$, or even to 1; the use of masks of such a radius is identical to the use of a neighborhoods of the same radius;

(2) the edge or line detector

$$F_b = \begin{cases} 1, \; if \; (g \leq T) \wedge (c > T), \\ 0, \; otherwise; \end{cases} \tag{1.33}$$

if the pixel (ξ, η) identified using this detector contains among the pixels covered by the detector mask at least one pixel of the interior identified by the use of detector (1.32), then pixel (ξ, η) is considered to be a boundary pixel. Otherwise, if there are no interior pixels in the neighborhood of $O_\varepsilon(\xi, \eta)$, then pixel (ξ, η) is identified as a line pixel;

(3) the object shadow detector (for a pronounced shadow)

$$F_s = \begin{cases} 1, \; if \; (g > T) \wedge (c > T), \\ 0, \; otherwise. \end{cases} \tag{1.34}$$

In comparison with analogs of the form (1.5), the application of the threshold detector (1.33) allows not only to detect the contrast pixels of an image $(c > T)$, but also to check for the presence of non-contrast pixels of object's interior in the neighborhood $(g \leq T)$, what is important for clipping residual noise and pronounced shadow pixels for reducing the levels of false and excessive segmentation.

At the same time, it should be understood that due to the use of a mask with a non-minimal radius and the presence of distortions in the brightness distribution of the image, the application of the detector (1.33) will lead to detection of not only the boundary pixels, but also of object's shadow and near-boundary pixels.

Meanwhile, the detector (1.33) has a limitation on its use. It cannot be used if in the neighborhood of the pixel (ξ, η) of the boundary (line) there can be simultaneously two pixels: very bright $(u_i > T)$, and very dark $\left(u_j < -T\right)$ which belong to two different objects.

In practice, this situation is extremely rare, since the radius of the mask is most often determined by the value 2.

If such a situation is possible, one should either return to using of the next classic type detector

$$F_b = \begin{cases} 1, \; if \; (c > T), \\ 0, \; otherwise; \end{cases} \tag{1.35}$$

or, instead of calculating the series (1.27), proceed to consideration of the following signless assessment of the contrast

$$u_i = |f(\xi, \eta) - f(\xi + M[1, i], \eta + M[2, i])|, i = 1, \ldots, n, \tag{1.36}$$

At this, the levels of contrast c and homogeneity g are calculated directly—by the values of the obtained contrast spectrum as follows

$$g = \min_i \{u_i\}_i, c = \max_i \{u_i\}_i, i = 1, \ldots, n. \tag{1.37}$$

The use of such simplifications will lead to a certain increase in the level of excessive segmentation of boundary pixels in comparison with the use of a detector (1.33). The first of these approaches (viz. (1.35)) is more computationally efficient. The second one allows us to additionally reduce the level of excessive segmentation of boundary pixels in comparison with the detector (1.35), what is better from the viewpoint of post-processing due to accelerating the time-consuming procedure of non-maximal suppression, which is used for thinning the boundary [29].

As the results of simulation and the conducted studies show, the use of the signless contrast estimate (1.36) compared to using the estimate (1.27) leads to a noticeable increase in the number of segmented shadow pixels. Therefore, unless otherwise specified, the use of the signed estimate (1.27) is assumed.

For additional filtering of shadow pixels, we use the contrast balance detector

$$Bal = \begin{cases} 1, \ (g/c) > T_b, \\ 0, \ otherwise, \end{cases} \tag{1.38}$$

where $CRB = (g/c)$ is the contrast balance coefficient, which takes relatively small values at the object boundary, since the homogeneity value of g after filtering and smoothing will tend to zero, and for strongly pronounced shadow pixels, on the contrary, relatively large values.

1.4.7 *Indicator Detector of Object Boundary*

Applying a threshold detector may lead to simultaneous segmentation of contrast boundary pixels of object and background. In general, these boundaries will go parallel to each other and, in the absence of a shadow, may be fragmentary linked to each other.

Suppose that in the neighborhood of the object boundary the background pixel brightness is distributed uniformly; in this situation, the indicator detector Cr_I can be used to filter falsely segmented boundary pixels of the background. The algorithm of this detector is intended just for the sign estimation of contrast; it operates as follows.

First, we calculate the sum S and the sum of modules Sm of only significant differences in brightness $\mathrm{u_i}$, $|\mathrm{u_i}| > T$, of the contrast spectrum U of the boundary pixel that was identified with the use of detector (1.33).

Further on, if $|S| = Sm$, then all significant differences in brightness are of the same type; in such a situation, to identify the pixel of the boundary, it suffices to check the fulfillment of the specified condition $S > 0$, or $S < 0$. In the first situation, all significant brightness differences u_i are positive, and in the second—negative. Therefore, in the first situation we have a light object on a dark background, and in the second—a dark object on a light background.

If the background is homogeneous, with the use of the indicator detector you can filter out all falsely segmented pixels of the background boundary. Otherwise, the value of the indicator function (the sign of the sum Sm) can be effectively used for adequate binding of pixels of the same type into a component of object/background boundary without mixing them with each other.

1.4.8 Algorithm of Statistical Identification of the Boundary

In non-trivial data analysis conditions, to improve the quality of segmentation using the detector (1.33), you can use additional information that is contained in the contrast distribution of pixel $U = \{u_i\}_{i=1,\dots,n}$ for the purposes of alternative (or additional) statistical classification of an image pixel.

The main idea of this approach consists in checking whether the brightness distribution in the neighborhood of $O_\varepsilon(\xi, \eta)$ corresponds the definition of boundary.

So, suppose that the series $U = \{u_i\}_{i=1,\dots,n}$ is ordered in ascending order

Select in it the series of non-positive values $U = \{u_i\}_{i=1,\dots,p}$ and the series of non-negative values $U = \{u_i\}_{i=q,\dots,n}$.

Find the dispersion of these series of brightness

$$\sigma_{(1)}^2 = \frac{1}{p}\sum_{i=1}^{p} u_i^2, \sigma_{(2)}^2 = \frac{1}{n-q+1}\sum_{i=q}^{n} u_i^2. \tag{1.39}$$

In such a situation, the simplest approach to detecting the boundary is to check the fulfillment of criterion (1.33), where instead of differential contrast estimates c and g, their analogues are used—the RMS deviations $\sigma_{(1)}$ and $\sigma_{(2)}$.

In order to minimize the effect of masking the extreme observations, it may be advisable in this criterion to move from considering the standard deviation to considering the variances $\sigma_{(1)}^2$, $\sigma_{(2)}^2$, since the quadratically enlarged extreme observations will be seen much better.

But this approach has one major drawback. The contrast estimates between object pixels may have different signs. Consequently, they will fall into different groups. To minimize this effect, it is advisable to do as follows.

First, the primary set of pixels is formed, which is determined by homogeneity in the sense that when constructing an adaptive filter, in the set U the given number of contrast values is revealed which have the minimal modulo. The primary set can also be built by applying a threshold detector to the set U for obtaining the limiting values p and q of the above-described indices; and these are those limits which define the elements of the primary set as a series of elements of the set U.

Subsequently, the primary set is constructed using one of these two rules, or both, depending on the learning outcomes.

Then, the dispersion of the remaining positive $(q + 1, \dots, n)$ and negative $(1, \dots, p - 1)$ parts of the distribution $U = \{u_i\}_{i=1,\dots,n}$ are estimated. If the dispersion of the negative part is smaller $\sigma_{(1)} < \sigma_{(2)}$ (at $\sigma_{(1)} < T$), then the primary set is attached to this part, otherwise (at $\sigma_{(2)} < T$) the primary set is attached to the positive part of the series. After that, the values of p and q are corrected, the variance (1.39) is calculated, and the boundary identification detector (1.33) is applied. If $\sigma_{(1)}, \sigma_{(2)} > T$, then the pixel is no longer considered.

For the purposes of subsequent analysis, it may be necessary to isolate the entire homogeneous part of the neighborhood. By default, such an operation will be performed according to the criterion (1.33) by joining the pixels to a homogeneous set one by one.

So, assume that to this moment a homogeneous set $U = \{u_i\}_{i=1,\ldots,p}$, $p < n$, is constructed with the dispersion

$$\sigma^2 = \frac{1}{p}\sum_{i=1}^{p} u_i^2 = \frac{1}{p}\sum_{i=1}^{p} (f(\xi,\eta) - f_i)^2. \tag{1.40}$$

Does the value u_{p+1} belong to this homogeneous set or not? If the condition is met

$$|u_{p+1}| > k \cdot \sigma, \tag{1.41}$$

then it is considered that the hypothesis re to the extremity is confirmed, and the observation u_{p+1} should not be included in the set $U = \{u_i\}_{i=1,\ldots,p}$ as atypical. Otherwise, we assume $p = p + 1$ and may transfer to the following check.

Without loss of generality we can assume that the verification of observation u_{q-1} for the set $U = \{u_i\}_{i=q,\ldots,n}$ is carried out similarly; at this, instead of condition (1.41), the following is checked

$$|u_{q-1}| > k \cdot \sigma. \tag{1.42}$$

In the expressions (1.41) and (1.42), we used the $k\sigma$-criterion because most of the well-known statistical criteria for excluding outliers (for example, the Tietjen-Moore and Grabbs tests) presume that the distribution under consideration obeys the normal distribution, whereas to reliably establish the type of the distribution law for a relatively small neighborhood is simply impossible.

Using the $k\sigma$-test is also convenient for another reason. Due to the possibility of changing the parameter k (together with the possibility of choosing the estimation rule—relative to a dispersion or rms), one can effectively tune into the features of the brightness distribution of objects and background.

As a result of formation of homogeneous set in the neighborhood, for the purpose of the subsequent construction of connected components of the boundary, it is advisable to construct a descriptor of boundary pixel that stores information about the homogeneous part of the neighborhood. In its simplest form, a descriptor is a matrix of pixel coordinates of the homogeneous part of the neighborhood.

Consequently, we arrive at the following algorithm for statistical identification of the boundary.

Step 1. Arrange the series $U = \{u_i\}_{i=1,\dots,n}$ in the ascending order.
Step 2. Select the homogeneous primary set H in U.
Step 3. Combine the set H with the negative or positive part of U and, if possible, determine the boundaries of p and q, and classify the pixel. The combined set is also be denoted H.
Step 4. If the pixel under test belongs to the boundary, we augment it while the set H remains homogeneous.
Step 5. Construct a descriptor for the set H.

In comparison with analogs, the construction of a descriptor is in many respects a decisive factor in the efficiency of constructing components of the object boundary, since the descriptor exhaustively characterizes intra-group pixel connections in the respective neighborhood.

To minimize the computational complexity, it is advisable to first apply a threshold detector of the form (1.33), possibly with a lowered threshold, and to select a relatively small number of potentially boundary pixels in the image; only after that a more time-consuming statistical estimation algorithm should be used for them, if necessary. And only on completion of these operations one is recommended to start pixel marking and construction of connected components of the boundary.

At this, in order to minimize the computational complexity, the descriptors are advisable to build only for the boundary pixels. For minimization of memory costs, it is advisable not to store the descriptors, but to build and immediately to use them as a part of the chosen method of constructing the boundary component; e.g. within the wave method.

Notice, that for the given neighborhood a homogeneous set of pixels and a descriptor can be constructed with the use of detector (1.33). In conditions of sufficiently high contrast, it is sufficient for this to mark the pixels of neighborhood, the absolute contrast values of which are less than the specified threshold value. However, this approach is much more coarse and suitable only for high contrast conditions.

1.4.9 Direct and Indirect Boundary Identification Strategies

Until now, the work considered a strategy of direct identification of the boundary, based on the use of object edge detectors. This strategy is good for relatively high contrast conditions, when the result of segmentation is characterized by quite low levels of false, redundant and partial segmentation.

In conditions of very low contrast, it may be advisable to modify and/or to supplement the direct strategy.

In this respect, perhaps the simplest thing is to select all the pixels of the boundary of one object, which can be easily implemented using the wave method in combination with the analysis of descriptors, and only then to proceed the following one. The advantage of this approach consists in absence of need to separate the closely allocated boundaries. To construct the required approximation of the object boundary component in the form of an 8-connected line one pixel wide, immediately after segmentation the boundary can be traced with elimination of boundary branches and loops [29].

In low contrast conditions, by analyzing the image pixel descriptors one can find the interior of the object, attach pixels of a homogeneous neighborhood to it (without directly detecting the boundary), and only then apply the respective method for detection of boundary pixels. At this, instead of (or in addition to) direct application of a boundary detection method, with the aim to improve the quality of the result you can apply the approach described above with the trace of the boundary.

The proposed models, algorithms and detectors constitute a spectrum of tools that can be used not only independently, but also in combination with existing models and algorithms in order to build composite strategies that allow us to effectively detect the internal and boundary pixels of objects and their boundaries by taking into account the features of applications.

1.5 Conclusion

As it shows the analysis of the current state of the issue, including the results of modeling and numerous experiments, the efficient boundary segmentation, ensuring a qualitative and productive construction of connected and closed boundary components of one pixel width, is possible only in the context of a system approach to detecting of object boundary pixels.

Considering the most common situation of the absence of distortions of a special type, in order to achieve high-quality segmentation of an object boundary it is advisable to adhere to the procedure proposed in the work, including the stages of noise filtering, smoothing and segmentation.

In this relation the filtering is used for quick suppression of extreme observations, while the smoothing stage following it is used for approaching the brightness distributions of image objects to the average levels.

An important feature of the proposed algorithms is filtering and smoothing in a homogeneous neighborhood. This approach allows us not to expand the shadow of objects, which contributes to a significant simplification of the subsequent image processing after segmentation.

Building and using models for constructing a homogeneous neighborhood of a pixel stands out through all the sections of the work. And this is no accident. Even an approximate preliminary pixel classification underlies the efficient processing of brightness distribution in a pixel neighborhood in terms of both filtering and segmentation.

At the same time the filtering and smoothing are considered optional stages of image processing. They should be applied in such a situation (and in such a volume), when distortions of the brightness distribution of objects do not allow you to effectively segment the boundaries of objects and lines.

In the course of presenting this work, it is shown that the construction of an edge detector (and detector mask) should begin with the building of an adequate boundary profile, taking into account the presence and parameters of the object shadow, since only under such conditions one can adequately construct a model of an edge detector and tune it to the features of the applied problem.

Based on the simulation results, the models of mask and edge detector are proposed, which allow one to take into account the topological, geometric and photometric characteristics of objects and the properties of their shadow in order to obtain an adequate estimate of the contrast of the object's boundary relative to the background, not to the shadow.

For improving the efficiency of combining the boundary's pixels into an 8-connected component, a model for constructing a homogeneous set of pixels in a neighborhood of the considered one has been proposed with the aim to present the results as a pixel descriptor. This model is the foundation for the subsequent efficient binding of pixels into a boundary component.

At this, the direct and indirect boundary segmentation strategies are described for various visual field object contrast conditions.

With the aim to provide a formalized tools for setting the requirements to the models, detectors and data processing algorithms, as well as to analyze the results of their use and to adjust the parameters according to the conditions of the application problem, the unified system of indicators and quality criteria is developed for boundary segmentation by template.

Based on the use of these criteria and indicators, the principles of tuning of the parameters of data processing models are described, including the choice of type and tuning the parameters of mask and edge detector.

The simulation results allow us to conclude that with respect to global estimation of segmentation quality, the law of error multiplication is fair: the presence of one significant brightness distortion in the original distribution of certain pixel leads to significant distortions of the contrast of all pixels in its neighborhood and, moreover, entails problems in all subsequent segmentation stages.

In this regard, the foundation of global quality optimization is based on the system approach which requires consideration of all segmentation stages and methods that can be applied at them, with the aim to reveal and eliminate significant brightness distortions at early stages, even before segmentation, for well-timed preventing of appearance of inadequate contrast estimates. It also improves the computational efficiency of the subsequent segmentation, both by using more simple and efficient detectors, and by reducing the number of boundary breaks and eliminating the unjustified growth of number of false border components.

This way, the implementation of the principles of a system approach and adaptive data processing allows one to achieve high quality boundary segmentation with time costs corresponding to the theoretical complexity of the task and acceptable for the considered applications.

References

1. Goodfellow I, Bengio Y (2016) Aaron Courville deep learning. MIT Press, 787p
2. Norvig P, Russell S (2016) Artificial intelligence: a modern approach, Global Edition. Pearson Education Limited, 1152p
3. Deng L, Yu D (2013) Deep learning methods and applications. Found Trends® Signal Process 7(3–4):197–387
4. Zhengbing H, Bodyanskiy Y, Tyshchenko O, Tkachov V (2017) Fuzzy clustering data arrays with omitted observations. Intell Syst Appl 9(7):24–33
5. Ruban I, Smelyakov K, Martovytskyi V, Pribylnov D, Bolohova N (2018) Method of neural network recognition of ground-based air objects. In: IEEE 9th International Conference on Dependable Systems, Services and Technologies (DESSERT), 24–27 May 2018. pp 589-592. https://doi.org/10.1109/dessert.2018.8409200
6. Smelyakov K, Pribylnov D, Martovytskyi V, Chupryna A (2018) Investigation of network infrastructure control parameters for effective intellectual analysis. In: IEEE 14th international conference on advanced trends in radioelecrtronics, telecommunications and computer engineering (TCSET), 20–24 Feb 2018. pp 983–986. https://doi.org/10.1109/tcset.2018.8336359
7. Smelyakov K, Chupryna A, Yeremenko D, Sakhon A, Polezhai V (2018) Braille character recognition based on neural networks. In: IEEE second international conference on Data Stream Mining & Processing (DSMP), 21–25 Aug 2018, pp 509–513
8. Steger C, Ulrich M, Wiedemann C (2018) Machine vision algorithms and applications, 2nd edn. Wiley, Hoboken, 516p
9. Corke P (2017) Robotics, vision and control, 2nd edn. Springer, Heidelberg, 693p
10. Gonzalez RC, Woods RE (2018) Digital image processing, 4th edn. Pearson/Prentice Hall, 1168p
11. Forsyth DA, Ponce J (2015) Computer vision: a modern approach, 2nd edn. Pearson Education Limited, 792p
12. Ionescu RT, Popescu M (2016) Knowledge transfer between computer vision and text mining. Springer, Heidelberg, 250p
13. http://www.neurotechnology.com
14. Rattani A, Roli F, Granger E (2015) Adaptive biometric systems. Springer, Heidelberg 134p
15. Marcel S, Nixon MS, Li SZ (2014) Handbook of biometric anti-spoofing. Springer, Heidelberg, 281p
16. Bowyer KW, Burge MJ (2016) Handbook of iris recognition, 2nd edn. Springer, Heidelberg, 568p
17. Schmalstieg D, Höllerer T (2016) Augmented reality: principles and practice. Addison-Wesley Professional, Boston, 528p
18. Prince JL, Links J (2015) Medical imaging signals and systems, 2nd edn. Pearson Education, NJ, USA
19. Toennies KD (2017) Guide to medical image analysis, 2nd edn. Springer, Heidelberg, 589p
20. Lu L, Zheng Y, Carneiro G, Yang L (2017) Deep learning and convolutional neural networks for medical image computing. Springer, Heidelberg, 326p
21. Batchelor BG (2012) Machine vision handbook. Springer, London, New York, p 2290
22. Beyerer J, León FP, Frese C (2016) Machine vision. Springer, Heidelberg, 798p
23. Klette R (2014) Concise computer vision. Springer Science & Business Media, 429p
24. Szeliski R (2011) Computer vision: algorithms and applications. Springer London Ltd., 812p
25. Prince SJD (2012) Computer vision: models, learning, and inference. Cambridge University Press, Cambridge, 598p
26. Nixon M, Aguado A (2012) Feature extraction and image processing for computer vision, 3rd edn. Academic Press is an imprint of Elsevier, 632p
27. Bishop CM (2011) Pattern recognition and machine learning. Springer, New York, 738p
28. Parker JR (2010) Algorithms for image processing and computer vision, 2nd edn. Wiley Publishing Inc., Hoboken, 504p

29. Sonka M, Hlavac V, Boyle R (2014) Image processing, analysis, and machine vision, 4th edn. Cengage Learning, 896p
30. Smelyakov K, Ruban I, Sandrkin D, Martovytskyi V, Romanenkov Y (2018) Search by image. New search engine service model. In: International scientific and practical conference «Problems of Infocommunications Science and Technology», Kharkiv, Ukraine, 9–12 Oct 2018, pp 181–186
31. Churyumov G, Tokarev V, Tkachov V, Partyka S (2018) Scenario of interaction of the mobile technical objects in the process of transmission of data streams in conditions of impacting the powerful electromagnetic field. In: 2018 IEEE second international conference on Data Stream Mining & Processing (DSMP), 21–25 Aug 2018 (Lviv, Ukraine), pp 183–186
32. Zaitseva E, Levashenko V (2017) Reliability analysis of multi-state system with application of multiple-valued logic. Int J Qual Reliab Manag 34(6):862–878
33. Lemeshko O, Arous K, Tariki N (2015) Effective solution for scalability and productivity improvement in fault-tolerant routing . In: Proceedings of second international IEEE conference «Problems of Infocommunications. Science and Technology» (PICS&T-2015), Kharkiv, Ukraine, 13–15 Oct 2015, pp 76–78
34. Demaagd K, Oliver A, Oostendorp N, Scott K (2012) Practical computer vision with SimpleCV. O'Reilly Media Inc., 254p
35. Kaehler A, Bradski GR (2017) Learning OpenCV 3. O'Reilly Media Inc., USA, 1024p
36. http://www.photo-war.com/photo/shop/5394.htm

Chapter 2
Swarm Methods of Image Segmentation

Igor Ruban and Hennadii Khudov

Abstract A distinctive feature of the thematic image segmentation of onboard opto-electronic surveillance systems is the search for rational solutions in the multidimensional space of alternatives. Searching for rational solutions in a multidimensional space of alternative is peculiarized by non-linearity, non-differentiation, multi-extremality, ravine surface, lack of the analytic expression of objective functions, high computational complexity, high dimensionality of the search space, and the complex topology of the region of acceptability. Finding a solution by exact methods of optimization is complicated by non-linearity, non-differentiation and lack of the analytic expression of the objective function. High computational complexity, high dimensionality of the search space, and the complex topology of the region of acceptability lead to wasting valuable time when using precise optimization methods. Currently, methods of searching for global extremum are being developed, which provide the convergence to the exact solution of the optimization problem, provide the optimal (minimum or maximum) value of the objective function. Such methods include meta-heuristic optimization methods, which, unlike classical optimization methods, can be used even if there is no information about the nature and properties of the objective function. Meta-heuristic methods have the following features: they manage the process of finding the optimal solution; they efficiently study the search space to find the optimal solution; they use simple local search procedures and complex training processes; they are approximate methods and, as a rule, non-deterministic; they take into account the probability of trapping in a limited search space; they are universal (solve various application problems); they use a priori information to find the optimal solution. Let us detail the use of the variety of meta-heuristic optimization methods (ant colony optimization (ACO) and artificial bee colony optimization (ABC) for the thematic image segmentation of onboard optical and electronic surveillance systems.

I. Ruban
Kharkiv National University of Radio Electronics, 14 Nauky Ave, Kharkiv 61166, Ukraine
e-mail: ihor.ruban@nure.ua

H. Khudov (✉)
Ivan Kozhedub Kharkiv National Air Force University, 77/79 Sumska Str, Kharkiv 61023, Ukraine
e-mail: 2345kh_hg@ukr.net

V. Mashtalir et al. (eds.), *Advances in Spatio-Temporal Segmentation of Visual Data*, Studies in Computational Intelligence 876,
https://doi.org/10.1007/978-3-030-35480-0_2

Keywords Image segmentation · Ant colony optimization · Artificial bee colony optimization · Onboard system · Optic-electronic surveillance · Object of interest · Fitness function

2.1 The ACO Method of the Thematic Segmentation of Images of Onboard Optical and Electronic Surveillance Systems

In the simplest case, the image segmentation can be represented as a set of the following areas of movements of agents (ants) (Fig. 2.1): route starting point (RSP), straight sections, route finishing point (RFP). Straight sections come across route turning points (RTP), where the direction of the agent movement changes. Hereafter, believe that the location of RSP, RFP and RTPs completely determine the route of the agent movement.

Moving along each of the route sections and taking turns in selected RTPs have certain troubles and requires some cost of resources, which makes one route more advantageous than another. Since there is a great variety of RTPs locations, the number of possible routes will be extremely large, which makes the selection of the route of movement difficult. Let us demonstrate how to build a route using the simplest ACO method.

The ACO method is based on the simulation of the natural mechanism a colony of ants (agents) use for finding the shortest way to the source of food. The self-organization of the system is provided by the low-level interaction of agents, and agents exchange only local information; to transfer information, agents use a special secretion, pheromone, which each agent deposits along the route. The following agent, who is near the route of the first one, feels pheromone and will, with a high

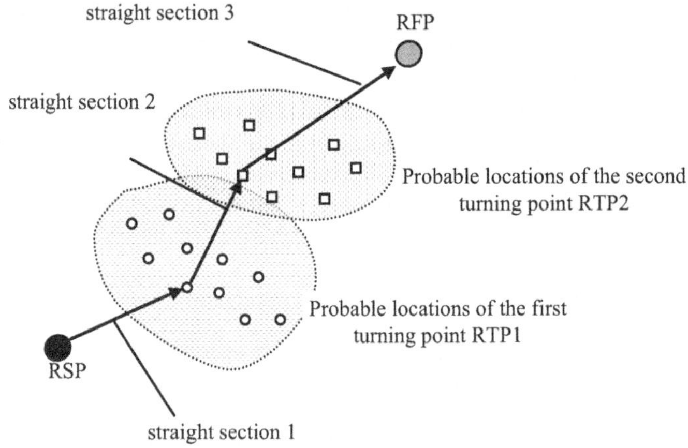

Fig. 2.1 A sample route of the agent movement while segmenting the image

degree of probability, go on along the route of the first agent, excreting the secretion (increasing the level of its concentration along the route). The higher the concentration of pheromone along the route, the greater the attractivity of this route for following agents. The distribution of pheromone in the environment is like a dynamic memory of a system. Each agent at a certain point in time perceives and changes one cell of this memory—the pheromone level in the vicinity of the point where the agent is located.

The concentration of pheromone deposited along the route is proportional to the attractivity (quality, efficiency) of the route. The more attractive the route, the greater the concentration of pheromone along it, as a result, the best routes are stored in the global memory of the colony of agents and are more likely to be selected by the following agents.

Over time, pheromone evaporates, providing feedback. As noted above, the concentration of pheromone will be gradually increased along attractive routes, and the pheromone evaporation rate is constant, so in some time unsuccessful routes will disappear, and more and more agents will travel only along successful routes. The use of feedback (evaporation) prevents the early convergence of solutions—the selection of the same suboptimal route by agents.

In the simplest ACO method, in each iteration of the iterative process, m agents search for a solution and renew pheromones on the selected route. While image segmenting, each m-th agent starts moving from RSP, sequentially come through the selected RTPs and finish moving in one of RFPs. RTP is selected from J probable ones on the basis of the probabilistic rule which determines $P_i^m(t)$ probability of the m-th agent coming in i-th RTP taking into consideration the attractivity of the i-th section of L_i route and the pheromone concentration F_i in this section at time t as follows (Eq. 2.1):

$$P_i^m(t) = \frac{F_i(t)^\alpha \cdot L_i^\beta}{\sum_{j=1}^{J} F_j(t)^\alpha \cdot L_j^\beta}, \tag{2.1}$$

where α and β are the parameters that determine the pheromone weight and the attractivity of the section ("greedy" method).

If $\alpha = 0$, agents come into the nearest RTP at each step and the ACO method of segmentation becomes a "greedy" method of the classical optimization theory. When $\beta = 0$, only the pheromone effect is taken into account, which will quickly lead to a suboptimal solution. Probabilities of RTP selection are also found according to Eq. (2.1). The selection is made according to the "roulette wheel" principle. It can be implemented, for example, by dividing a segment with S length into J parts with the length proportional to $P_i^m(t)$, generating a random number uniformly distributed in the interval [0, S] and selecting RTP depending on which part of the segment S by its ordinal number a random number falls.

Assume, that the attractivity of a section along L_i route in the ACO method is inversely proportional to the efforts for passing the section, that is (Eq. 2.2):

$$L_i = \frac{1}{D_i}, \tag{2.2}$$

where D_i is the length of the i-th section of the route.

At the beginning of the iterative process, the amount of pheromone in the route sections is assumed to be the same and equal to little number F_0. After each iteration, the pheromone concentration in the sections selected by agents is renewed according to Eq. (2.3):

$$F_i(t+1) = (1-\rho)F_i(t) + \sum_{m=1}^{M} \Delta F_i^m, \tag{2.3}$$

where $\rho \in [0, 1]$ is the rate of pheromone evaporation; ΔF_i^m is the pheromone concentration in the i-th section of the route of the m-th agent.

As a result of a certain number of iterations, the most attractive routes are selected according to the maximum concentration of pheromone. Pheromones on unattractive routes gradually dry out and unattractive routes disappear.

The efficiency of the ACO method will be checked using a test case. To visualize the results in all examples, the rectangular coordinate system was used where RSP, RFP and RTPs are in the same horizontal plane. The attractivity of the route sections was calculated according to Eq. (2.2). The results of calculations are presented in Fig. 2.2.

Source data:

- a number of RSP $N_{RSP} = 1$;
- a number of RFP $N_{RSP}= 1$;
- a number of probable first RTP (RTP1) $N_{RTP1} = 20$;
- a number of probable second RTP (RTP2) $N_{RTP2} = 20$ (RST, RFP, RTP1 and RTP2 locations are shown in Fig. 2.2);
- method "greediness" $\beta = 1$;
- pheromone weight $\alpha = 2$;
- a rate of pheromone evaporation $\rho = 10^{-3}$;
- a number of method iterations $N = 400$;
- a number of agents in an iteration $m = 10$;
- the initial amount of pheromone $F_0= 10^{-2}$.

As a result of N iterations, using the ACO method of image segmentation, the agent route is shown in Fig. 2.2 as a bold solid line. The dashed line shows the optimal route of agents during segmenting the image (reference segmentation). The analysis of Fig. 2.2a indicates that the route of agents resulted from the use of the ACO method, is slightly different from the optimal one shown in Fig. 2.2a as the dashed line. However, as it is shown in Fig. 2.2b, the method can be implemented with unsuccessful results, which indicates the need for further studies of the possibility of applying the advanced ACO method to solve the problem of segmentation.

The advanced ACO method of segmentation is the development of the classical ACO method. It is peculiarized by the fact that only the best agents increase the level

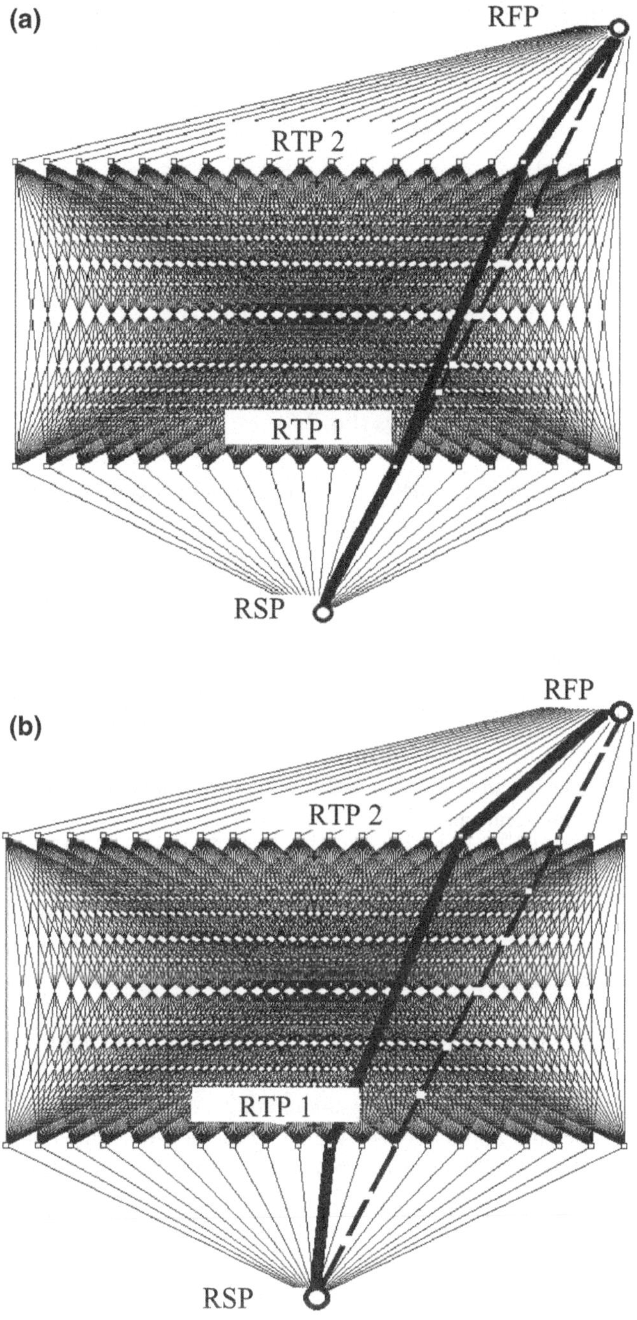

Fig. 2.2 The results of the ACO method of image segmentation for the source data of the test case: **a** the route of the agents slightly differs from the optimal route; **b** the route of agents is significantly different from the optimal route

of pheromone along their routes and that the level of pheromone along the routes is limited. The pheromone level along the routes can be renewed according to Eq. (2.4):

$$F_i(t+1) = \left[(1-\rho)F_i(t) + \Delta F_i^{best}\right]_{F_{\min}}^{F_{\max}}, \tag{2.4}$$

where $F_{\max}$ and $F_{\min}$ are the upper and lower limits of the pheromone level; $[x]_b^a$ is an operator determined by Eq. (2.5):

$$[x]_b^a = \begin{cases} a, \text{ if } x > a \\ b, \text{ if } x < b \\ x, \text{ otherwise,} \end{cases} \tag{2.5}$$

and ΔF_i^{best} is determined as follows (Eq. 2.6):

$$\Delta F_i^{best} = \begin{cases} \frac{1}{L_{best}}, & \text{when i---the best route at the iteration;} \\ 0, & \text{otherwise,} \end{cases} \tag{2.6}$$

and L_{best} is the length of the best agent route. It can be either the best route found in the current iteration—iteration best, L_{ib} or the best solution that has been found since the beginning of the method operation—best-so-far, L_{bs}.

The functional check of the advanced evolutionary method was checked using test case data. In contrast to the classical ACO method, when the advanced ACO method was used during the image segmentation, the optimal route of agents was found in all implementations. Figure 2.3 sequentially shows the operation of the method after 40, 80, 200 and 400 iterations.

Figure 2.3 shows that after 40 iterations, the level of pheromone along all the routes is still slightly different from the initial level F_0, however, the agents have already marked two best routes with pheromones (more bolded lines). After 80 iterations the level of pheromone along all the routes, except for the best ones, significantly reduced due to evaporation (the corresponding lines have become thinner), the three best routes are clearly distinguished, between which there will be a "controversy" from now forward. After 200 iterations, the best route is already clearly seen although several routes are still in the state of "controversy"; after 400 iterations, the best route (which is the optimal one) is obviously dominating.

Let us consider the application of the ACO method to segmenting the image obtained from the onboard optoelectronic surveillance system [1–3]. Figure 2.4 shows the source image obtained from the onboard optoelectronic surveillance system. The colour image made by the onboard optoelectronic surveillance system will be considered as the source one (Fig. 2.4). The image dimensions are 868 × 847 pixels, the image is presented in the RGB colour space.

The result of the source image segmentation (Fig. 2.4) by the ACO method is given in Fig. 2.5.

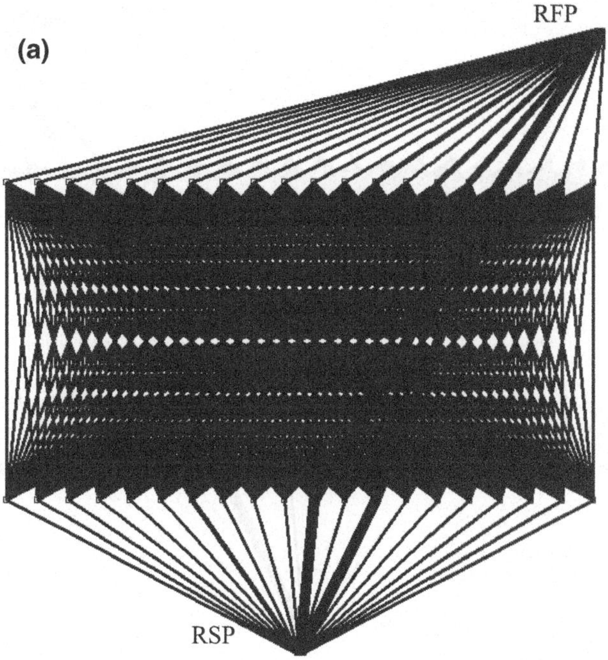

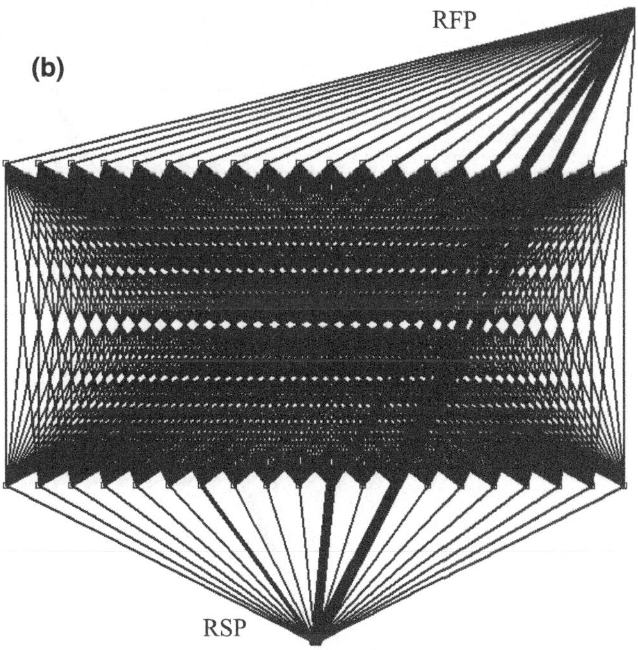

Fig. 2.3 The results of the operation of the advanced evolutionary method of the image segmentation for the source data of the test case: **a** after 40 iterations; **b** after 80 iterations; **c** after 200 iterations; **d** after 400 iterations

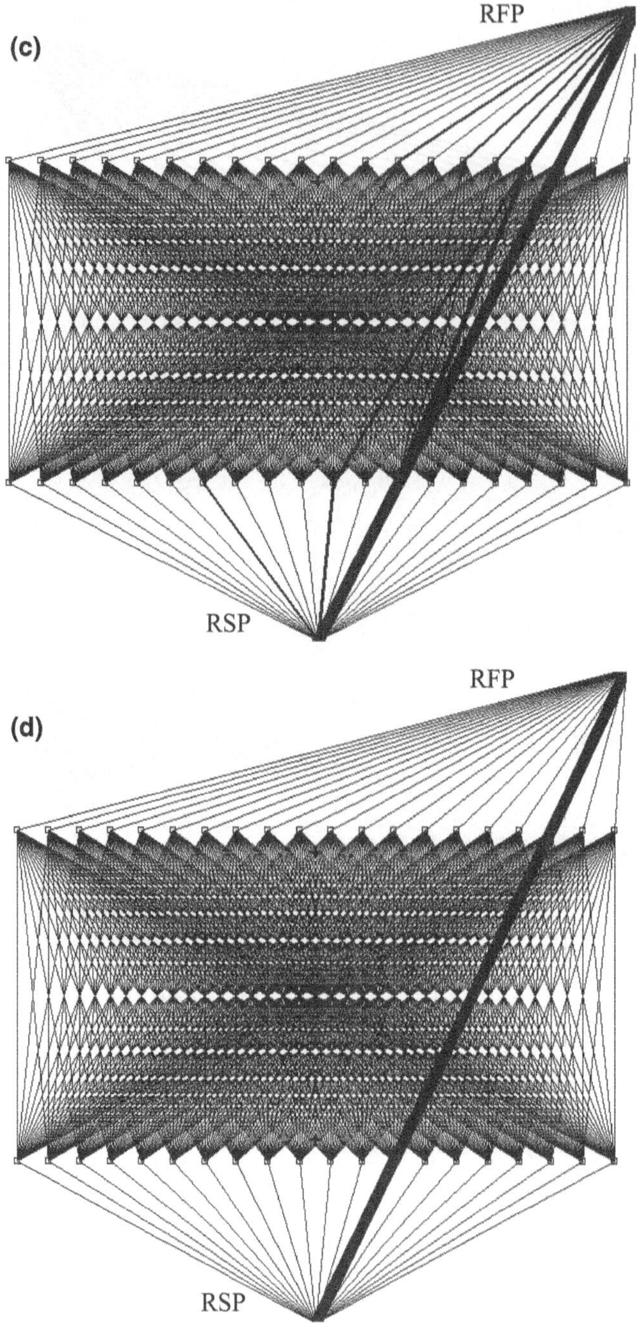

Fig. 2.3 (continued)

Fig. 2.4 The source image [1–3]

In the segmented image (Fig. 2.5), some objects of interest can be identified in the future, for example (Fig. 2.6):

- containers with oil or fuel for airplanes;
- planes that escaped the strike;
- planes damaged or destroyed, etc.

The comparison of Figs. 2.4 and 2.6 shows that the contours of the main objects of interest marked in Fig. 2.6 coincide with the boundaries of the objects in the original image (Fig. 2.4).

The decryption of these objects of interest, recognition, thematic classification, etc., is the subject of further research and lies beyond this work.

A large number of marked contours of small objects should be noted in the segmented image. An example of such an area with a large number of small objects is shown in Fig. 2.7 and marked like an ellipse.

Fig. 2.5 The result of the source image segmentation (Fig. 2.4) by the ACO method

Similar areas can be seen in other parts of the segmented image. The direction of further research is to reduce the areas with a large number of small objects to decrypt objects of interest in a more efficient way.

Thus, the probability of using the simplest and advanced ACO method to segment the image obtained from the onboard optoelectronic surveillance system has been considered in the Section.

The functional check of the simplest ACO method has indicated errors in segmenting. The functional check of the advanced ACO method using the data of the test case has testified the optimal way of agents while segmenting the image in all implementations.

Experimental studies of the segmentation of optoelectronic images have confirmed that the advanced ACO method is efficient. In the segmented image, probable objects of interest were marked as an example, namely containers with oil or fuel for airplanes; planes that escaped the strike; planes damaged or destroyed, etc. It has been established that the marked contours of the main objects of interest coincide

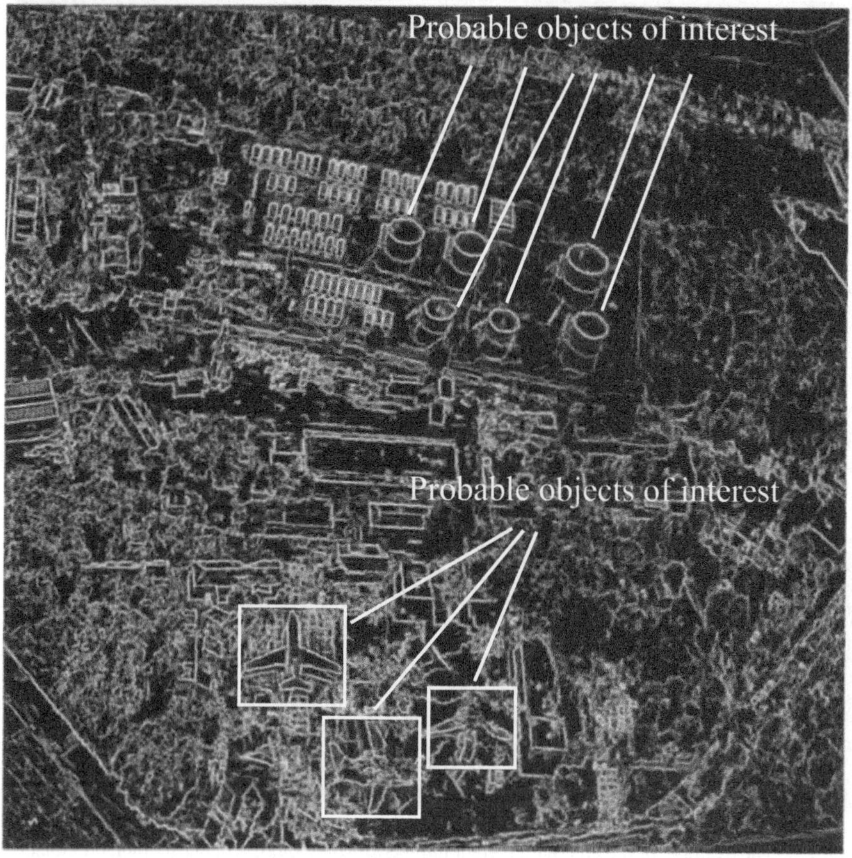

Fig. 2.6 Probable objects of interest in the segmented image

with the boundaries of the objects in the source image. The efficiency of the ACO method has been assessed visually.

One of the important advantages of the ACO method is the rapid adaptation to the conditions of variable light and a number of objects of interest of different types and classes that are in the image. Under these conditions, the ACO method of segmentation quickly finds a new optimal solution.

The main drawbacks of the advanced ACO method of segmentation are:

- a large number of marked contours of small objects in the segmented image;
- the need for a larger computing resource.

The advanced ACO method of segmentation can be used:

- in software and hardware complexes for processing images obtained from onboard optoelectronic surveillance systems (airborne, space-based);
- while improving technologies and means of processing optoelectronic images;

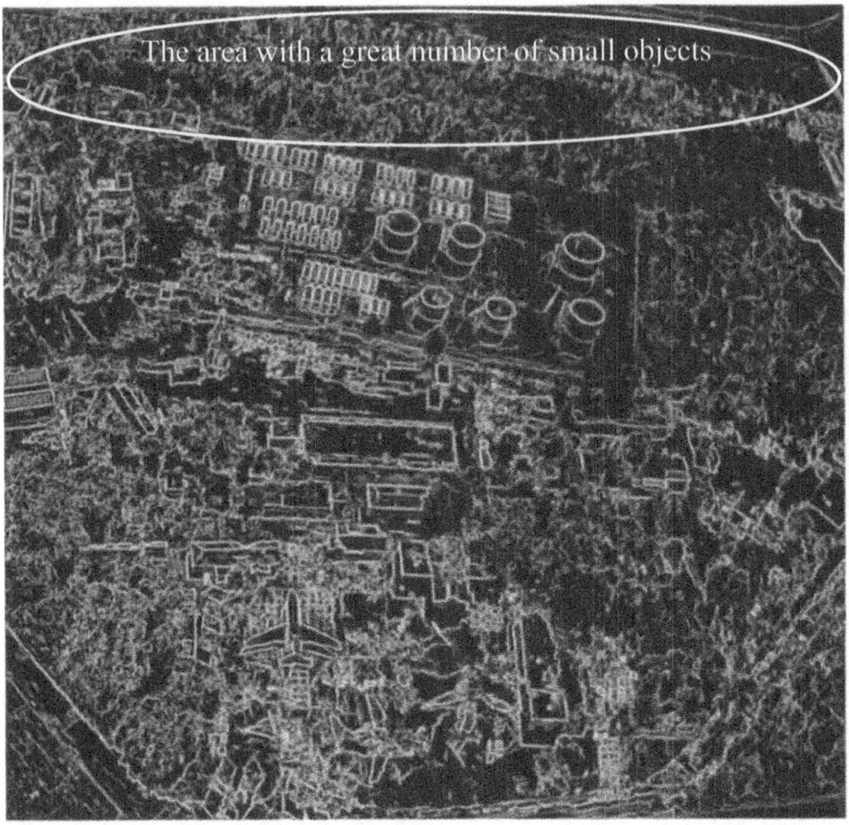

Fig. 2.7 The example of the area with a great number of small objects in the segmented image

- while processing optoelectronic images in optoelectronic stations of onboard complexes of aircraft protection systems and other safety-critical systems [4].

2.2 The ABC Method of the Thematic Segmentation of Images of Onboard Optoelectronic Surveillance Systems

The ABC method is based on the simulation of the behaviour of bees when they are harvesting nectar [3]. A swarm of bees sends several scouts in random directions to find nectar. Having returned, the scouts report about the areas where nectar was found and other bees fly there. In this case, the more nectar in the area, the more bees are sent to the areas. But the bees might deviate from the desired flight direction.

On returning, bees exchange information and new scout-bees and employed bees are sent.

The ABC method involves performing the following sequence of actions:

1. Determining the location of the nectar source.
2. Initializing the starting points for the bee-scouts.
3. Searching for new sources by employed bees.
4. Determining the objective function.
5. Repeating steps 1–4 until the decision stops improving.
6. Checking the termination condition. Termination criteria can include achieving the maximum permissible number of iterations, finding the necessary solution, the lack of a significant improvement in the value of the objective function within a certain number of iterations.
7. Memorizing the best source of nectar.

2.2.1 Mathematical Statement of the Task of Thematic Segmentation of an Optoelectronic Image

The segmentation of a source optoelectronic image (OEI) $f(x,y)$ is expression (2.7)

$$f(x, y) \rightarrow fs(x, y), \tag{2.7}$$

where $f(x,y)$ is the source OEI; $fs(x,y)$ is the segmented image.

Segmenting the source OEI $f(x,y)$ according to expression (2.7) involves imaging pixels (points) of the source image with coordinates (x, y) to a certain environment of features and introducing a metric (proximity measure) to this environment of features. The brightness in a certain colour space of pixels with coordinates (x, y) in the image $f(x, y)$ is used as a feature. And the brightness features such as discontinuity and uniformity are used. A metric is the distance between pixels in the colour space and the segmentation of the source image, according to (2.7), provides for segmenting $f(x,y)$ into B_i, which meets conditions (2.8):

$$\begin{cases} \bigcup\limits_{i=1}^{K} B_i = B; \\ B_i \cap B_j = \emptyset, \text{ for } i \neq j; \ \forall i, j = \overline{1, K}; \\ LP(B_i) = 1; \ \forall i = \overline{1, K}; \\ LP(B_i \cap B_j) = 0, \text{ for } i \neq j; \ \forall i, j = \overline{1, K}. \end{cases} \tag{2.8}$$

where B: $B = \{B_1, B_2, \ldots, B_K\}$ are segments in the image $fs(x,y)$; K is a number of segments B_i, $(i = 1, 2, \ldots, K)$; LP is the predicate determined on B, that takes the value "1—true—if and only if any pair of points from each segment of B_i satisfies some criterion of homogeneity—expression (2.9):

$$LP(B_i) = \begin{cases} 1, & if\ f(x_1, y_1) = \cdots = f(x_M, y_M); \\ 0, & otherwise, \end{cases} \quad (2.9)$$

where $(x_m, y_m) \in B_i$; $m = 1, 2, \ldots, M$; M is a number of points in the segment B_i, or

$$LP(B_i) = \begin{cases} 1, & for|f(x_m, y_m) - f(x_l, y_l)| \le T; \\ 0, & others, \end{cases} \quad (2.10)$$

where (x_m, y_m), (x_1, y_1) are random points of the segment B_i; T is a certain preset threshold.

The thematic segmentation of OEIs of onboard optoelectronic surveillance systems results in separating the image into artificial objects (military objects) and natural objects (background or scenery), that is, a number of segments in a segmented image is $K = 2$.

2.2.2 Summary of the Method of the Thematic Segmentation of Images of Onboard Optoelectronic Surveillance Systems

For the OEI thematic segmentation and separating the image into artificial and natural objects, the optimal threshold of the thematic segmentation should be determined. To determine the optimal threshold of the thematic segmentation, a swarm method of the artificial bee colony will be used.

To switch over from the formal description of the ABC swarm method application to the thematic segmentation of the images of onboard optoelectronic surveillance systems, a set of nectar sources should be presented as the solution search space, and the amount of nectar as a criterion of the optimization problem (objective function) of the ABC swarm method. At each step, among the positions of agents, n_b of the best positions is selected among the positions of agents by the value of the objective function, n_g of the best so called "selected" (perspective) positions is also selected among others. The positions are selected as follows: among the two closest positions, the one with the worst value of the objective function is discarded and instead of it the position of the second agent that meets the condition is selected.

The marked positions (a set of best N^b positions and a set of perspective N^g positions) are memorized and c^b of agents is sent to the vicinity of each best position at the following step, and c^g of agents is sent to the vicinity of each perspective position. An agent that is sent to the vicinity of the position gets at a random point inside the position. In a two-dimensional space, for example, the vicinity of the position centered at the coordinates (x, y) is the area $([x - rx; x + rx], [y - ry; y + ry])$, where rx and ry are the method parameters. The only rx coefficient can be used in all dimensions or $\boldsymbol{RX}$ vector. Scout-agents consisting of n_s are sent to random positions within the entire search space at each iteration. The method ends when a specified number of iterations has been reached or a satisfactory solution has been obtained or when some time has passed. The sequence of operations of the method can be represented as follows.

1. Sending n^s scout-agents to random points and calculating the value of the objective function in these points.
2. Selecting N^b best and N^g perspective positions from the found ones taking into consideration intersections. If a number of found positions is less than $(N^b + N^g)$, all positions are selected taking into consideration intersections.
3. Sending c^b agents to the vicinity of each best position and c^g agents to the vicinity of each perspective position with the calculation of the objective function.
4. Sending n^s scout-agents.
5. If the condition for completing the method is fulfilled, the method ends, otherwise step 2 is repeated. The best solution obtained from all iterations will be the result of the method. The generalized description of the ABC swarm method can be presented by expression (2.11):

$$\text{ABC} = \{S, M, A, P, In, Out\}, \tag{2.11}$$

where S is a set of swarm agents; M is the object for exchanging experience among S agents; A is the rules of the swarm method (creation, behaviour, modification of agents); P is the parameters that are used in A rules (heuristic coefficients); In, Out are the input and output of the swarm method (the interaction with the external environment and control system is carried out).

The communication scheme of the swarm with the external environment is given in Fig. 2.8.

The following designations are used: $In = \{In_1, In_{fb}\}$, In_1 are the inputs where the objective function $\varphi(\mathbf{X}_{ij})$ and constraints $G(X)$ are fed, $\mathbf{X} = (x, y)$ is the agent position vector; $\mathbf{X}_{ij}(x_{ij}, y_{ij})$ is the vector of position of agents at the j-th iteration in the i-th point of the image, In_{fb} is the feedback input, $Out = \{Out_1, Out_{fb}\}$, Out_1 are outputs for the best found solution $\mathbf{X}_{final}^{best}$ and optimal value of the objective function

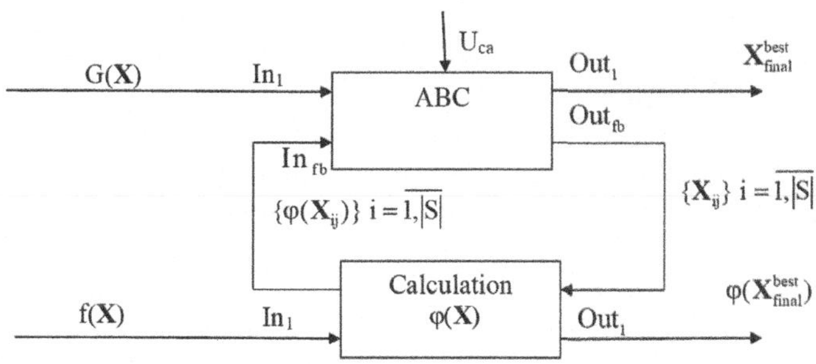

Fig. 2.8 The communication scheme of the swarm with the external environment

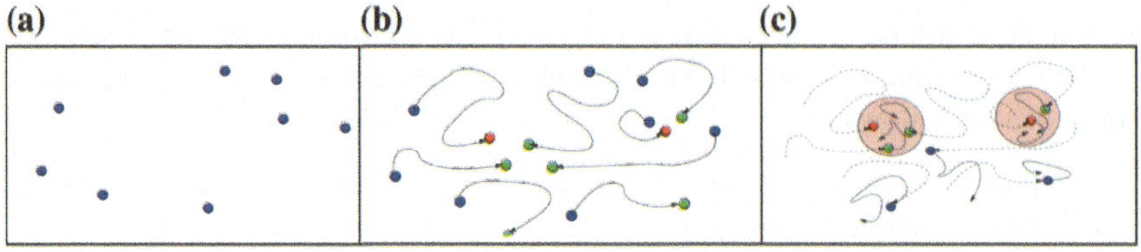

Fig. 2.9 Positions of the agents of the swarm method of the thematic segmentation: **a** initializing starting positions, **b** the migration of agents, **c** determining the optimal position

$\varphi(\mathbf{X}^{best}_{final})$, Out_{fb} is the feedback output, U_{ca} is the control action (start, stop, number of agents, method parameters).

A set of agents $S = \{s_1, s_2, \ldots, s_{|S|}\}$,$|S|$ is a number of agents. At the j-th iteration, the i-th agent is characterized by the state $s_{ij} = \{\mathbf{X}_{ij}(x_{ij}, y_{ij})\}$. The object for exchanging experience among agents is the list of the best and perspective positions (expression 2.12):

$$M = \{N^b_{ij}, N^g_{kj}\}, \tag{2.12}$$

where $i = 1, 2, \ldots, n^b$, $k = 1, 2, \ldots, n^g$; N^b_{ij} is a number of the best positions for the i-th best agent at the j-th iteration; N^g_{kj} is a number of perspective positions for the k-th perspective agent at the j-th iteration.

The operation of the swarm method of the OEI thematic segmentation is presented as follows. The positions of the agents of the swarm method of the thematic segmentation are given in Fig. 2.9.

1. Initializing starting positions of agents in the image (Fig. 2.9a). The condition is fulfilled only for n^s scout-agents at the first iteration ($j = 1$) (expression 2.13):

$$\mathbf{X}_{i1} = rand(f(\mathbf{X})), \tag{2.13}$$

where $\mathbf{X}_{i1}(x_{i1}, y_{i1})$ is the vector of the positions of agents at the first iteration; $rand(f(\mathbf{X}))$ is the random number generator; $I = 1, \ldots, n^s$.

2. Determining the objective function $\varphi(\mathbf{X}_{ij})$ of each i-th agent ($i = 1, 2, \ldots, |S|$) at the current j-th iteration.
3. Developing the best and perspective positions of agents taking into consideration the values of the objective function $\varphi(\mathbf{X}_{ij})$ of each i-th agent (expression 2.12).
4. The migration of agents (Fig. 2.9b). After developing the best and perspective positions of agents in the vicinity of these positions, employed agents are sent: c^b employed agents are sent to the vicinity of each best position, c^g employed agents are sent to the vicinity of each perspective position. The positions of all employed agents are determined by expressions (2.14)–(2.15):

$$\mathbf{X}_{(i-1)cb+kj} = N^b_{ij-1} + Rnd \cdot rad \tag{2.14}$$

where $\mathbf{X}_{(i-1)cb+kj}$ is the vector of positions of employed agents which are sent to the best positions at the *j*-th iteration; N^b_{ij-1} is the best position of the *i*-th agent at (*j*-1)-th iteration; $i = 1, \ldots, n^b, k = 1, \ldots, c^b$; n^b is a number of the best positions at the *j*-th iteration; c^b is a number of agents which are sent (migrate) to the best positions; *Rnd* is a random number; *rad* is the coefficient that determines the scattering agents when they are sent to the best and perspective positions;

$$\mathbf{X}_{n^b c^b+(i-1)c^b+k\,j} = N^g_{ij-1} + Rnd \cdot rad \tag{2.15}$$

where $\mathbf{X}_{n^b c^b+(i-1)c^b+k\,j}$ is the vector of positions of employed agents that are sent to the perspective positions at the *j*-th iteration; N^g_{ij} is the perspective position of the *i*-th agent at the *j*-th iteration; $i = 1, \ldots, n^g$, $k = 1, \ldots, c^g$; n^g is a number of perspective positions at the *j*-th iteration; c^g is a number of scout-agents which are sent (migrate) to perspective positions.

Scout-agents are sent to the positions whose coordinates are random values and that are evenly distributed over the entire permissible range of values (over the entire image) (expression 2.16):

$$\mathbf{X}_{n^b c^b+n^g c^g+i\,j} = rand(f(\mathbf{X})), \tag{2.16}$$

where $i = 1, \ldots, n^s$, n^s is a number of scout-agents at the *j*-th iteration.

5. Coefficients (parameters) of the method that are used in expressions (2.14)–(2.15) are included in the vector of method coefficients $P = \{n^s, n^b, n^g, c^b, c^g, rad, rx, ry\}$.
6. The initial data of the method include the determination of the optimal position of agents (Fig. 2.9c) which ensures the minimum or maximum value of the objective function $\varphi(\mathbf{X}_{ij})$ and the optimal threshold value for the OEI thematic segmentation.

Thus, the general scheme of the method of the thematic segmentation of the OEIs of onboard optoelectronic surveillance which is based on the ABC swarm method is given in Fig. 2.10.

The behaviour of agents of the ABC method is based on self-organization—a set of dynamic mechanisms according to which the system is globally regulated due to the interaction of components at the lower level without direct interaction among these components. The main components of the self-organization of agents of the ABC method in the segmentation of the OEIs are as follows:

- *positive feedback*, which is achieved by performing simple behavioural empirical techniques that provide finding solutions. Positive feedback lies in the fact that agents, basing on the information obtained from other agents, start moving to the indicated sources of resources;

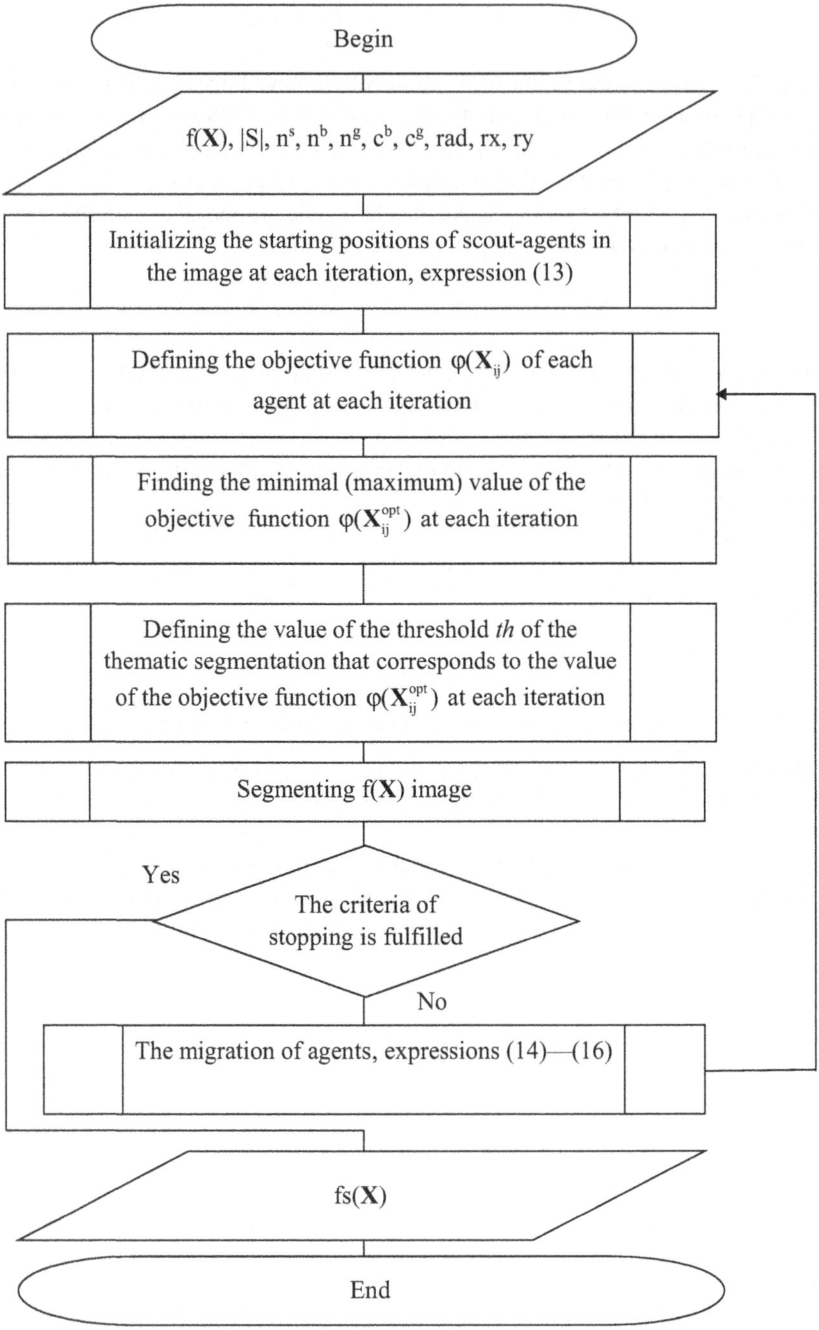

Fig. 2.10 The general scheme of the method of thematic segmentation of the OEIs of onboard optoelectronic surveillance systems

- *negative feedback*, which counterbalances positive feedback, which leads to the stabilization of the collective behaviour. Negative feedback lies in the fact that an agent, basing on the information obtained from other agents, can decide that the position found by this agent is worse;
- *the instability of positive feedback:* the behaviour of agents is based on stochastic rules, so random deviations in the solutions are the basis for finding new solutions. The agent that has missed the course can find new, undiscovered positions and bring other agents;
- *the requirement of multiple interactions among agents*, which results in new best solutions. The multiple interactions of the ABC method lie in the fact that the information about the position found by a single agent is available to all other agents.

Besides, the self-organization of the agents of the ABC method while segmenting OEIs is characterized by certain properties:

- dynamism, that is finding solutions requires a constant interaction among agents and the environment. This interaction provides positive feedback, due to which certain groups are created, while negative feedback reduces the effect of direct feedback;
- emergence, that is properties of self-organizing systems are more complex than the properties of individual agents included in the system. Such properties are the result of a nonlinear combination of interactions among agents;
- non-linear interactions among agents lead to branching in the behaviour of self-organizing systems. Due to this, qualitative changes in the collective behaviour of agents are achieved and new sustainable solutions are found;
- *multi-stability,* in other words, the system can reach different stable states for given set parameters depending on the initial state and random deviations.

To apply the method (Fig. 2.10) to the thematic segmentations of OEIs, the objective function $\varphi(\mathbf{X}_{ij})$ should be selected and the optimization task should be formulated to determine the optimal value of the threshold of the OEI thematic segmentation.

2.2.3 Selecting the Objective Function of the ABC Swarm Method for Segmenting Optoelectronic Images

To solve the optimization task and find optimal solutions in the multidimensional space of alternatives in general and to determine the optimal value of the threshold *th* in particular, the basic requirements for the objective function $\varphi(\mathbf{X}_{ij})$ are:

- meeting the condition of the adequacy of the problem that is being solved;
- it should not be too "sharp" (the problem of a "bottleneck");
- minimum computational resource should be used;
- minimum number of local optimums.

To substantiate the choice of the objective function of the method of thematic segmentation, let us consider model images in the case where an object (the internal square) and a background are separated and not separated (Fig. 2.11).

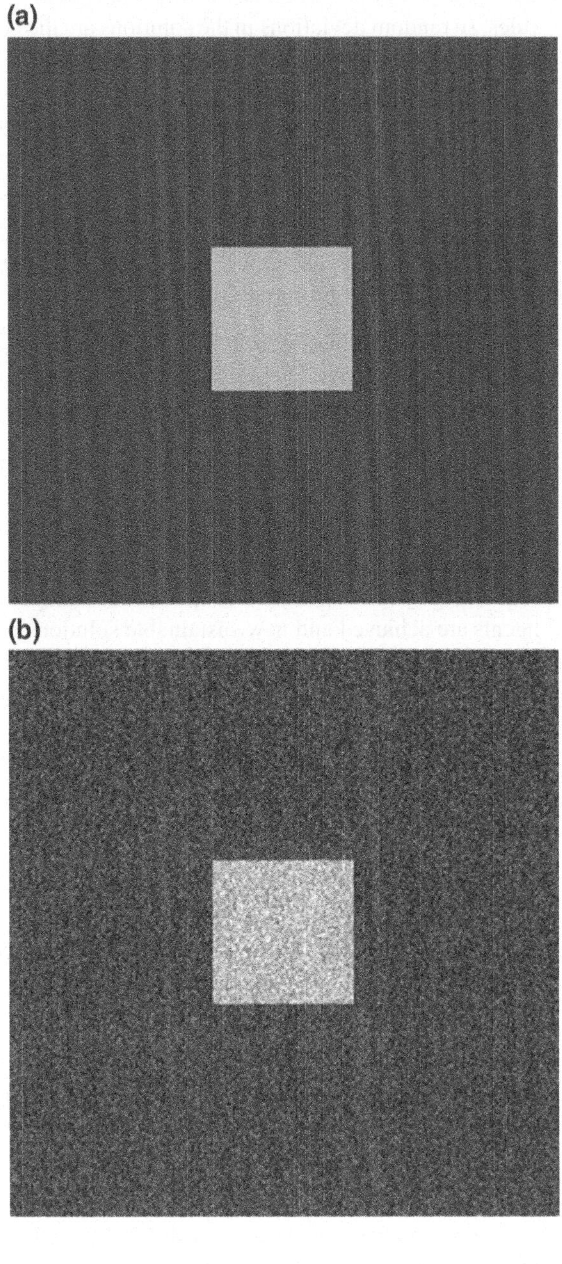

Fig. 2.11 Model images of an object and a background: **a** an object and a background are separated, **b** an object and a background are not separated

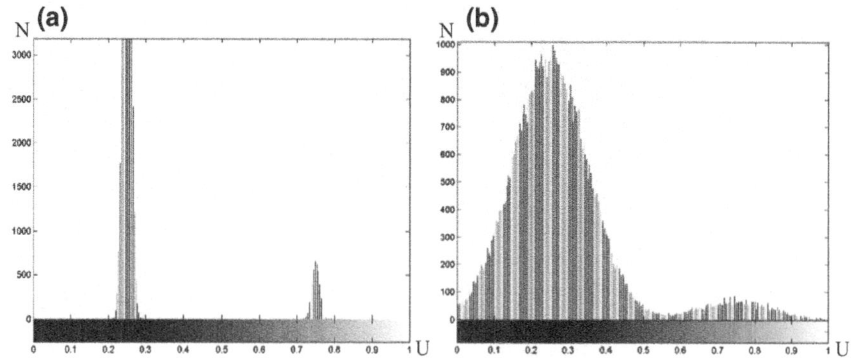

Fig. 2.12 The histograms of the distribution of model images brightness: **a** an object and a background are separated, **b** an object and a background are not separated

The following is taken into account: the background size significantly exceeds the size of the object, the brightness of the pixels of the object and the background is distributed according to the normal law with the parameters (Fig. 2.12):

Then, for the case when an object and a background are separated (Fig. 2.11a), let us binarize the image (Fig. 2.11a) and consider the dependence of the value of the dispersion of the intensity of subthreshold $D_1(th)$ and after-threshold $D_2(th)$ pixels as the function of the threshold *th*. For threshold extreme values $D_1(th = 0) = 0$, $D_2(th = 0) = D_0$, $D_1(th = 1) = D_0$, $D_2(th = 1) = 0$, where D_0 is the dispersion of the model image (Fig. 2.11a).

To determine the expression for the dispersion of the model image D_0, let us calculate the mathematical expectation of the model image according to expression (2.17):

$$\mu = \frac{\tilde{s}_1\mu_1 + \tilde{s}_2\mu_2}{\tilde{s}_1 + \tilde{s}_2}, \tag{2.17}$$

where $\tilde{s}_1$, $\tilde{s}_2$ are the plane (a number of pixels) of an object and a background in the model image (Fig. 2.11a).

The value of D_0 can be easily determined using expression (2.18):

$$D_0 = \frac{\tilde{s}_1\mu_1^2 + \tilde{s}_2\mu_2^2}{\tilde{s}_1 + \tilde{s}_2} - \mu^2. \tag{2.18}$$

When the distribution of the pixel intensity of an object and a background are separated (Fig. 2.12a), condition (2.19) should be fulfilled:

$$(\mu_1 - 3\sigma_1) > (\mu_2 + 3\sigma_2). \tag{2.19}$$

Then,

$$D_1((\mu_2 + 3\sigma_2) < th < (\mu_1 - 3\sigma_1)) = \sigma_2^2, \tag{2.20}$$

$$D_2((\mu_2 + 3\sigma_2) < th < (\mu_1 - 3\sigma_1)) = \sigma_1^2. \tag{2.21}$$

Let us consider function $D(th) = D_1(th) + D_2(th)$, which has the physical meaning of the sum of the variances of the intensity of subthreshold and after-threshold pixels. We obtain:

$$D(th = 0) = D(th = 1) = D_0, \tag{2.22}$$

$$D((\mu_2 + 3\sigma_2) < th < (\mu_1 - 3\sigma_1)) = \sigma_1^2 + \sigma_2^2 < D_0, \tag{2.23}$$

that is, within the interval of values of threshold *th*, where an object is separated from a background, $D(th) = \min$.

Figure 2.13 shows function D(*th*) for the model image in Fig. 2.11a, when an object and a background are separated. When an object and a background are not separated (Fig. 2.11b), the quantitative analysis is more complex and is not given in the paper. Nevertheless, the quality pattern of function $D(th)$ does not change and is given in Fig. 2.14. Therefore, while using the swarm method of the thematic segmentation,

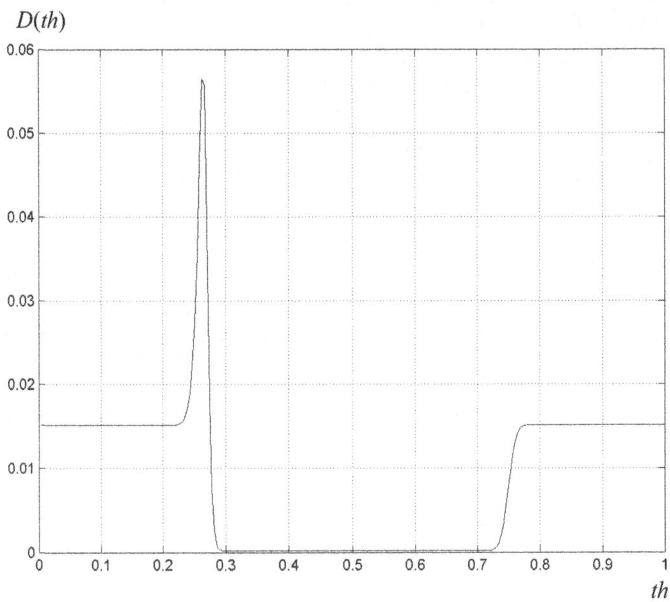

Fig. 2.13 The pattern of function $D(th) = D_1(th) + D_2(th)$ for the model image in Fig. 2.11a

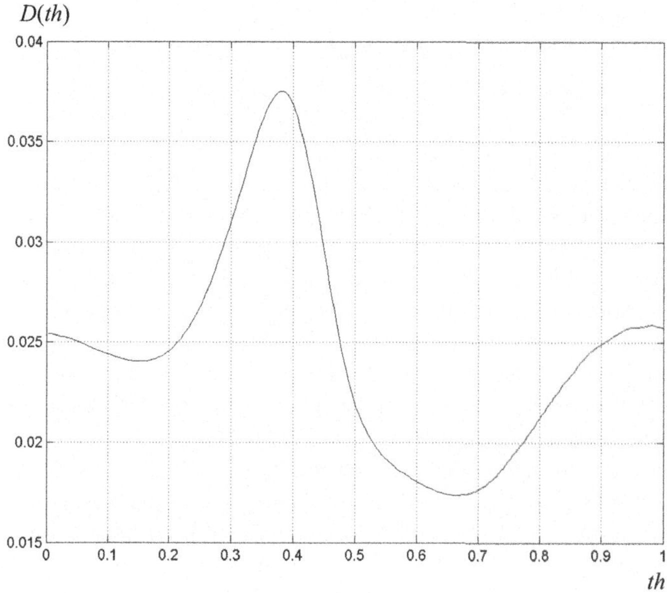

Fig. 2.14 The pattern of function $D(th) = D_1(th) + D_2(th)$ for the model image in Fig. 2.11b

function $D(th)$ should be chosen as an objective function as this function has the physical meaning of the sum of the variances of the segments of the segmented image.

In general, the optimization problem of selecting the threshold of the OEI thematic segmentation at each j th iteration is presented by expressions (2.24)–(2.30) and is to minimize the objective function (2.24) at each j-th iteration taking into consideration definitions and constraints (2.25)–(2.29).

$$D(th) = \sum_{j=1}^{K} D_j(th) \rightarrow \min; \tag{2.24}$$

$$D_j(th) = \frac{1}{N_{1j}(th) - 1} \sum_{k=1}^{N_{1j}(th)} \left| U_{kj} - \mu_j \right|^2 - \text{for } U_{kj} > th; \tag{2.25}$$

$$D_j(th) = \frac{1}{N_{0j}(th) - 1} \sum_{k=1}^{N_{0j}(th)} \left| U_{kj} - \mu_j \right|^2 - \text{for } U_{kj} \le th; \tag{2.26}$$

$$\mu_j(th) = \frac{1}{N_{1j}(th)} \sum_{k=1}^{N_{1j}(th)} U_{kj} - \text{for } U_{kj} > th; \tag{2.27}$$

$$\mu_j(th) = \frac{1}{N_{0j}(th)} \sum_{k=1}^{N_{0j}(th)} U_{kj} - \text{for } U_{kj} \leq th; \tag{2.28}$$

$$0 \leq th \leq U_{\max}, \tag{2.29}$$

where K is a number of segments; $N_{0j}(th)$ is a number of pixels in the j-th segment whose brightness is less than or equal to the value of the threshold level th; $N_{1j}(th)$ is a number of pixels in the j-th segment whose brightness is greater than the value of the threshold level th; U_{kj} is the brightness of the k-th pixel in the j-th segment; μ_j (th) is the mathematical expectation of the brightness in the j-th segment; $U_{\max}$ is the maximal value of the brightness of pixels in the image.

So, taking into consideration the definition of the objective function (expressions 2.24–2.29), the scheme of the thematic segmentation of images of onboard optoelectronic surveillance systems is shown in Fig. 2.15.

A colour image of the onboard optoelectronic surveillance system will be considered as the source (Fig. 2.4). Let us view the brightness channels Red (R), Green (G), Blue (B) of the colour space of the source OEI RGB representation (Fig. 2.4).

Figures 2.16, 2.17 and 2.18 show the histograms of the brightness distribution in each channel of brightness (the dependence of U brightness on N number of pixels for channels R, G, B is presented). The analysis of histograms indicates (Figs. 2.16, 2.17 and 2.18) that they are not unimodal, which makes the selection of the optimal threshold of the thematic segmentation more difficult.

For the thematic segmentation of the colour image, the method is suggested (Fig. 2.19) which includes:

- entering initial data: $f(X)$, $|S|$, n^s, n^b, n^g, c^b, c^g, rad, rx, ry;
- defining a colour channel in the source image $f(X)$: $f_R(X), f_G(X), f_B(X)$ (where $f_R(X), f_G(X), f_B(X)$ are images in colour channels Red, Green, Blue);
- defining the brightness channel in each colour channel of the source image: $f_R(X)$, $f_G(X), f_B(X)$;
- thematic segmenting the image in each colour channel by the method presented in Fig. 2.14 and obtaining segmented images in each colour channel: $fs_R(X), fs_G(X)$, $fs_B(X)$ (where $fs_R(X), fs_G(X), fs_B(X)$ are segmented images in colour channels Red, Green, Blue);
- transitioning to the RGB colour model (combining colour channels);
- obtaining segmented colour image $fs(X)$.

Figures 2.20, 2.21 and 2.22 show the images of the brightness channel of each colour channel of the RGB colour space. Thematic segmenting in each colour channel is carried out by the ABC swarm method (Fig. 2.15). The iterative process of finding optimal thresholds for thematic segmentation in the OEI colour channels is presented in Fig. 2.23, there are 20 iterations. Figure 2.23 shows the location of the local and global minima of the objective function at each stage of iteration. The analysis of Fig. 2.23 shows that the greater number of iterations, the greater number of agents are located in the area of the minimal value of the objective function.

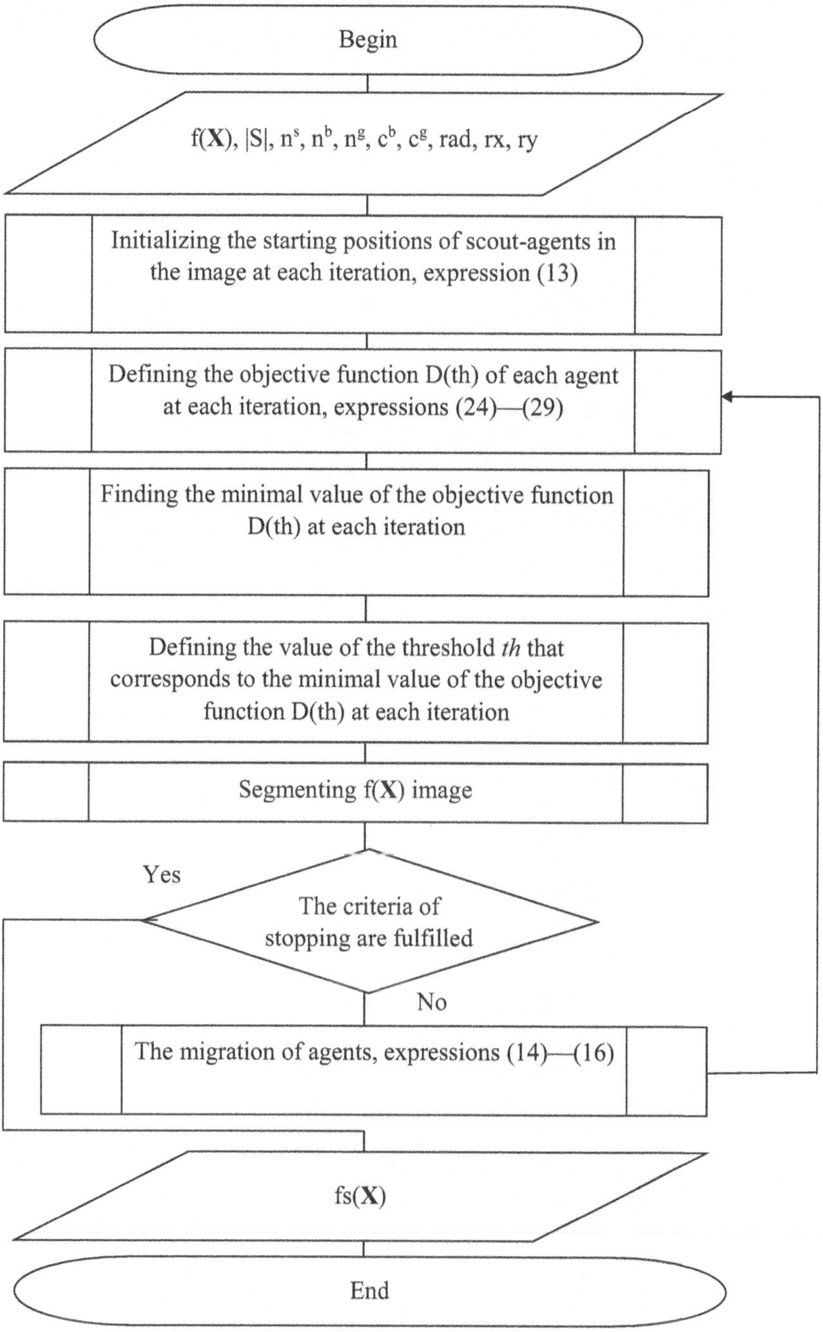

Fig. 2.15 The scheme of the method of thematic segmentation of the OEIs of onboard optoelectronic surveillance systems

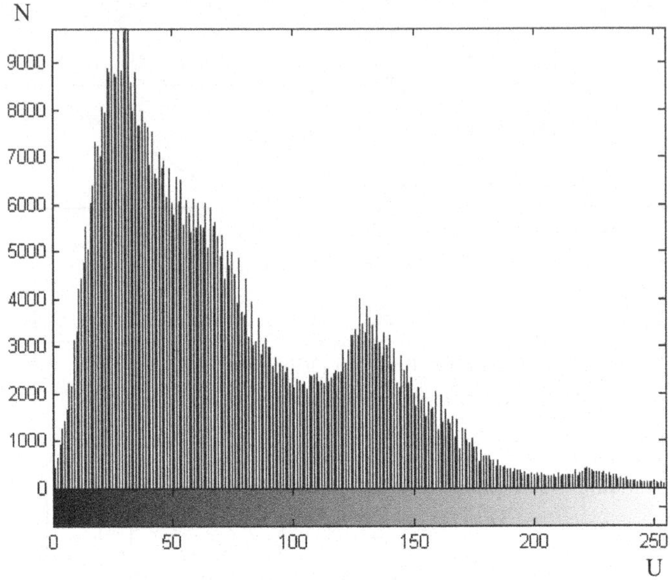

Fig. 2.16 The histograms of the brightness distribution for the Red channel

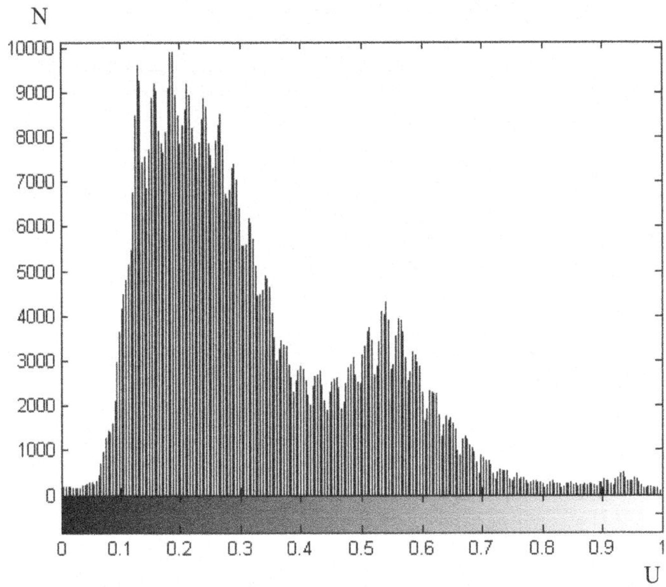

Fig. 2.17 The histograms of the brightness distribution for the Green channel

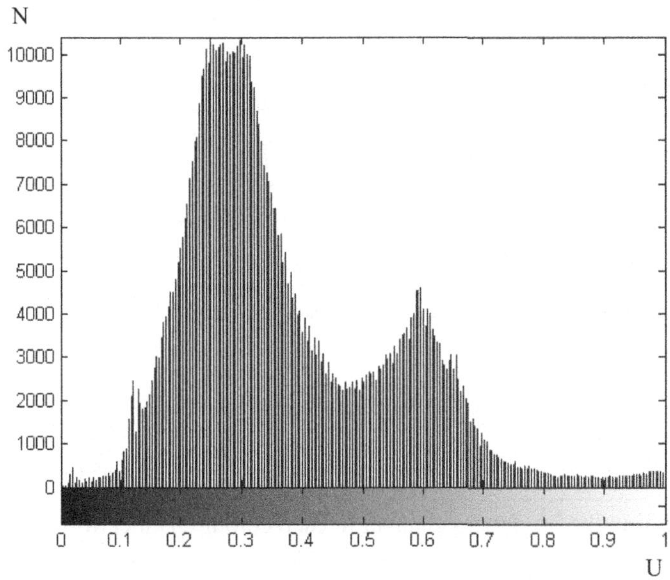

Fig. 2.18 The histograms of the brightness distribution for the Blue channel

Figures 2.24, 2.25 and 2.26 show the examples of constructing the objective function $D(th)$ for different brightness channels (Red, Green, Blue). The value of the optimal threshold th of the OEI thematic segmentation is defined for each brightness channel.

Transitioning to the RGB colour model is performed by joining colour channels using the rule of colour mixing and the laws of mathematical logic (Table 2.1).

Figure 2.27 shows the three-dimensional visualization of the source image (Fig. 2.4) that is represented as a cloud of points in the RGB colour space of the appropriate brightness.

Sectional views that correspond to the optimal values of the threshold level of each colour channel are shown in the three-dimensional RGB colour space in Fig. 2.28. These sectional views are R = 0.4089; G = 0.3978; B = 0.4837.

The result of the thematic segmentation of the source OEI (Fig. 2.4) is shown in Fig. 2.29.

The analysis of the results of the thematic segmentation (Fig. 2.29) by the visual assessment of brightness shows that it became possible to conduct a thematic segmentation of the colour image of the onboard optoelectronic surveillance system and to identify the objects of reconnaissance—airplanes, oil storages, structures and others.

To assess the visual brightness of the segmented image, its binarization should be carried out. Physically, this means, that in the three-dimensional RGB colour space of brightness of pixels, which are located in the right top farther parallelepiped, take on a value of “1”, all others—“0”. The binarized image is presented in Fig. 2.30.

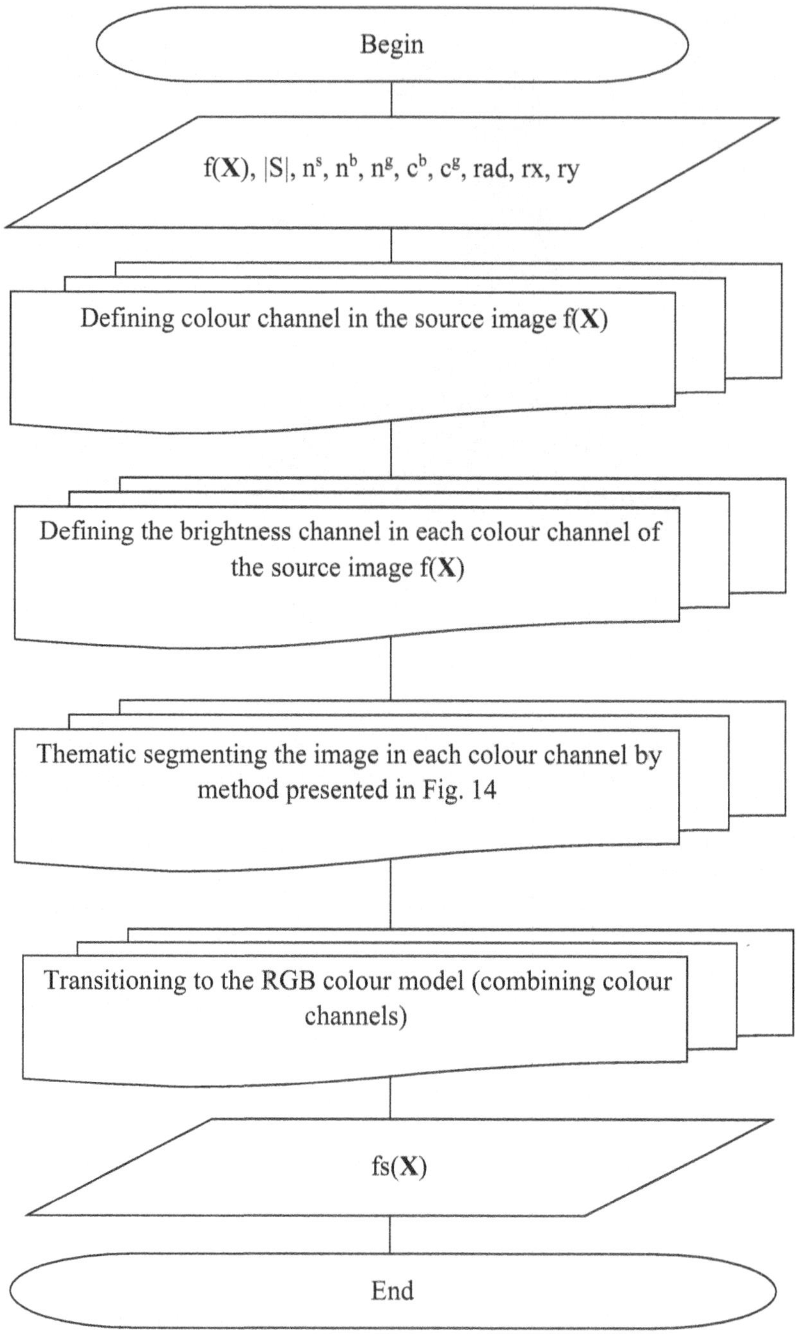

Fig. 2.19 The scheme of the thematic segmenting of the colour image (RGB colour space) of the onboard optoelectronic surveillance system

Fig. 2.20 The image in the brightness channel of the Red colour channel

The visual comparison of Figs. 2.30 and 2.29 indicates the need for additional binarization of the segmented image.

2.3 The Quality Indicators of the Thematic Segmentation of Images of Onboard Optoelectronic Surveillance Systems

Existing quality indicators of the segmentation are known to have the following drawbacks [5]:

- the results of segmentation that are the best in the opinion of experts have a greater percentage of misclassified pixels;
- the location of erroneous pixels relative to the corresponding segment is not taken into consideration, that is why an error at the boundary and in the centre of a segment should be fined differently;
- the difference in the importance of individual sections of the image for segmentation is not taken into consideration, that is errors should have different weights for different segments of the image;
- there is no information about the class of pixels, which makes the greatest error;

Fig. 2.21 The image in the brightness channel of the Green colour channel

- small ("littered") areas in the segmented image are not taken into consideration.

To assess the quality of the OEI thematic segmentation, the thematic segmentation errors of the first and second kind will be used as well as such information indicators as the Kullback–Leibler divergence and the entropy of a segmented image.

The thematic segmentation errors of the first and second kind are determined when the maximum likelihood criterion is used, which results from the generalized criterion of mean risk [5–7]. The thematic segmentation errors of the first kind $\alpha_1(f(\mathbf{X}), fs(\mathbf{X}))$ are determined by expression (2.30) [5–7]:

$$\alpha_1(f(\mathbf{X}), fs(\mathbf{X})) = \frac{1}{NM} \sum_{\mathbf{X} \in Z_f} fs(\mathbf{X}), \tag{2.30}$$

where $Z_f = \{\mathbf{X} | f(\mathbf{X}) = 0\}$ is a set of pixels that do not belong to the objects of interest in the source image $f(X)$; (NxM) is a number of pixels in the image.

The thematic segmentation errors of the second kind $\beta_2(f(\mathbf{X}), fs(\mathbf{X}))$ are determined by expressions (2.31) [5–7]:

$$\beta_2(f(\mathbf{X}), fs(\mathbf{X})) = \frac{1}{NM} \sum_{\mathbf{X} \in Z_{fs}} f(\mathbf{X}), \tag{2.31}$$

Fig. 2.22 The image in the brightness channel of the Blue colour channel

where $Z_{fs} = \{\mathbf{X} | fs(\mathbf{X}) = 0\}$ is a set of pixels that do not belong to the objects of interest in the segmented image $fs(X)$.

When moving to the plane of the objects of interest and background, the errors of the first and second kinds are calculated by expressions (2.32)–(2.33)

$$\alpha_1(f(\mathbf{X}), fs(\mathbf{X})) = \frac{S_1(fs(\mathbf{X}))}{S_2(f(\mathbf{X}))}, \tag{2.32}$$

$$\beta_2(f(\mathbf{X}), fs(\mathbf{X})) = 1 - \frac{S_3(fs(\mathbf{X}))}{S_4(f(\mathbf{X}))}, \tag{2.33}$$

where $S_1(fs(X))$ is the area of the background sections that erroneously assigned to the objects of interest in the segmented image $fs(X)$; $S_2(f(X))$ is the area of the background sections of the source image $f(X)$; $S_3(fs(X)f(X))$ is the area of correctly segmented objects of interest; $S_4(f(X))$ is the area of the objects of interest in the source image $f(X)$.

The errors of the first and second kind are not related to the amount of information that is presented in the segmented image as compared to the source one. That is why to assess the segmentation quality, the information indicator—the Kullback–Leibler divergence $K(p_\xi, p_\eta)$ will be introduced—expression (2.34) [3, 8]:

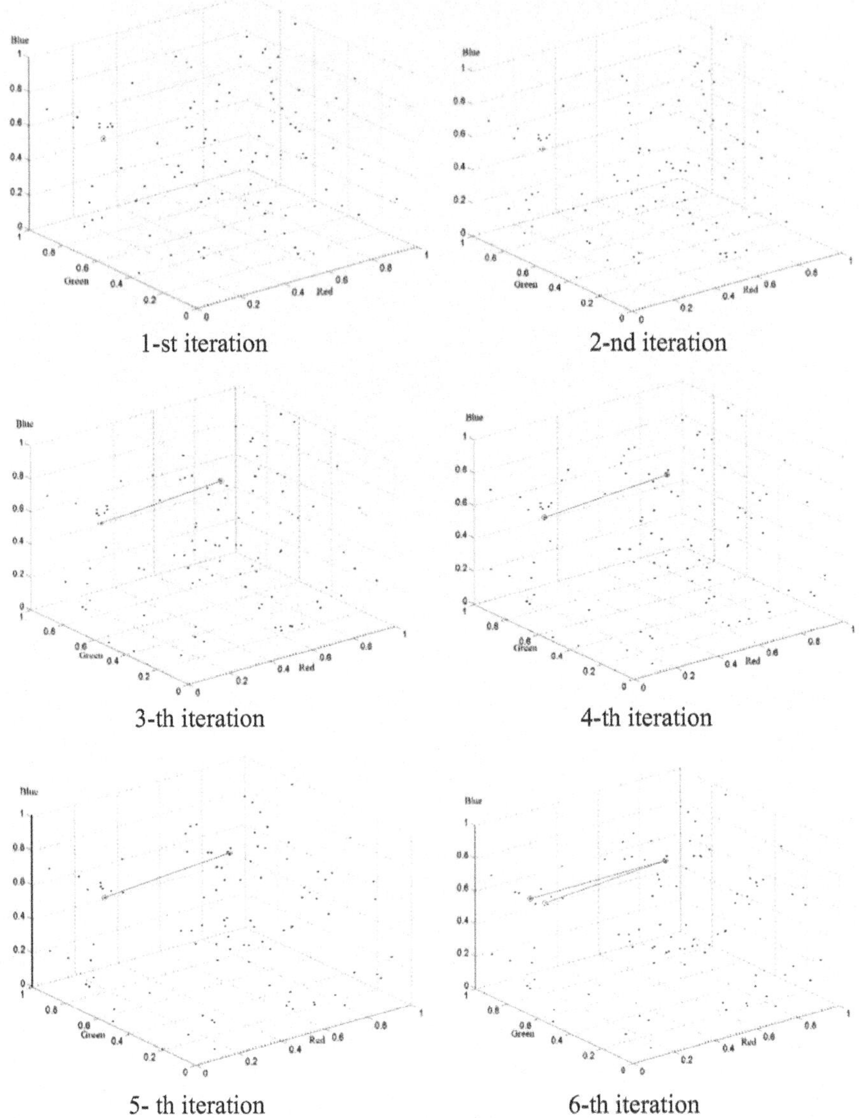

Fig. 2.23 The iterative process of finding optimal thresholds of the thematic segmenting in the OEI colour channels, page 1

$$K(p_\xi, p_\eta) = \int\limits_{R^2} p_\xi(U) \log_2 \frac{p_\xi(U)}{p_\eta(U)} dU, \tag{2.34}$$

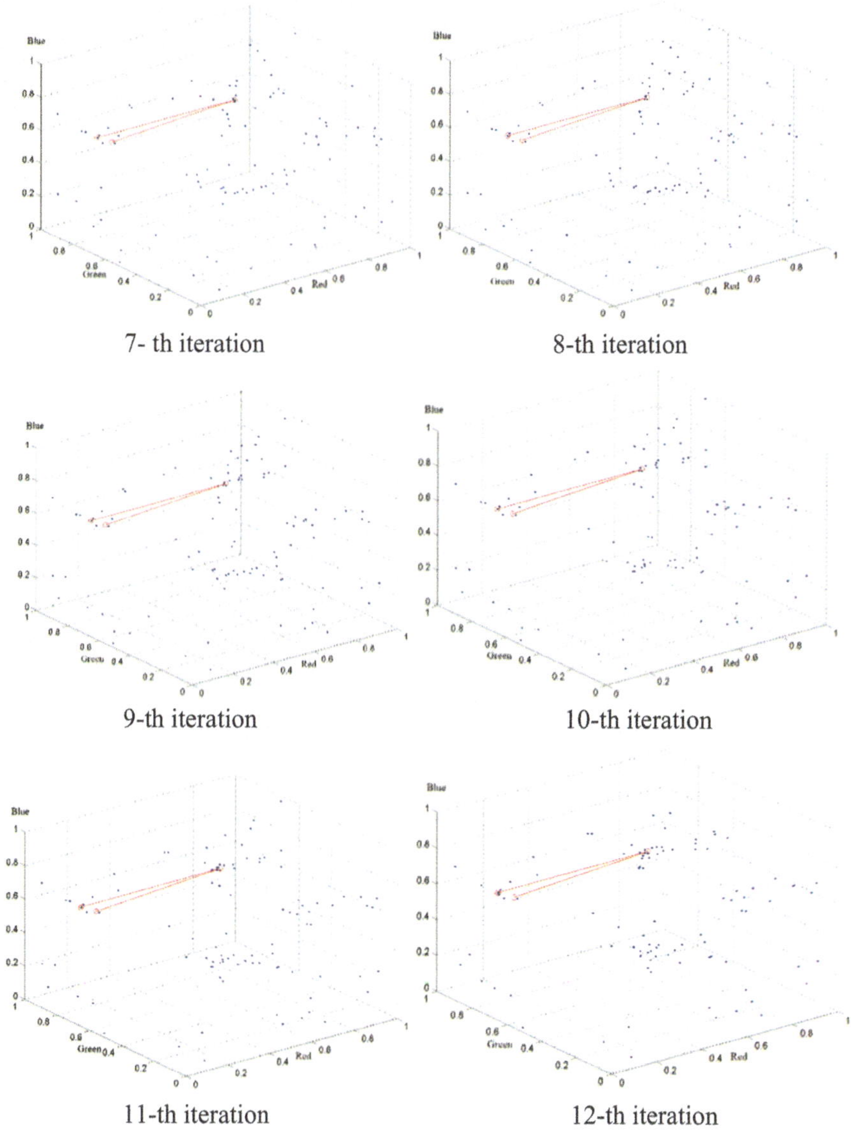

Fig. 2.23 (continued)

where $p_\xi(U)$ is the density of the brightness distribution in the source image; $p_\eta(U)$ is the density of the brightness distribution in the segmented image; R^2 is the area of the image.

The information indicator Kullback–Leibler divergence $K(p_\xi, p_\eta)$ enables assessing the information loss while thematic segmenting OEIs. Let p_i—probability of pixel brightness entering the i-th segment be introduced ($\sum_{i=1}^{N} p_i = 1$, N is

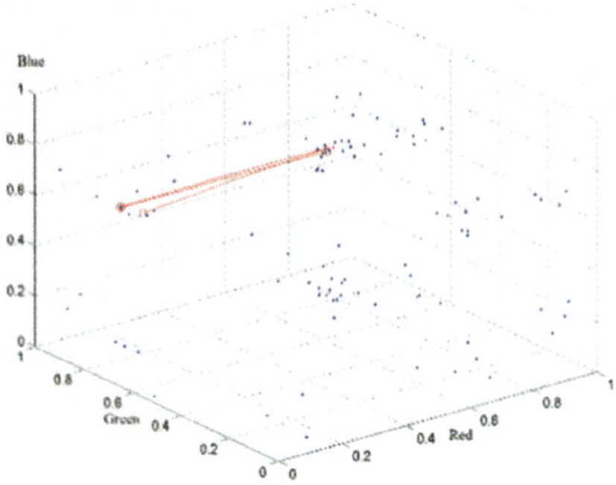

13-th iteration

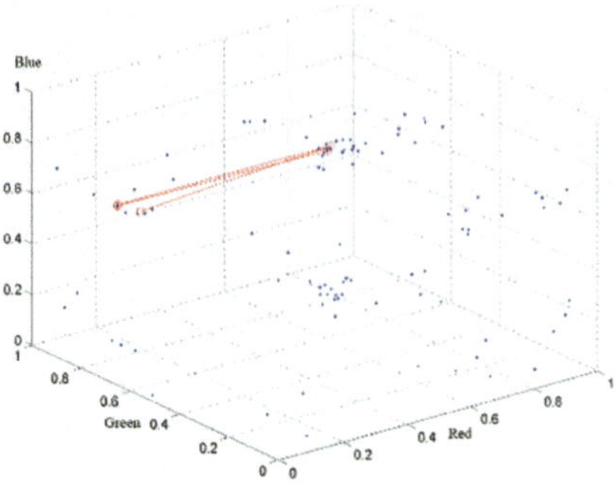

14-th iteration

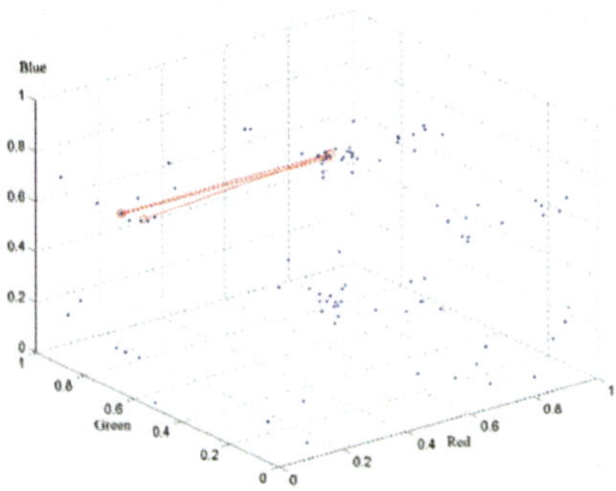

15-th iteration

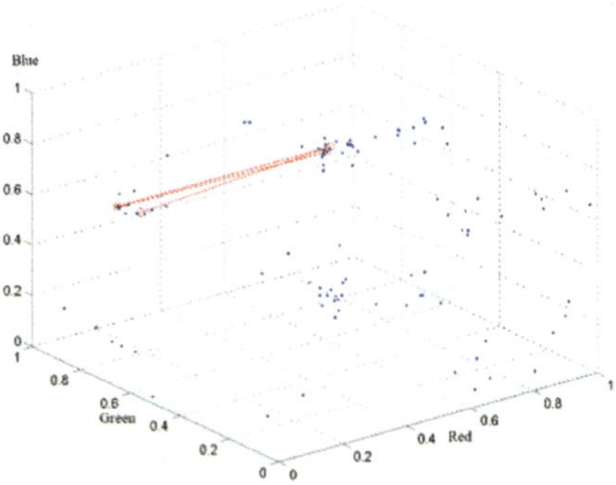

16-th iteration

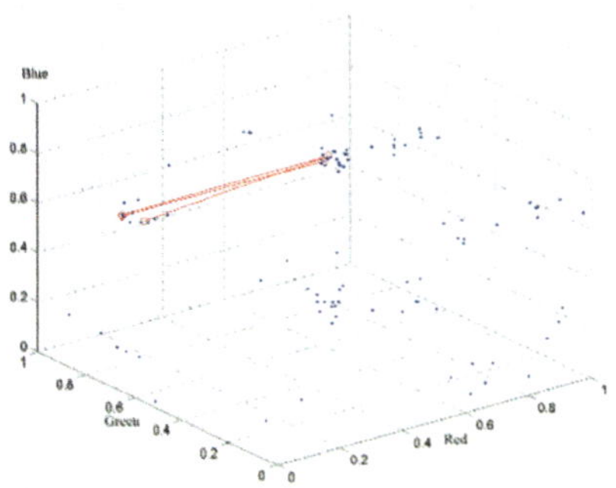

17-th iteration

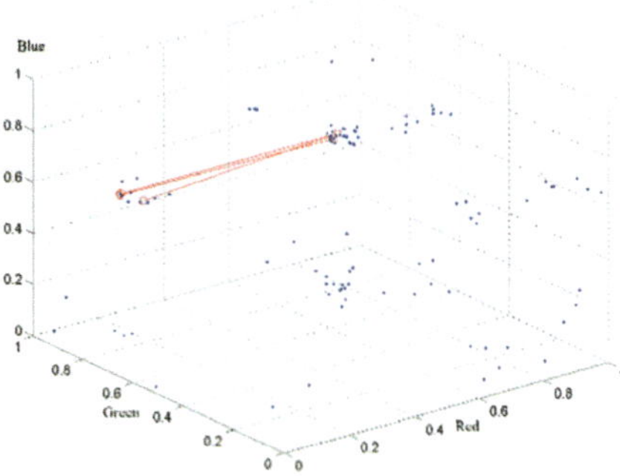

18-th iteration

Fig. 2.23 (continued)

a number of segments). If p_i is the probability for the i-th segment of the source image and q_i is the probability for the i-th segment of the segmented image, the Kullback–Leibler divergence can be calculated by expression (2.35):

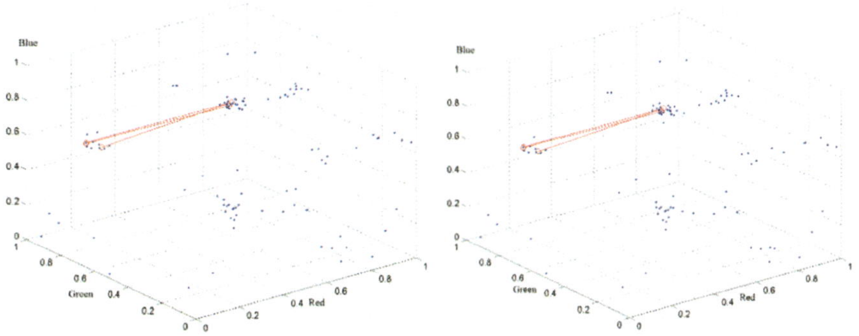

19-th iteration 20-th iteration

Fig. 2.23 (continued)

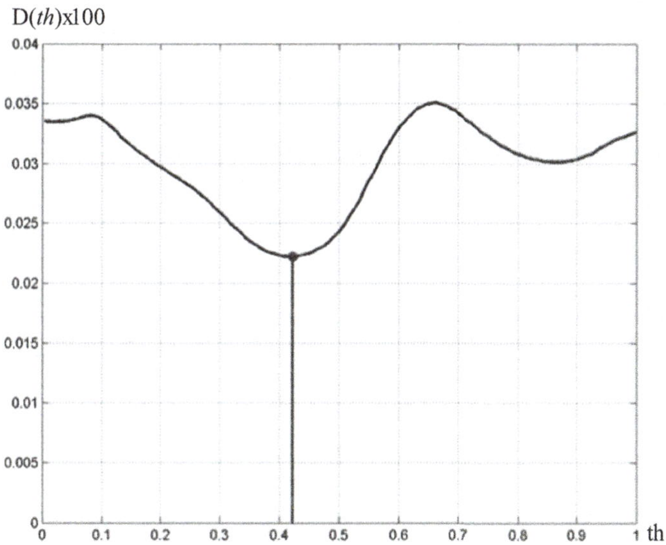

Fig. 2.24 Defining the optimal value of the threshold level for the brightness channel Red

$$K(p_\xi, p_\eta) = \sum_{i=1}^{N} p_i \log_2 \frac{p_i}{q_i}. \tag{2.35}$$

In the work, along with the information indicator—the Kullback–Leibler divergence, another information indicator—the entropy of the segmented image will be used [3]:

$$H = -\sum_{c} p_c \log_2 p_c, \tag{2.36}$$

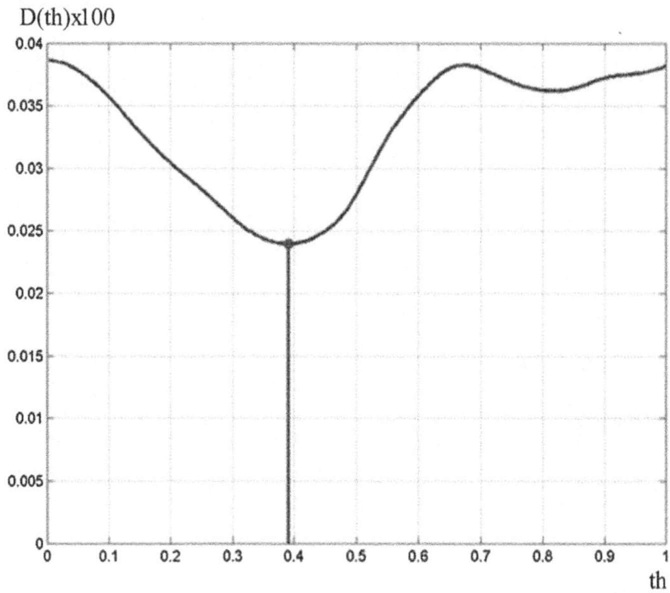

Fig. 2.25 Defining the optimal value of the threshold level for the brightness channel Green

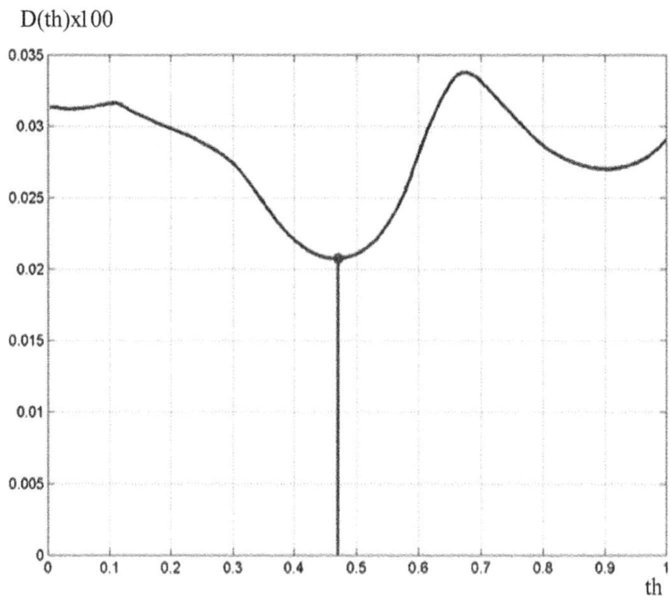

Fig. 2.26 Defining the optimal value of the threshold level for the brightness channel Blue

Table 2.1 The rules of combining colour channels of the segmented image

Red	Green	Blue	The result of combining colour channels	
0	0	0		Black
0	0	1		Blue
0	1	0		Green
0	1	1		Cyan
1	0	0		Red
1	0	1		Magenta
1	1	0		Yellow
1	1	1		White

where p_c is the probability of c discoloration in the image, which is calculated by the image brightness histogram.

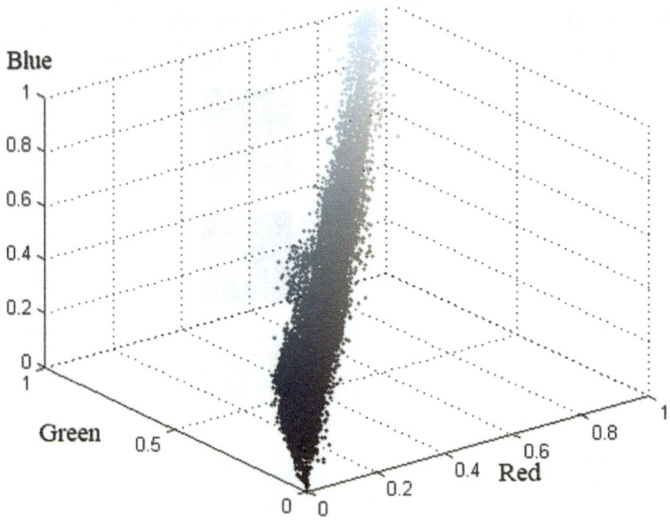

Fig. 2.27 The source image (Fig. 2.4) presented as a cloud of points in the RGB colour space

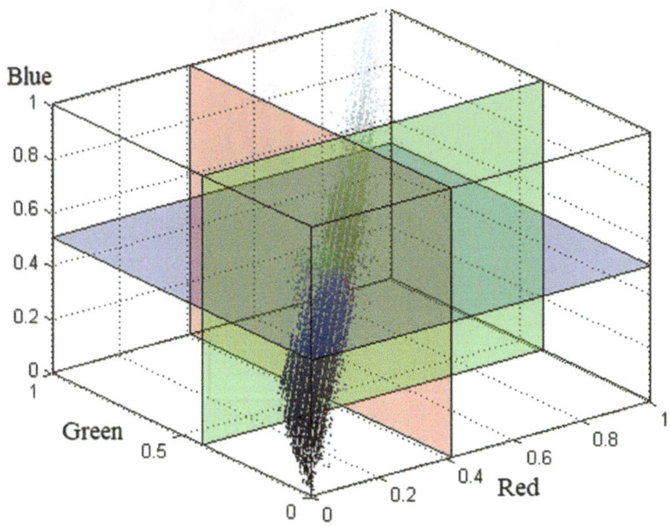

Fig. 2.28 Sectional views that correspond to the optimal values of the threshold level of each colour channel

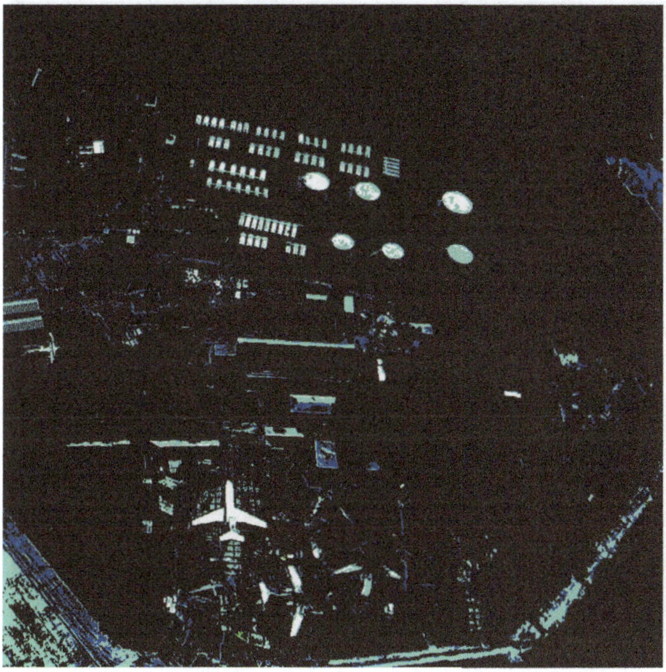

Fig. 2.29 The result of thematic segmenting the source colour OEI

2.4 The Comparative Quality Assessment of the ACO Method, the ABC Method and the Known Methods of the Thematic Segmentation of the Images of Onboard Optoelectronic Surveillance Systems

As the known methods of the OEI thematic segmentation, the following ones are selected:

- *Random forest* method [9–11];
- *Otsu* method [9, 10];
- *Canny* method [12, 13];
- *k-means* method [9–11, 14, 15].

The methods will be compared under the conditions of the effect of the basic confounding factors: OEI rotating, OEI zooming, noise impact (Additive White Gaussian Noise (AWGN), "salt and pepper", speckle) in OEIs. The methods will be compared by visual assessing, calculating the errors of the first and second kind, the Kullback–Leibler divergence and the segmented image entropy.

Figures 2.31, 2.32, 2.33 and 2.34 shows the results of the OEI thematic segmentation, namely Fig. 2.31 shows the segmentation by Otsu method; Fig. 2.32—by Canny method; Fig. 2.33—by *k*-means method ($k = 5$) with different values of classes *k*;

Fig. 2.30 The result of thematic segmenting the source reconnaissance colour image

Fig. 2.34—by the Random forest method. The visual assessment of the quality of images segmented by the known methods (Figs. 2.31, 2.32, 2.33and 2.34), ACO method (Fig. 2.5) and ABC method (Fig. 2.30) testifies to better segmentation by the ACO method and the ABC method.

Table 2.2 shows the results of assessing the error values of the first and second kind of the OEI segmentation without confounding when the OEI is rotated and zoomed. Figure 2.35 shows the assessment of the Kullback–Leibler divergence depending on the factor of zooming. Figure 2.36 shows the dependence of the segmented image entropy on the root-mean-square deviation σ_n of the AWGN intensity.

The numbers in Figs. 2.35 and 2.36 correspond to the methods in the following way:

- 1—ABC method;
- 2—ACO method;
- 3—*Canny* method;
- 4—*Random forest* method;
- 5—*Otsu* method.

The analysis of Table 2.2 and Figs. 2.35, 2.36 shows that the use of the ACO method and ABC method for the thematic segmentation will enable increasing the quality of the OEI processing, namely:

Fig. 2.31 The OEI segmented by the Otsu method

- decreasing the segmentation errors of the first and second kind by 4–13%;
- increasing the quality of the segmented image without the impact of confounding factors (the gain in the value of information indicators: the Kullback–Leibler divergence—from 3 to 12%, the segmented image entropy—from 5 to 13%);
- increasing the quality of the segmented image under the impact of confounding factors (the gain in the value of information indicators: the Kullback–Leibler divergence—from 5 to 14%, the segmented image entropy—from 7 to 16%).

2.5 Conclusion

For segmentation of the images of onboard systems of optoelectronic observation, the rowing methods—ant method and artificial bee colony method are developed. The choice of rowing methods for segmentation is due to their main advantages, namely: managing the process of finding the optimal solution; effective research into search space to find the optimal solution; using simple local search procedures and complex learning processes; the methods are approximate and, as a rule, non-deterministic; taking into account the possibility of trapping in a limited search space; universality.

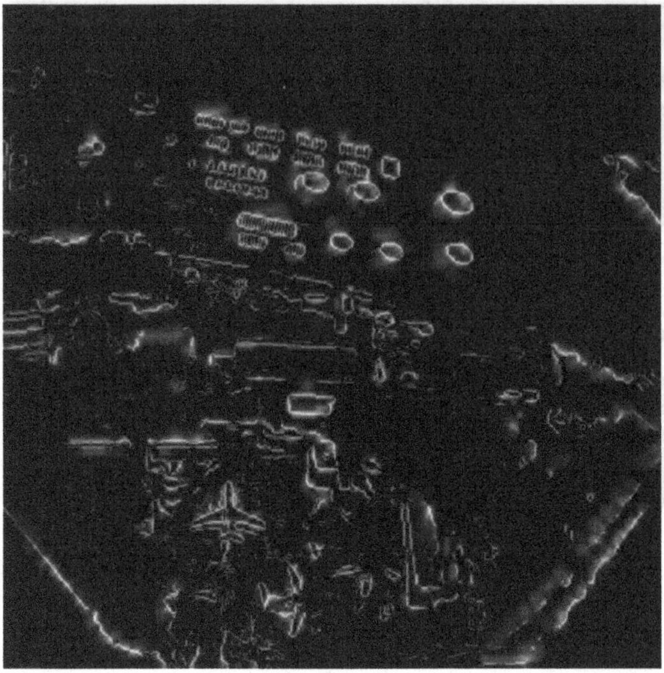

Fig. 2.32 The OEI segmented by the Canny method

A simpler ant method of segmentation of the on-board optical-electronic observation system is developed. A practical verification of theoretical calculations concerning the application of a simpler ant method to segmentation of optoelectronic image is carried out. Practical results have shown that there are errors in segmentation.

The advanced ant method of segmentation of the on-board system of optoelectronic observation, in which only the best agents increase the level of pheromone on their routes, has been developed. The test of the efficiency of the advanced ant method using the data of the control sample showed the optimal way of agents in the segmentation of the image in all implementations. Experimental studies of the segmentation of the optoelectronic image confirmed the performance of the advanced ant method. In a segmented image, as an example, possible objects of interest are identified. It was found that the selected contours of the main objects of interest coincide with the boundaries of objects in the original image. The evaluation of the effectiveness of the ant method is carried out in a visual way.

It is proved that one of the important advantages of the ant method is rapid adaptation in the conditions of the variable light and the presence of a plurality of objects of interest of various types and classes in the image. Under these conditions, the ant segmentation method quickly finds a new optimal solution. The main disadvantages of the advanced ant method of segmentation are also noted.

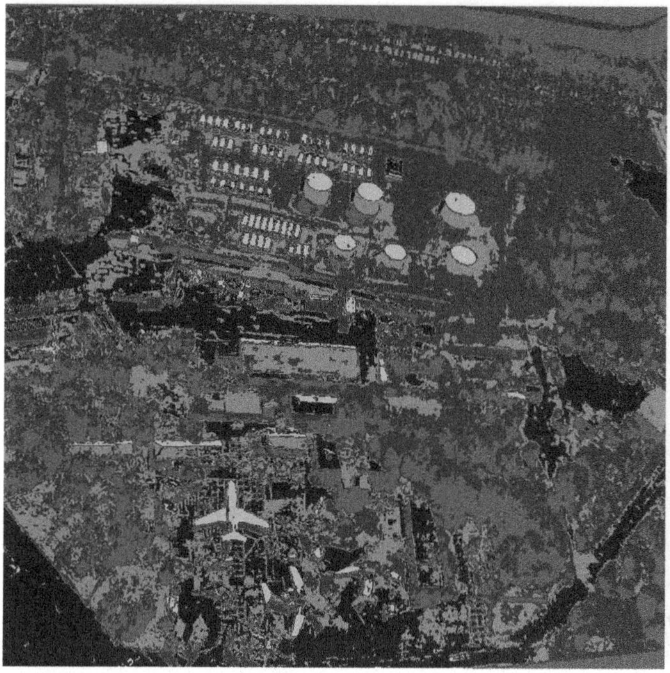

Fig. 2.33 The OEI segmented by the k-means method ($k = 5$)

An artificial bee colony method has been developed to determine the segmentation threshold for separating the image into objects of interest and the background. The essence of the method of artificial bee colony consists in determining the initial positions of the agents, their migration, determining the conditions of the stopping of the iterative process by the criterion of the minimum of the target function and determining the optimal value of the threshold level. The generalized diagram of the method of segmentation of the images of onboard systems of optoelectronic observation, based on the method of artificial bee colony, is presented.

The choice of the target function for segmentation of the images of onboard systems of optoelectronic observation is substantiated. It is proved that as a target function it is advisable to choose the sum of the dispersion of the brightness of segments of a segmented image. The objective function for typical model images is presented and the optimization problem of choosing the threshold of segmentation of the optoelectronic image is formulated. The optimization problem is solved by the method of iterative calculations.

Experimental researches on the segmentation of the optoelectronic image by artificial bee colony were conducted. Experimental studies have confirmed the efficiency of the developed segmentation method, which is based on the method of artificial bee colony. An estimation of the quality of segmentation of the optical-electronic image of the on-board monitoring system is carried out. For the quantitative assessment of

Fig. 2.34 The OEI segmented by the Random forest method

Table 2.2 Assessing the errors of the first and second kind of the OEI thematic segmentation by various methods of the thematic segmentation

The method of thematic segmentation	Without distortions		Rotation		Change of scale	
	α_1 (%)	β_2 (%)	α_1 (%)	β_2 (%)	α_1 (%)	β_2 (%)
Otsu	18.3	25.7	20.34	26.66	19.01	26.33
Canny	10.5	14.3	15.76	18.66	13.61	16.55
k-means (k = 2)	18.5	25.8	21.55	28.89	20.77	27.7
k-means (k = 3)	17.3	23.7	20.99	27.8	20.33	27.1
k-means (k = 4)	15.5	20.3	19.33	25.55	18.88	25.13
k-means (k = 5)	14.1	16.08	18.11	24.22	17.66	24.01
Random forest	15.9	17.76	19.12	24.66	18.12	24.88
ACO method	4.15	6.84	8.23	10.89	8.01	10.66
ABC method	2.7	4.55	3.45	5.66	3.66	5.69

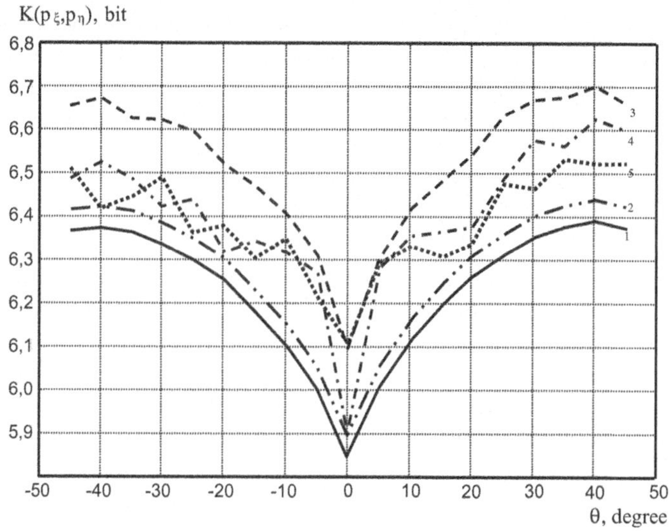

Fig. 2.35 Assessing the Kullback–Leibler divergence when the OEI is rotated

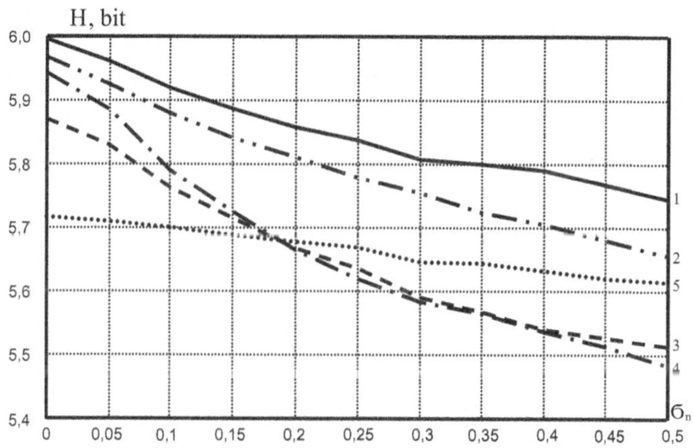

Fig. 2.36 Assessing the entropy of the segmented OEI under the AWGN effect

segmentation quality, the calculation of segmentation errors of the first and second kind has been carried out, the information indicators—the distance of Kulbak-Leibler and the entropy of the segmented image.

It is proved that application of the developed methods to segmentation of images of onboard systems of optoelectronic observation will allow:

- to reduce segmentation errors of the first and second kind by 4–13%;

- to improve the quality of the segmented image without the influence of distorting factors (gain in the meaning of information indicators: Kulback-Leibler distance—by 3–12%, entropy of the segmented image—by 5–13%);
- to improve the quality of the segmented image under the influence of distorting factors (the gain in the meaning of information indicators: Kulback-Leibler distance—by 5–14%, the entropy of the segmented image—by 7–16%).

The following factors support the validity and validity of the results:

- the use of theoretically substantiated and tested in practice methods of analysis and synthesis of complex information systems, decomposition of multilevel modeling, mathematical methods of image comparison, methods of digital image processing, probability theory methods, mathematical statistics, rooted intelligence, data clustering, evolutionary calculations, mathematical modeling, mathematical logic, methods of theory of optimization, analytical and empirical methods of comparative research;
- good coincidence of theoretical calculations and results of image processing.

References

1. Ruban I, Khudov H, Khudov V, Khizhnyak I, Makoveichuk O (2017) Segmentation of the images obtained from onboard optoelectronic surveillance systems by the evolutionary method. East-Eur J Enterp Technol 5/9)(89):49–57
2. Ruban I, Khudov V, Khudov H, Khizhnyak I (2017) An improved method for segmentation of a multiscale sequence of optoelectronic images. In: Problems of infocommunications science and technology: 4th international scientific-practical conference October, 10–13, 2017: thesis of reports. Kharkiv, 2017. pp 137–141
3. Ruban I, Khudov V, Khudov H, Khizhnyak I (2018) A swarm method for segmentation of images obtained from on-board optoelectronic surveillance systems. In: Problems of infocommunications science and technology: 5th international scientific-practical conference, October, 9–12, 2018: thesis of reports. Kharkiv, 2018. https://easychair.org/proceedings/paper.cgi?paper=434107;track=206520;a=19749082
4. Kharchenko VS, Illiashenko OA, Kovalenko AA, Sklyar VV, Boyarchuk AV (2014) Security informed safety assessment of NPP I&C systems: Gap-Imeca technique. Int Conf Nucl Eng, Proc, ICONE, 2014., vol 3. https://doi.org/10.1115/ICONE22-31175
5. Zhenbao L, Sicong T, Shuhui B, Hao Z (2013) New evaluation metrics for mesh segmentation. Comput & Graph 37:553–564
6. Dhanachandra N and Chanu Y (2017) A survey on image segmentation methods using clustering technigues. Eur J Eng Res Sci 2(1):15–20
7. Sharma N, Aggarwal L (2010) Automated medical image segmentation techniques. J Med Phys 35(1):3–14
8. Zhai H, Liu Y, Chen W (2012) Applying minimum-risk criterion to stochastic hub location problems. Procedia Eng 29:2313–2321
9. Faroogue MY, Raeen MS (2014) Latest trends on image segmentation schemes. Int J Adv Res Comput Sci Softw Eng 4(10):792–795
10. Choudhary R, Gupta R (2017) Recent trends and techniques in image enhancement using differential evolution—a survey. Int J Adv Res Comput Sci Softw Eng 7(4):106–112

11. Levashenko V, Zaitseva E, Puuronen S (2007) Fuzzy classifier based on fuzzy decision tree. In: Proceeding of the EUROCON 2007—The international conference on computer as a tool, pp 823–827
12. Canny JF (1986) A computational approach to edge detection . IEEE Trans Pattern Anal Mach Intell 8:679–698
13. Kumar B, Kumar D, Jarial SK (2017) A review on artificial bee colony algorithms and their applications to data clustering. Cybern Inf Technol 17(3):3–28
14. Ayman E-B, Jiang X, Suru JS (2016) Biomedical image segmentation: advances and trends. CRC Press, US, 546 p
15. Waseem K (2013) Image segmentation techniques: a survey. J Image Graph 1(4):166–170

Chapter 3
Spatio-Temporal Data Interpretation Based on Perceptional Model

Anatolii Kargin and Tetyana Petrenko

Abstract This chapter describes approach to interpretation of the heterogeneous data from sensors based on a new perception model, that implement cognitive functions as abstraction of data, tracking context and switching attention. Based on fuzzy L-R numbers, the knowledge presentation and rules engine inference are introduced. The resilient and interoperability of the perception model is shown in two examples of interpretation heterogeneous spatio-temporal data from sensors about the situation at the intersection and about the navigable path between landmarks of the robot route.

Keywords Internet of things · Smart rules engine · Fuzzy L-R numbers · Abstraction inference · Cognitive robotics · Context · Attention · Prototype · Hierarchical structured fact

3.1 Interpretation of Heterogeneous Data from Sensors in Mobile Smart Machines

3.1.1 Mobile Smart Machines Design Problems

In the epoch of the fourth digital revolution (Industry 4.0) are actual the solutions of automation tasks based on the concepts of the Internet of Things (IoT) [1, 2], Embedded Artificial Intelligence (EAI) [3–5] and Smart Machines (SM) [6, 7].

For example, the plans for the implementation of the strategic development of the railway industry [8], as one of the leaders in the implementation of Industry 4.0 technologies, include a wide range of solutions based on the concepts of IoT, EAI and SM. Created smart rail cars and smart trains [9–11]. Leading automotive

A. Kargin (✉) · T. Petrenko
Ukrainian State University of Railway Transport, 7 Feierbakh Square, Kharkiv 61050, Ukraine
e-mail: kargin@kart.edu.ua

T. Petrenko
e-mail: petrenko_tg@kart.edu.ua

V. Mashtalir et al. (eds.), *Advances in Spatio-Temporal Segmentation of Visual Data*, Studies in Computational Intelligence 876,
https://doi.org/10.1007/978-3-030-35480-0_3

companies are working to create and improve smart cars [12]. The common task of these solutions is to control the movement of SM along a given route. There are two possible approaches to solving this problem. The first is the preliminary ordering of the environment in which the SM operates, and the second approach is the creation of intelligent control systems that support the execution of the task in open-ended environment not designed specifically for SM [13]. The first approach is applied in production automatic lines, processes of servicing automated warehouses, in auto driving systems of electric vehicles of metros. The second approach is preferable when, in automation projects, the costs of environmental streamlining are significant, or the environment ordering is completely impossible due to the specificity of the SM application conditions. For example, in the case of a Mobile SM (MSM) crossing a traffic intersection or a pedestrian crossing, safety cannot be ensured only by ordering the environment. For such applications, MSM should have intelligence enough, for example, to avoid obstacles to prevent collisions [14–17]. The MSMs control relies on the Global Situation Model (GSM) [8], which is formed based on heterogeneous data from various sources in real time. First, it is data on the internal state of MSM, such as the technical condition of machine units: position and orientation in space, speed, and others. Secondly, it is data on the state of the MSM environment from video sensors, sound sensors, ultrasound and infrared sensors; data from sensors that monitor weather conditions, geographical location and many others. Video data in this case is one of the important sources of information.

In the concept of the IoT, as well as in the concept of SM, machines are considered in mutual relation with each other and with other Internet of things objects [11, 18, 19]. SMs communicate with each other via the Machine-to-Machine (M2M) protocol, exchanging data for use in management and decision-making tasks. Relevance GSM is supported by data flow from SMs. Monitoring environmental, economic, technological, production situations [20–22] is an example in which GSM plays an important role. Particularly relevant are systems in which, through monitoring, the safety of facilities is achieved [23]. In such systems, increased demands are placed on reliability characteristics. The solution is possible due to the introduction of additional requirements for models and methods of information processing, software and hardware [24]. Based on the IoT concept, another way to improve reliability is proposed. It is based on obtaining additional clarifying information using autonomous robots as the kind of MSMs [25]. Monitoring in this case is organized as follows. The main components of the monitoring system are stationary distributed in space, and MSMs (robots, drones) move in this space in real time [25, 26]. Data from sensors received from MSMs do not duplicate information from stationary devices, including SMs, but complement the description of the situation, which increases its reliability. This method of monitoring is economical in terms of costs for large-scale distributed objects. The stationary components of the system localize the state of the monitoring object at a general level, and MSMs perform the clarification of the situation. Examples of using the MSM group in conjunction with stationary components to obtain additional clarifying information are video surveillance systems, security systems and military systems. In the monitoring systems of fire-hazardous situations in the room, some parameters are centrally controlled with the help of permanently installed sensors (smoke, temperature, flame sensors, etc.). To eliminate equipment false alarms or false alarms that are caused by other causes, additional information from MSMs is used. A greater amount of information coming from the place of localization of a dangerous situation can significantly reduce the number of false alarms arising due to the influence of the environment [27, 28]. This is true for security, and for many other special purpose systems. So, for security systems, the place of unauthorized entry (the area where the motion sensor has triggered) is registered, and the identification of the type of violation requires additional information at the location. This information can be obtained through MSMs, which are equipped with a set of sensors, primarily video sensors, and are able to move to the area where the violation was located.

The functions that MSM performs are as follows:

1. Movement in the monitoring space from the park position to the specified position along the route. The MSM is controlled under noisy and disordered environments based on heterogeneous information from the sensors.
2. Monitoring the environment in the current position. Observation is performed using the sensor sets that MSM is equipped with. The observation process is distributed in space and in time, which is necessary for calculating the dynamic properties of the environment and is associated with search tasks.
3. Formation of a Local Situation Model (LSM), which is built on data from sensors and is a fragment of GSM.

The functions performed by the stationary monitoring components are as follows:

1. When signs appear that are characteristic of any violations of the normal situation, the MSM position is localized and a decision is made on how to obtain additional information about the state of the environment. Further, the parameters of the situation and the route to reach the target position are specified.
2. Update GSM based on the additionally received information in the form of LSMs received from MSMs and relevant information in previous positions of MSMs.

It is not always possible to obtain an MSM control solution by analyzing LSM and, moreover, GSM, presented directly based on primary data from sensors. Difficulties are caused by the need to analyze many prototypes of situations that should be constructed in the form of specific sets of heterogeneous data from sensors [29] from various sources in real time.

Analyzing high-dimensional data and real-time events is a common and urgent task for both SM and IoT applications. The problem of complexity today combines many different areas of research.

To determine the reaction to events in MSM and IoT systems, a mechanism that is built on rule processing (Rule based Engine, RE) is widely used [30–35]. Rules are processed based on various models, such as decision trees, data and flow diagrams, spatio-temporal segmentation of complex processes, Apache Spark space, business process management, and Bayesian networks [32]. For the representation of individual objects and their interaction in the IoT system, patterns of various levels of complexity were developed. The largest information companies offer their services for the accumulation and processing of IoT data (Amazon, IBM, Google). The service for building RE processing rules is one of the main parts of the IoT system. Naturally, there is a need to create framework that allow to solve these problems. For example, a successful Waylay project [31] has been created, which, thanks to the satisfaction of the RE requirements, is the most universal platform for managing events in IoT systems based on a rule mechanism.

Requirements to the RE need be defined by the characteristics of the SM and IoT applications.

First, the RE must support event handling in real time from a large variety of devices with different communication protocols.

Secondly, the RE must provide scaling and flexible control rules based on the needs of a particular system, the possibility of combining online event analytics with offline analytics.

Third, since the SM and IoT systems have significant uncertainty, the event handling rules in such systems must take into account the incompleteness and uncertainty of the data.

Different approaches are proposed and implemented aimed at meeting the specified requirements.

The problem of combining data from different sources is solved by granulating the input data into terms and using the terms of different data sources in one rule [36]. This approach is supported by both models of production rules and fuzzy systems with rules containing terms of linguistic variables [37, 38].

The problem of scaling and flexible configuration of rules is potentially embedded in the concept of rule-based systems. Traditional production model decides this problem. Production model supports the organization of knowledge in independent portions in the form of rules and facts [32, 35].

Fuzzy systems do not solve the problem of scaling and flexible configuration of rules. Using rules with linguistic variables does not allow modifying or supplementing the rule base without changing existing rules. This is due to the need to use the full rule base on the entire set of input data for correct output [38].

The problem of uncertainty and incompleteness of information symbolic calculations, which include these production models, do not cope [35, 37]. Therefore, approaches based on fuzzy systems [38, 39], the certainty factor [35, 37] or Bayesian reasoning [31] are used. The latter model works well with applications that interact with external data accumulated in cloud storage. For SMs, implemented in the form of intelligent robots or embedded cognitive systems [14, 40, 41], besides the probability it is necessary to take into account the incompleteness of the data and their fuzzy nature.

However, there are unresolved RE problems that do not allow to fully consider the features of the domain of SM and IoT applications [29, 34]. These are:

1. Dynamic changes in the environment cause difficulties in adapting the RE.
2. The problem of combining data from different devices and sensors.
3. RE do not work well in conditions of uncertainty and incomplete information.

4. The large dimension of the solution search space limits the possibility of a comprehensive analysis of the situation during the formation of the reaction.

The most promising approaches for the integrated solution of these problems are considered in cognitive robotics [37, 40–42]. The model of processing of heterogeneous data arriving in real time from several sources, including sensors, is divided into three main tasks: perception, movement planning and movement implementation, which is directly reduced to control.

The model of perception in cognitive psychology [43–45] solves the problem of interpreting data from sensory systems, which involve such psychological processes as meaning and relationships, the use of prior experience and context. The formalization of these psychological processes is considered in artificial intelligence as a representation of knowledge and a conclusion on knowledge by the method of generalization and abstraction.

In the perception model created by wildlife, the problems of the dimension of the input data are most effectively solved.

This section deals with the task of interpreting heterogeneous data (the first stage of information processing in MSM), which comes in real time from several sources, including video sensors. The result of data interpretation is the LSM of MSM, which is used in the subsequent stages of planning and control. The task of spatio-temporal segments data interpretation is solved based on a perception model borrowed from the cognitive sciences. The perception model is constructed on the rules based system concept which an inference mechanism supports two processes: (1) generalization and abstraction; (2) context and attention. This solution allows designing the MSM control system based on the RE [29]. The RE implementation provides a resilient and interoperability system with a flexible architecture [2].

3.1.2 *Abstraction of Data from Sensors*

As mentioned earlier, the perception is the first stage of the data processing at the process of control or decision making. This requires that the perception model is compatible and automatically aggregated with models of subsequent data processing steps. This integration can be performed based on the Lotfi A. Zadeh conception—From Numbers to Words [46]. The perception model presented in this section is a further development of the theory of Computing With Words (CWW) and Perceptual Computing (PC) [47, 48]. In this paper, in accordance with the approaches of cognitive psychology [43–45], the perception of data from sensors is the process of abstraction from data. The main components of model of perception of data from sensors are shown in Fig. 3.1.

Since the perception model is based on the CWW principles, data processing is performed on the basis of knowledge. Externally, domain knowledge is represented by the ontology in form of prototypes of the semantic network. Internally, knowledge representation model parameters take into account dynamic properties of the domain,

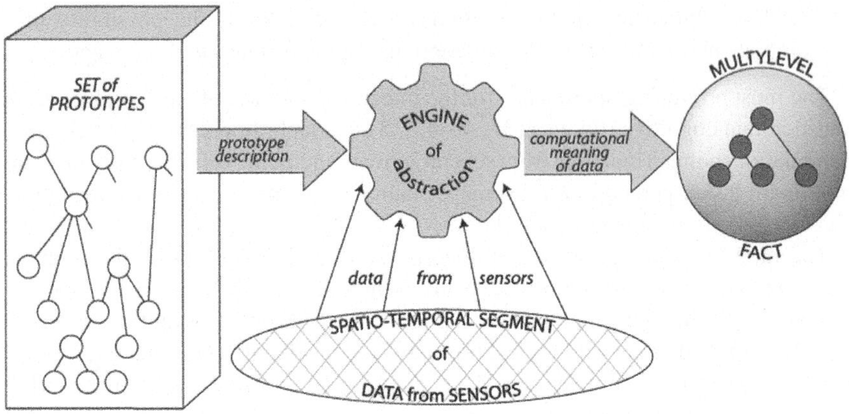

Fig. 3.1 Perception model components

incompleteness and uncertainty knowledge and relevance associated with the aging of the data from sensors. Knowledge representation model serves as a bridge between measurement-based information (spatio-temporal segmented of data from sensors) and perception-based information (fact).

Fact is computational meaning of data from sensors. Internally, the meaning of a fact is the Fact Fuzzy Characteristic (FFC) which expresses the degree of agreement of data from sensors with the certainty semantic network prototype. Externally, the meaning of a fact is expressed in natural language through its definition. The definition of a fact is given on the base of semantic network through its connections with other facts of the lower levels of abstraction.

Thus, the essence of the abstraction process of the perception model is the computation of the meaning of a spatio-temporal segmented data represented at different levels of abstraction. The Engine of Abstraction (EoA) implements the abstraction process (Fig. 3.1).

In this section, the computational meaning of a spatio-temporal segment of data will be considered as a result of input data interpretation. Since the applications of smart machines are considered, this interpretation will be presented as a *situation* semantic description. The dimension of the semantic presentation of the situation is significantly reduced compared with the presentation at the level of input data due to abstraction.

Consider the place of the perception model at the data processing chain in the MSM control system. At the first stage, the EoA performs abstraction from data from sensors. The abstraction result is the set of multi-level facts. The situation description obtained at the first stage as a set of facts is used at the second stage of data processing for the production of control decisions. In the MSM control system, the control decision is also presented in abstract form. At the third stage of data processing, the problem of specification is solved: a set of elementary control actions is assigned.

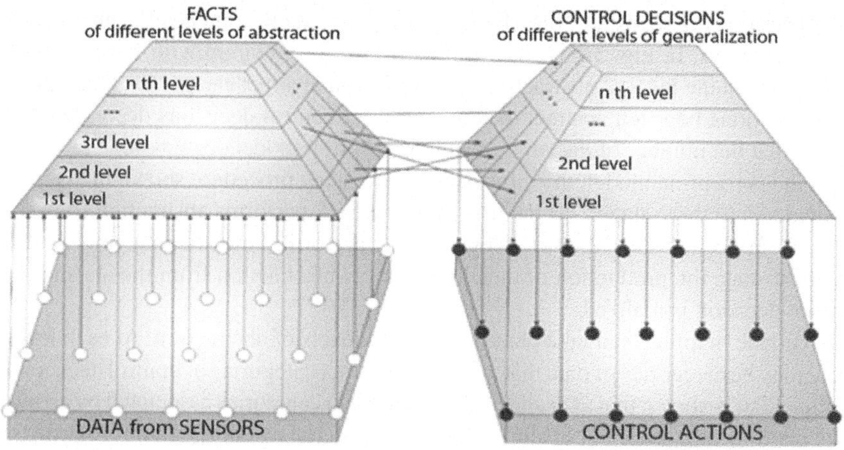

Fig. 3.2 Diagram of control decision making based on model of abstraction from data

Such an information processing scheme was proposed in [49, 50] and summarized in [51]. Figure 3.2 shows the information processing scheme of discussed model.

We illustrate the above scheme with the following example. MSM following a route must cross an unregulated crossroads. The MSM is equipped with an Ultrasonic Sensor (US) for the distance to an obstacle, which is mounted on a turntable platform. In addition, MSM is equipped with a video sensor and a sound sensor, which provide information about the state of the environment.

Being directly at the crossroads, MSM first scans the carriageway to the left of itself, then straight ahead and then to the right of itself. The result of viewing is the primary data from sensors (Fig. 3.2) about the distance to obstacles, audio and video information about the environment. It is not possible in all cases to decide to continue the movement directly based on only these input data. Difficulties are caused by the need to analyze many input data templates. Moreover, the situation at the intersection is dynamic and additional information is needed to decide (re-viewing of any direction). It can be seen, that the problem of decision making is complicated by the dimension of the problem and the incompleteness of information.

The person solves this problem by abstracting the data from the sensors. Thus, for the example under consideration, person uses the verbal presentation of the situation at the abstract level as a fact "an object is slowly approaching from the left at a great distance, there are no objects at all directly, and an object that is removed and creates a high level noise is very close to it". Then this knowledge is related to a certain class of situations, for example, "not very dangerous for crossing the intersection". Description of the data set from sensors in the form of a set of abstract categories in the verbal form "an object is slowly approaching from the left at a great distance", "there are no objects directly", "the object is very close to the right and creates a high noise level", "not very dangerous" there is a semantic representation of the situation. Each of these concepts describing a situation is considered as a structured fact of some level of abstraction. The above facts have a different level of abstraction from input data. In Fig. 3.2 is shown in the form of a pyramid with flat, that with an increase in the level of abstraction, the number of abstract categories, the number of connections between abstract data and abstract control decisions decrease. This means that the number of decision rules of perception model decreases.

In the perception model, the information processing procedure shown in Fig. 3.2, is divided into three stages. At the first stage the EoA performs abstraction from the data. At the second stage the fuzzy logic system performs heuristic mapping. And at the third stage the production of the elementary control action from the abstraction control decision is realized.

Heuristic mapping can take place at different levels of abstraction. At each level, there may be categories of data that associate abstract categories of control decisions. In Fig. 3.2 it is shown that at each level there are data categories associated by arrows with control categories. For example, for the considered example "at the crossroads" the category "there are no objects" refers to the first level of abstraction and may have a direct connection with the categories of controls "it is possible to cross the intersection". At the same time, the category "not very dangerous situation at the intersection" belongs to the seventh abstraction level and it can be associated with the same control category. The relationship between the abstract data category and the abstract control category is characterized by a degree of certainty. Such a solution reduces the dimension of the problem compared to the task of comparing specific data from sensors and a elementary control action.

At the third stage of processing information, in Fig. 3.2 it is indicated by an arrow connecting the pyramid with flat "control decisions" and the parallelepiped "control actions", the elementary control action is produced from the set representing the category. Control categories of different levels are chains of control actions of different levels of generalization. For example, "drive straight through the intersection", "drive through the intersection with a right turn" and so on. Each of the listed generalized controls represents a sequence of control actions of the lower level.

For the perception model, the key issue is that the domain knowledge and the facts have multi-level organization. This allow the different levels of abstraction to be used by control. This provides high efficiency of rules based control systems, as for the representation of knowledge, accumulated knowledge bases in the form of semantic networks [37], conceptual graphs [51], classification schemes and ontologies [52, 53] can be used.

The next equally important question is how to formalize Lotfi A. Zadeh bridge between numerical data from sensors and words description? The process of abstraction from numerical data is multi-step and includes the following types of abstraction [51]:

1. *Quantitative abstraction* from data. This is restrictions on the numerical data from the sensors based on the requirements for the solution accuracy.
2. *Definitive abstraction*. An abstract concept in form of nature language words is associated with each quantitative restriction, which reflects this quantitative restriction on the semantic level.

3. *Abstraction by generalization.* Semantic network and ontologies is example of this type of abstraction.

Researches in the field of granular computing are dedicated to solving the first two steps (quantitative and definitive) of process of abstraction. The concept of an informational granule, as a portion of knowledge, was introduced in [46] and discussed in [54–58]. In these works, the informational granule is considered not as a separate isolated unit, but as an aggregate organized in the form of a multilevel hierarchical structure. At different levels of the structure there are granules representing knowledge of different levels of abstraction. It is stated that the granules of different levels are interconnected "according to the principle of the bridge". Concretization of the "bridge" reveals the method for constructing top-level granules based on lower-level granules (bottom-up technology) or lower level granules based on top-level granules (top-down technology). However, the "bridge principle" is not generally formalized and is considered more from a philosophical point of view than from a practical one.

The organization of the nervous system of animals, which explicitly has a multilevel organization, confirms the bottom-up technology of sensory data processing. The structure is implemented using neurons organized as a network. Directly sensory neurons of the nervous system of animals through axons are connected with the following neurons of the network. For any neuron there is a chain of neurons in the network, connecting it with sensory neurons. For two neurons in the chain of the ith and jth, $i > j$, the neuron i is farther from the sensory neurons than the jth. It is known from biology that the ith neuron is activated by a larger group of sensory neurons than the jth one. Neuron ith has a higher level of generalization than jth. Nature in the nervous system of living beings has explicitly implemented a multi-level hierarchical structure for processing sensory data: in neurons of level i, the nervous system stores more generalized and higher levels of abstraction of knowledge about the surrounding world than in neurons of level j.

To build a perception model for embedded artificial intelligence or embedded cognitive systems [42, 59], it is necessary to answer the fundamental question. Where, at what level of abstraction will the boundary between numerical data and its verbal description? The aggregation of two different models of representation the numerical data and the semantic knowledge is performed using the head-on wave technology. The granulating data from sensors is performed "bottom-up". The granulating knowledge is performed "from top to bottom". Counter-granulation of data and knowledge continues until such level as knowledge granule become a direct verbal definition of the data granule. In perception model this level named as zero level.

The mapping of numerical data from sensors into zero-level granules is the task of the first stage of quantitative abstraction. At the second stage of the definitive abstraction, the input data, represented by zero-level granules, is assigned their meaning in verbal form. Starting from the third step, abstraction is performed on the basis of the first, second, and so on levels of granules, representing generalized semantic knowledge. The result of granulation is a multi-level hierarchical structure, which is used to represent both domain knowledge and the results of abstraction in the form of multilevel facts (Fig. 3.3).

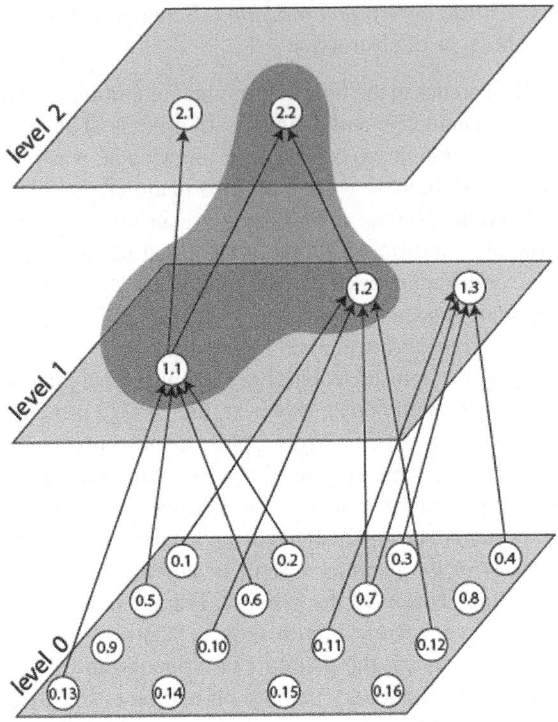

Fig. 3.3 The base granular multi-level hierarchical structure, which is used to represent both domain knowledge and facts

Suppose as an example in Fig. 3.3 shows the structure of the facts. For example, fact at the first level ***1.1*** represents a knowledge portion in the form of a semantic definition, which corresponds to the prototype of the data segment. The prototype is determined on the data from different sensors. In the figure, four rows of zero-level granules reflect the granulation of data from four sources. The range of possible data values from each source is divided into four granules. Thus, semantic knowledge of fact ***1.1*** describes a prototype of data built on granules ***0.2***, ***0.5***, ***0.6*** and ***0.13***. The fact of the second level, for example, **2.2** already has only semantic links. A portion of knowledge about the meaning of fact **2.2** is a generalized semantic knowledge about facts ***1.1*** and ***1.2*** of the first level of abstraction. Fact **2.2** of the second level is semantically defined by means of two portions of knowledge of the first level, which represent facts ***1.1*** and ***1.2***. The latter, in turn, are defined semantically by portions of knowledge about the zero-level granules ***0.2***, ***0.5***, ***0.6***, ***0.13*** and ***0.1***, ***0.7***, ***0.10***, ***0.12***. And the granules of the zero-level are determined by prototypes through certain data values from the sensors. Thanks to such a chain of links, the facts of the upper levels, presented semantically in the form of abstract concepts, give a semantic description to an arbitrary set of data segments. The process of abstracting from the bottom up, using data from sensors and a semantic representation of the domain, is the process of interpreting data from sensors in a perceptual model.

The perception model, in addition to EoA for interpretation of data from sensors, also includes Context Tracking Engine (CTE) and Switching Attention Engine (SAE). In cognitive psychology [45, 59], the nature of context and attention is considered as an interaction mediated by fields of different origin: electric, electromagnetic, magnetic, the concentration of certain molecules in a solution. In the brain of animals and humans, the connection between neurons that are so far removed in space that axons and dendrites themselves cannot have contact is carried out through fields. A pair or more of neurons, which should form a portion of knowledge of a higher level of abstraction but located in areas of the brain that are distant from each other, are united by context. In perception model, the context highlights and combines a subset of knowledge granules (concepts) of domain knowledge into one segment, which we will call the Context Segment (CS). A CS is a dynamically time-varying subset of domain knowledge. The CTE changes the CS in the space of granular structure in accordance with the logic of the problem being solved. A similar approach is used in this section and for the attention model. The Segment of Attention (AS) is understood as a certain subset of granular structure space. The SAE changes AS in accordance with the appearance of certain events in the environment.

Thus, there are two ways of organizing the interaction between the knowledge granules: through direct connections and through context and attention. If we now assume that Fig. 3.3 represents a granular structure of domain knowledge, then the CS depicted in dark color covers knowledge granules ***1.1***, ***1.2*** and ***2.1***. Context is related granules ***1.1***, ***1.2*** and ***2.1***.

Section 3.2 describes the Basic Perception Model (BPM). A model of the knowledge and fact presentation is introduced. A granular knowledge representation aggregates the verbally knowledge portion (concept) and an expert's knowledge about the characteristics of concept definition by prototype. Following shows the mapping segment of data from sensors into the set of structured multilevel facts. A computational model of abstraction is introduced, and an EoA is described.

Section 3.3 describes the Extended Perception Model (EPM). The model includes the tracking context and switching attention engines. A model for presenting contextual knowledge and a mechanism for tracking of the CS in real time are given. It is shown how the EPM performs the movement controlling of MSM along a given route.

Section 3.4 provides two examples of the EPM for MSM applications. The first is devoted to the interpretation of data from sensors about the situation at the crossroads using the BPM. The results of step-by-step modeling of a dynamic situation and its presentation in verbal form are given. The second example illustrates the use of the EPM in the SLAM task – interpreting the MSM motion path when is needed to identify the geometric shape of MSM path.

3.2 Perception Model

Knowledge-based systems are considered from two points of view. First, from the position of knowledge representation, it is necessary to answer the question: in what form are the domain knowledge formalized? In the perception model, knowledge is divided into two types: domain knowledge and knowledge representing the computational meaning of data. Secondly, from the standpoint of knowledge processing, it is required to show an EoA computational model.

Domain knowledge representation are considered in Sects. 3.2.1 and 3.2.2. The EoA computational model is considered in Sects. 3.2.3 and 3.2.4.

3.2.1 Granular Knowledge Representation

The perception model Knowledge Base (KB) contains knowledge about various essences and processes of the MSM environment. To formalize knowledge about these components, need to know in which environment MSM operates. The following classification of environment models is known [37, 60]:

1. Static or dynamic.
2. Deterministic or stochastic.
3. Open to change or enclosed.
4. Fully observable or partially observable.

If the environment state does not change during the execution of the MSM task, then we are talking about a static environment. If the environment state changes during this time, then it is necessary to consider the dynamic model. In this case knowledge representation model must take into account time changes (delays) and information aging parameters.

For a deterministic environment, causal dependences [60] are known in the form of some mappings. Knowledge about these dependencies can be represented by models based on classical set theory and logic. However, if information about the environment is not enough, not all causal relationships are known. In this case, probabilistic models or the certainty factor apparatus can be used [35, 37].

For enclosed environment, the set of possible data sources is constant and not replenished throughout the entire period of MSM operation. An open environment requires the introduction of new knowledge into the KB and use of data from new sources. Therefore, the model must be open and adapt to changing environment.

If the hypothesis of a fully observable environment is accepted, then the formalism is based on the apparatus of crisp set and crisp mapping. The assumption of a partially observable environment implies the need to consider two questions. There are uncertainty of knowledge and uncertainty of data [38].

Lack of full knowledge of the domain create an uncertainty of knowledge. Three sources of uncertainty are taken into account: (1) incomplete knowledge; (2) the expert's confidence in the definition of the concept; (3) aging of information about the components by which the definition of the abstract concept is given. For each of these three factors, the corresponding parameters are introduced into the knowledge representation model of EoA.

Insufficient or inaccurate information about a given situation generates data uncertainty. Sensors have limited resolution and are subject to interference. The EoA uses data from different sources (sensors). This requires data processing on different physical scales, which is an additional source of uncertainty. This problem is overcome by granulating data from sensors and presenting them with a fuzzy characteristic. The formalization can build on fuzzy sets [38, 39].

The perception model considered in this section deals with knowledge about the environment of the following type: dynamic, partially observable, stochastic, open to change. To take into account the peculiarities of this type of environment, the following characteristics of a knowledge portion about the environment essences are introduced. This is an assessment of certainty, a temporary delay, the information aging rate and the information completeness

$$< a, b, \nu, g > . \tag{3.1}$$

A knowledge portion can to characterize an environment essence with varying certainty. The quantitative assessment of certainty based on the Stanford theory [35, 37] of Certainty Factor (CF). In (3.1), the parameter $-1.0 \leq a \leq +1.0$ is the certainty. It is a numerical estimate of the certainty that an entity has some property. If it is necessary to indicate the presence or absence of a property to determine an essence, then the certainty of a knowledge portion about a property is estimated by a number close to $a = +1.0$ or vice versa $a = -1.0$, respectively. When, with equal certainty, it can be said that an essence can either have a property or not, then the value is $a \approx 0.0$.

The characteristic b in (3.1) is a dynamic characteristic of the properties of the essence. The dynamics characteristic $0 \leq b < \infty$ (time interval) is used in the definition of processes and time events. For example, one property should appear earlier by b time units relative to another. The characteristic b may describe the frequency of occurrence of an essence property.

The information aging rate $0.0 \leq \nu \leq 1.0$ indicates how fast the relevance of information about the essence is lost. If the essence is a rapidly occurring environmental process, then the aging rate is close to unity $\nu = 1.0$. For static essence, the aging rate is $\nu = 0.0$. Aging information is proportional to the time interval Δt, during which there is no opportunity to observe the essence.

An estimate of the information completeness $0.0 \leq g \leq 1.0$ is a characteristic of the sufficiency of one such knowledge portion to determine the essence. When the definition of an essence is given by enumeration of its representatives (instances), then for the knowledge portion describing a specific instance of the entity, $g = 1.0$. A knowledge portion is enough to make a conclusion about the essence when observing an instance. In the case when an entity definition needs a certain set of properties, for example, n, then the characteristics of information completeness for individual portion of knowledge must satisfy the condition $g_1 + g_2 + \ldots + g_n \geq 1.0$.

Thus, the essence is represented by a set of portions of knowledge about its properties. Such a presentation will be called a prototype. In the perception model, prototype consists of a set of Knowledge Granules (KGs). The KG represents concept [53] as knowledge, which is expressed by some notion at deferent abstraction levels. Before introducing a formal definition of a prototype model, consider an example of the knowledge representation.

Consider MSM, which, before making a decision about the passage of an unregulated crossroads, must build a GMS based on the data from the sensors. The example is implemented on the hardware resources of the IoT and SM research and development polygon [23]. Wheel robots are used as MSM, and humanoid robots are used as pedestrians. The US, sound sensor and video sensor are input data sources of MSM. We consider the knowledge representation about the distance to some object on the road adjacent to the crossroads. Zero-level granules are introduced to represent a portion of knowledge about road sections where any objects are located. For this, the possible distances are divided into intervals (sections). Each section is associated with the zero-level granular with the identifier ***0.i***. The meaning of the knowledge portion associated with this granular is the following: "Some object is in the *i*th section". In Fig. 3.4 in square brackets shows the values of the interval (boundaries of the section in centimeters), and inside the shaded circle is the identifier of the corresponding granular. The identifier begins with a number corresponding to the knowledge abstraction level. The abstraction by generalization starts at zero level. The zero-level KG, for example, ***0.i*** represents the data from sensor by a knowledge portion "some object is in the *i*th section", if the US data belongs to the *i*th interval (Fig. 3.4).

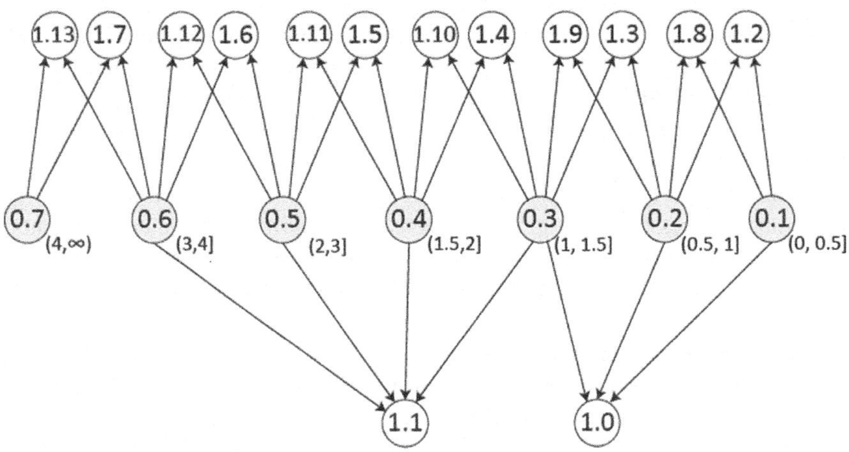

Fig. 3.4 Facts of the zero and first levels represent knowledge about the situation fragment at the intersection

At the first level of abstraction, data from sensors are represented by knowledge portions “The object is close” (KG identifier is ***1.0*** in Fig. 3.4.), “The object is far” (KG identifier is ***1.1***), “The object has moved from section 1 to section 2” (KG identifier is ***1.2***), “The object has moved from section 2 to section 3” (KG identifier is ***1.3***), and so on, if the object is removed from the crossroads. An object approach to the crossroads is described by other knowledge portions: “The object has moved from section 2 to section 1” (KG identifier is ***1.8***), “The object has moved from section 3 to section 2” (KG identifier is ***1.9***), and so on, “The object has moved from section 7 to section 6 “(KG identifier is ***1.13***). In Fig. 3.4 definition of the first level KG through the KGs of the zero-level are graphically shown as arcs. The following generalization levels of the data from the sensors will be given later. The model for presenting a KG by a knowledge portion is given in (3.2)

$$<N, know, \{<M_i, (a_i, b_i, v_i, g_i)>, M_i \in \Omega_N\}>, \quad (3.2)$$

where N is a KG identifier; *know* is a verbal description of a knowledge portion, which is an abstract concept; $\Omega_N = \{M_i\}_{i=1,2,\ldots,k}$ is a set of the lower levels KGs, which are used in the definition of the concept of *know*; M_i is a low-level KG identifier; (a_i, b_i, v_i, g_i) are the parameters (3.1) of the knowledge portion of the KG M_i.

For example, in Fig. 3.4 for the KG with identifier ***1.0***, the knowledge portion *know* = *object to be close* is a concept of domain. This concept is given by using knowledge portions of KG $\Omega_{1.0} = \{\boldsymbol{0.1}, \boldsymbol{0.2}, \boldsymbol{0.3}\}$: “Some object is in the 1st section”, “Some object is in the 2nd section”, “Some object is in the 3rd section”. In accordance with (3.2), the knowledge representation of KG ***1.0*** will take the form

$$<\mathbf{1.0}, \textit{object to be close}, \{<\mathbf{0.1}, (1.0, 0, 0.0, 1.0)>, <\mathbf{0.2}, (0.8, 0, 0.0, 1.0)>, <\mathbf{0.3}, (0.6, 0, 0.0, 1.0)>\}>. \quad (3.3)$$

With the help of the certainty parameter a, dynamic parameter $b = 0$, aging rate $v = 0.0$ and information completeness parameter $g = 1.0$ through three knowledge portions of KGs ***0.1***, ***0.2*** and ***0.3*** the generalized knowledge about the following situation is presented. The object is close when “the object is in the first section” or “the object is in the second section” or “the object is in the third section”. Each of the three knowledge portions describes a concrete instance of the concept *an object to be close*, therefore $g = 1.0$. Knowledge of the object location on at least one of the sections is enough to make a conclusion that the object is close. The knowledge portions in (3.3) are estimated by the expert values $a_1 = 1.0$, $a_2 = 0.8$, $a_3 = 0.6$. The expert expressed confidence that to define the concept *object to be close* enough to know that the object is in one of the three sections. It is enough to know only at the current time and does not require knowledge about its dynamics. Therefore, the dynamics parameter $b = 0$. Despite the dynamic domain is considered, the aging rate $v = 0.0$, because the frequency of updating the input data from the sensors is sufficient compared to the object moving speed.

3.2.2 Domain Knowledge Representation Based on Semantic Network

Original information about the domain for building a knowledge base is expert knowledge in the form of ontology, conceptual graph, semantic network, and others [35, 37, 51–53, 61]. Figure 3.5 shows an example of mapping a knowledge semantic representation to a multileveled prototypes presented by a set of parametrized knowledge granules.

The shown semantic network represents a fragment of knowledge about a safe situation in one of the directions at the crossroads. These are situations when there is no object on the road section adjacent to the crossroads (in Fig. 3.5 it is indicated as a circle with the number 1.14 inside) or close to the crossroads, an object slowly approaches and this is accompanied by a traditional noise level (the left branch in Fig. 3.5 is indicated by a circle with the number 4.0 inside). The right side of the Fig. 3.5 shows a graphical illustration of the presentation of the same knowledge in the form of a multileveled prototypes of situation. On it in the circles there are numbers of the corresponding KGs (1.14 and 4.0) representing this knowledge. Figure 3.5 shows the definition of not all the KGs of the first and second levels, but only those used to determine KGs 4.0 and 5.0. This fragment of knowledge representation is fully compatible with the fragment shown in Fig. 3.4. The KGs with the same identifiers in Figs. 3.4 and 3.5 represent one portion of knowledge. For example, 1.0, 1.1, 1.2 and others.

To present the prototype of the situation, we will use four basic relationships used in building semantic networks and ontologies based on them. There are: object–property (*consist_of*); whole–part (*part_of*); genus–species (*is_a*); action object–action–subject of action (*before*). The first three relationships are used in traditional

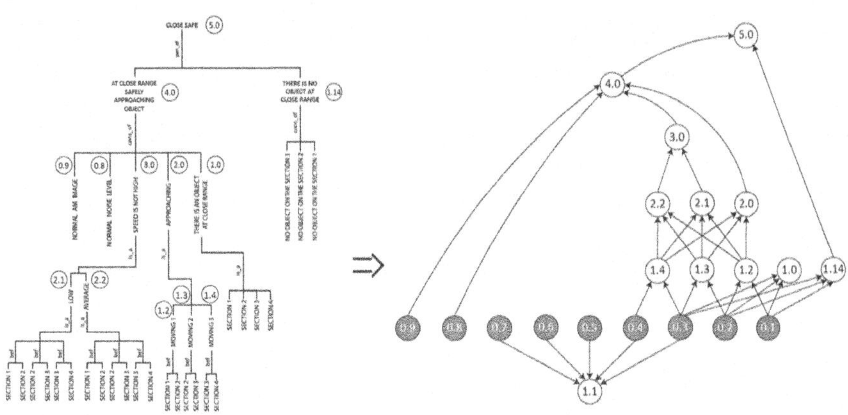

Fig. 3.5 Mapping the knowledge semantic representation to a multileveled prototypes presented by a set of parametrized knowledge granules

semantic network models of knowledge representation. The last relationship is introduced to represent knowledge of the processes passing in time including time events and actions such as "action object–action–subject of action". Any of these four relationships can be presented by model (3.2). One of them is a particular case of (3.2) and is specified by parameterization.

Consider these possibilities. The relationships *is_a* and *consist_of* are used to define concepts that do not describe any processes that take place in time. Therefore, in (3.2) the dynamics parameter is $b = 0$, which indicates on the current time. The *is_a* relationship is used to define a concept by listing representatives or instances belonging to a certain class. Portions of knowledge about any representative is sufficient to determine the essence as a whole. Therefore, the information completeness parameter in (3.2) is $g = 1.0$. If the definition using the *consist_of* relationship does not contain a description of at least one portion of knowledge, the definition will be incomplete. These arguments are the justification for choosing in (3.2) the completeness parameter value $g < 1.0$. For example, the concept is defined by the relationship *consist_of* through three portions of knowledge, as KG ***1.14*** in Fig. 3.5. In this case, for all three knowledge portions $g_i = 0.33$. However, it is not necessary that the completeness parameter be the same for all knowledge portions, it is important that the condition $g_1 + g_2 + g_3 \approx 1$ be fulfilled.

The relationship *before* specifies an ordered in time sequence of changes in the properties of an essence. The parameter b in (3.2) indicates the time interval between two consecutive properties changes. In the semantic network in Fig. 3.5 using the *before* relationship given the definition of events: *the object has moved slowly from section 2 to section 1* (KG ***1.2***); *the object has moved slowly from section 3 to section 2* (KG ***1.3***). Identifiers in brackets correspond to the KGs in Fig. 3.5. Different values of the dynamics parameter b allow to specify events: *the object has moved at an average speed from section 2 to section 1*; (KG ***1.2'***) *the object moved quickly from section 3 to section 2* (KG ***1.2''***). These knowledge portions are used in the concept definition of the second level of generalization: the *object approaches* (KG ***2.0***), *the object approaches slowly* (KG ***2.1***), *the object approaches quickly* (KG ***2.2***). An even higher level of generalization is represented by the knowledge portion *the object approaches with an acceleration*.

With the help of the *part_of* relationship, it is possible to define the entity as a whole through knowledge portions about parts with indication of the loss of certainty. Loss of relevance associated with the aging of information. Consider the following situation. Separate parts of a whole entity are not simultaneously available for observation. Portions of knowledge arrive randomly and distributed in time. This is the most common case when heterogeneous data using from various sources. Data is received asynchronously at different time. For example, visual, sound and data of infrared or ultrasonic sensors. At the time of using the knowledge represented by the *part_of* relation, different portions of knowledge are characterized by different time delays Δt from the moment of observation of the corresponding parts of the entity. This is the partial loss of relevance of observations caused by the aging of the data. The quantitative assessment of the influence of aging data on relevance is given by the aging rate parameter v. It should be noted that the time delay Δt since the last observation was not indicated in the knowledge representation (3.2). This characteristic relates to the EoA computational model. The time delay Δt should not be confused with the characteristic of the dynamics b, which is included in the representation (3.2). The dynamics characteristic b is the time interval from the moment when properties of the observed part is being changed. Figure 3.6 illustrates the difference between the time delay Δt elapsed since the last observation and the time interval b since the change in the observed property. Time changes of the FFC parameter α are shown in the Fig. 3.6. Bold dots denote α values, circles on the time axis are the moments of data arrival from a source, crosses are the time of a FFC changes. In Fig. 3.6a the case is given when the source data is received regularly at each time. For all time points the delays from the last observation is $\Delta t = 0$. The time interval since the α change is, for example, five for the 8th time point. In Fig. 3.6b there is a case

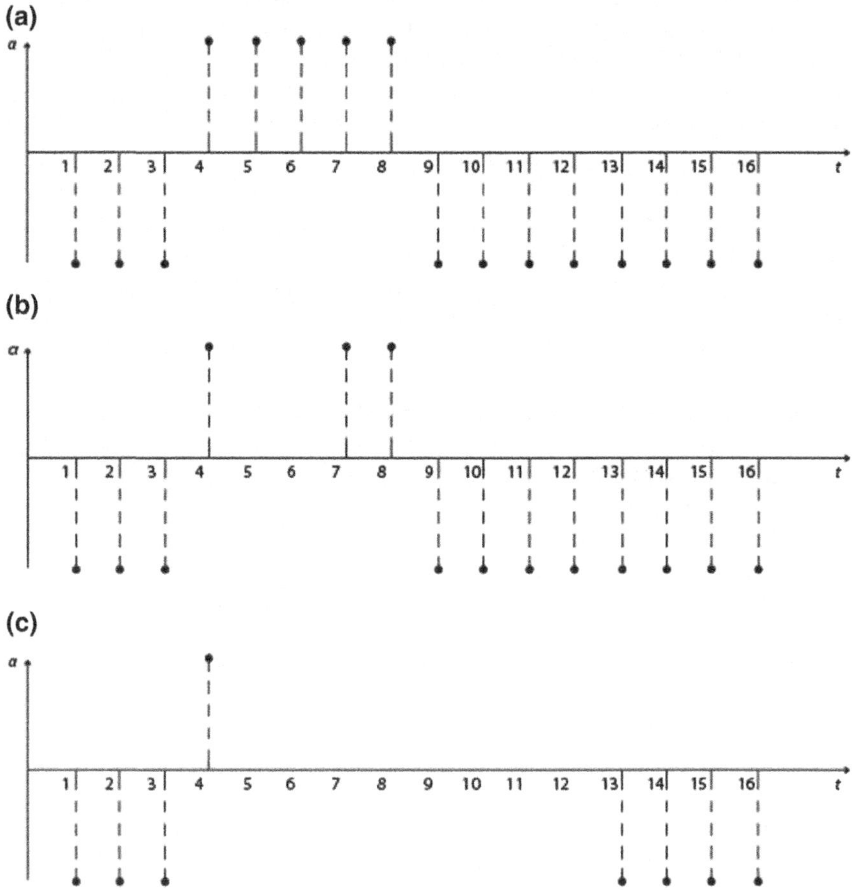

Fig. 3.6 Graphic illustration of the difference between the time interval Δt since the last observation and the time interval b since the change of the observed property

when data from the same source were absent (there was no observation) on the 5th and 6th time points. For these moments, the delays are, respectively, $\Delta t = 1$ and $\Delta t = 2$ time units. The data change interval is still five for the 8th time point. In Fig. 3.6c long delays in obtaining updated data from the source lead to a distortion of the time interval since the α change. It is equal to 9 time units. The lack of information from the source leads to a loss of relevance of previously obtained data.

In conclusion, the semantic knowledge of Fig. 3.5 is presented in the EoA prototype form (3.2).

1. <***0.1***, *some object is on the 1st section*, {<>}>;
2. <***0.2***, *some object is on the 2nd section*, {<>}>;

   ...

7. <***0.7***, *some object is on the 2th section*, {<>}>;
8. <***0.8***, *the sound of a slowly approaching vehicle*, {<>}>;
9. <***0.9***, *standard vehicle image*, {<>}>;
10. <***1.0***, *the object is close*, {< ***0.1***, (1.0, 0, 0.0, 1.0)>, <***0.2***, (0.8, 0, 0.0, 1.0)>, <***0.3***, (0.6, 0.0, 1.0)>}>;
11. <***1.1***, *the object is far away*, {<***0.4***, (0.8, 0, 0.0, 1.0)>, <***0.5***, (1.0, 0, 0.0, 1.0)>, <***0.6***, (1, 0, 0.0, 1.0)>}>;
12. <***1.2***, *the object has moved slowly from section 2 to section 1*, {<***0.1***, (1.0, 0, 0.15, 0.5)>, <***0.2***, (1.0, 2, 0.15, 0.5)>}>;
13. <***1.3***, *the object has moved slowly from section 3 to section 2*, {<***0.2***, (1.0, 0, 0.15, 0.5)>, <***0.3***, (1.0, 2, 0.15, 0.5)>}>;
14. <***1.4***, *the object has moved slowly from section 4 to section 3*, {<***0.3***, (1.0, 0, 0.15, 0.5)>, <***0.4***, (1.0, 2, 0.15, 0.5)>}>;
15. <**1.2′**, *the object has moved at an average speed from section 2 to section 1*, {<***0.1***, (1.0, 0, 0.15, 0.5)>, <***0.2***, (1.0, 12, 0.15, 0.5)>}>;
16. <***1.3′***, *the object has moved at an average speed from section 3 to section 2*, {<***0.2***, (1.0, 0, 0.15, 0.5)>, <***0.3***, (1.0, 12, 0.15, 0.5)>}>;
17. <**1.4′**, *the object has moved at an average speed from section 4 to section 3*, {<***0.3***, (1.0, 0, 0.15, 0.5)>, <***0.4***, (1.0, 12, 0.15, 0.5)>}>;
18. <***1.5***, *the object has moved slowly from section 5 to section 4*, {<***0.4***, (1.0, 0, 0.15, 0.5)>, <***0.5***, (1.0, 2, 0.15, 0.5)>}>;
19. <**1.6**, *the object has moved slowly from section 6 to section 5*, {<***0.5***, (1.0, 0, 0.15, 0.5)>, <***0.6***, (1.0, 2, 0.15, 0.5)>}>;
20. <***1.7***, *the object has moved slowly from section 7 to section 6*, {<***0.6***, (1.0, 0, 0.15, 0.5)>, <***0.7***, (1.0, 2, 0.15, 0.5)>}>;

   ...

21. <***1.14***, *there is no object near*, {<***0.1***, (−1.0, 0, 0.0, 0.33)>, <***0.2***, (−1.0, 0, 0.0, 0.33)>, <***0.3***, (−1.0, 0, 0.0, 0.33)>}>;
22. <**2.0**, *the object is approaching*, {<**1.2**, (1.0, 0, 0.0, 0.17)>, <***1.3***, (1.0, 0, 0.0, 0.17)>, <***1.4***, (1.0, 0, 0.0, 0.17)>, <***1.2′***, (1.0, 0, 0.0, 0.17)>, <***1.3′***, (1.0, 0, 0.0, 0.17)>, <***1.4′***, (1.0, 0, 0.0, 0.17)>}>;
23. <***2.1***, *the object is approaching slowly*, {<***1.2***, (1.0, 0, 0.0, 0.33)>, <**1.3**, (1.0, 0, 0.0, 0.33)>, <***1.4***, (1.0, 0, 0.0, 0.33)>}>;

24. <***2.2***, *object approaching at medium speed*, {<***1.2′***, (1.0, 0, 0.0, 0.33)>, <***1.3′***, (1.0, 0, 0.0, 0.33)>, <***1.4′***, (1.0, 0, 0.0, 0.33)>}>;
25. <***3.0***, *speed is not high*, {<***2.1***, (1.0, 0, 0.0, 0.5)>, <***2.2***, (1.0, 0, 0.0, 0.5)>}>;
26. <***4.0***, *object is approaching safely at close range*, {<***1.0***, (1.0, 0, 0.0, 0.25)>, <***2.0***, (1.0, 0, 0.0, 0.25)>, <***3.0***, (1.0, 0, 0.0, 0.25)>, <***0.8***, (0.8, *t*, 0.25, 0.35)>, <***0.9***, (0.8, *t*, 0.25, 0.35)>}>;
27. <***5.0***, *at close range the situation is safe*,

$$\{<\mathbf{4.0}, (1.0, 0, 0.0, 0.5)>, <\mathbf{1.10}, (1.0, 0, 0.0, 0.5)>\}>. \tag{3.4}$$

In (3.4), in the definition of KGs ***0.1***, ***0.2***, …, ***0.7*** there are no references to the KGs of the underlying levels and parameters. These are zero-level KGs, so there are no references in their definition.

3.2.3 Computational Fact Model: Generalization

A fact, is computational meaning of data from sensors. Externally, the meaning of a fact is expressed in natural language which is given in the definition of the KG as parameter *know* in (3.2). Internally, the meaning of a fact is the FFC which expresses the degree of agreement of KG with data from sensors. In other words, the FFC characterizes the certainty with which the input data satisfies the pattern presented in the form of knowledge (3.2). The Computational Fact Model (CFM) determines the FFC parameters. For the fact presented in the form (3.2), the CFM has k inputs and one output. The output of the model is a FFC, which is a meaning the input data of fact. The inputs of the CFM are the FFC of the lower levels facts $\{M_i\}_{i=1,2,\ldots,k}$, which are used in the definition of the concept *know* in (3.2). To obtain the FFC at the model output according to the FFC values at the inputs, the computational model uses prototypes knowledge in the form (3.2).

The FFC is based on the Certainty Factor (CF) model ($0.0 \le cf \le +1.0$). The CF model based on Stanford's theory of the certainty factor [35, 37]. In our work, CF is an integral fuzzy characteristic, taking into account the dynamic properties, stochastic nature, partial observability and aging of the input fact data.

We are proposed to use fuzzy L-R number [38, 62] as a FFC. For this, the fuzzy L-R numbers with three parameters are entered: certainty $-1.0 \le \alpha \le +1.0$, dynamic t_R and aging t_L.

$$\mathbf{X} : \{x | m_{\mathbf{X}}(x), \quad \forall x \in [-1, +1]\} \tag{3.5}$$

with Gaussian L-R function

$$\begin{aligned} m^L_{\mathbf{X}}(x) &= \exp(-(x-\alpha)^2/2t_L^2), \quad \forall x \in [-1, \alpha]; \\ m^R_{\mathbf{X}}(x) &= \exp(-(x-\alpha)^2/2t_R^2), \quad \forall x \in [\alpha, +1]. \end{aligned} \tag{3.6}$$

The parameters in (3.6) are the crisp numbers: certainty α and time intervals t_L and t_R. The first time interval characterizes aging due to the absence of input data for t_L time units. The second time interval t_R characterizes the dynamics of data changes.

A shot notation of the fuzzy L-R number (3.5), (3.6) is used in the form:

$$X = (\alpha, t_L, t_R). \tag{3.7}$$

Based on the FFC (3.5), (3.6), the integral numerical certainty estimate is determined. The fact CF cf is calculated by the equation:

$$cf = \alpha \cdot k_t;$$

$$k_t = 1 - \nu \cdot \frac{\sum_{\forall x \in [-1,\alpha]} m_{\mathbf{X}}^L(x) + \sum_{\forall x \in (\alpha,+1]} m_{\mathbf{X}}^R(x)}{Card([-1,+1])}. \tag{3.8}$$

The coefficient k_t in (3.8) takes into account the influence of both parameters of dynamics (t_R) and aging (t_L) on the CF, where ν is the normalized aging rate of the data.

The Fig. 3.7 shows two variants of FFC for which $k_a < k_b$, where k_a, k_b are the coefficients of CF, shown in Fig. 3.7a, b, respectively. With decreasing k, the cf tends to zero. This means that there is no certainty that the input data can be interpreted by a knowledge portion *know*, and that the input data cannot be interpreted by knowledge portion *know*.

Input variables of the CFM are fuzzy characteristics of facts. For the fact N in (3.2), the input variables are the set of FFCs $\boldsymbol{X}_{M1}, \boldsymbol{X}_{M2},\ldots, \boldsymbol{X}_{Mk}$, output variable is FFC $\boldsymbol{X}_N$. For example, for fact ***4.0*** in Fig. 3.5, which is determined through facts ***1.0, 2.0, 3.0, 0.8, 0.9*** the CFM has, respectively, five input variables, namely, FFCs $\boldsymbol{X}_{1.0}, \boldsymbol{X}_{2.0}, \boldsymbol{X}_{3.0}, \boldsymbol{X}_{0.8}, \boldsymbol{X}_{0.9}$. The output variable of the CFM is the desired fuzzy characteristic of the fact $\boldsymbol{X}_{4.0}$. The CFM is the foundation of EoA. To find $\boldsymbol{X}_N$, the CFM performs fuzzy calculations on fuzzy numbers $\boldsymbol{X}_{M2},\ldots, \boldsymbol{X}_{Mk}$. Finding $\boldsymbol{X}_N$ by the values of $\boldsymbol{X}_{M2},\ldots, \boldsymbol{X}_{Mk}$ will be denoted as a rule.

$$IF(\{<M_i, (a_i, b_i, \nu_i, g_i,)>, \quad \forall M_i \in \Omega_N\})\ THEN\ X_N, \ (\mathbf{X}_N). \tag{3.9}$$

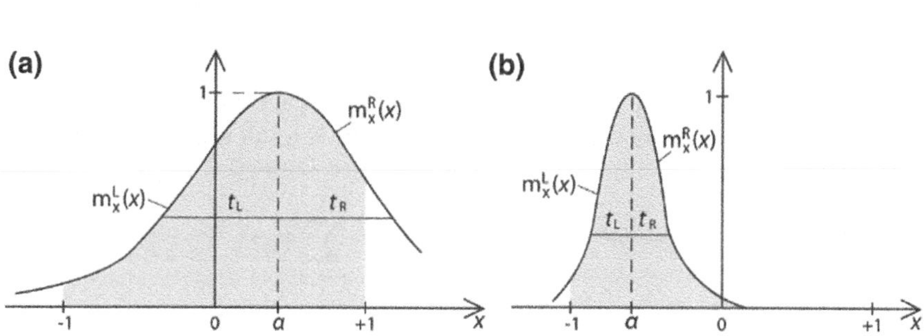

Fig. 3.7 Two variants of FFC for which $k_a < k_b$

The essence of the CFM calculations is the comparison of the prototype (3.2) with the input variables. The result of the comparison is the FFC, which should be considered as a fuzzy estimate of the closeness degree of the situation description given by the input variables and the prototype in the *IF* field of rule (3.9). The CFM computational procedure is divided into three phases: *matching* is pairwise comparison of the prototype, presented in the *IF* field of the rules (3.9), with the FFC of the corresponding input variable; *aggregation* of similarity estimates obtained in the first phase for all input variables; *actualization* the FFC parameters of the X_N in (3.9).

Matching is the operation of comparing two fuzzy L-R numbers: $\boldsymbol{X}_i = (a_i, t_{Li} = t, t_{Ri} = b_i)$, obtained from the rule (3.9), and the FFC $\mathbf{X}_i^{in} = (\alpha_i, t_{Li}, t_{Ri})$, corresponding to the CFM input. The result of the comparison is the new L-R number $\boldsymbol{X}'_i$. It is a fuzzy certainty that the compared fuzzy numbers are close. Calculations on this and following phases are based on the Zadeh generalization principle [63].

$$\mathbf{X}'_i = (\alpha'_i, t'_{Li} = t_{Li}, t'_{Ri} = t_{Ri}), \tag{3.10}$$

where

$$\alpha'_i = \exp(-\nu_i \cdot t_{Li}) \cdot MIN(\Delta_i^1, \Delta_i^2); \tag{3.11}$$

$$\Delta_i^1 = 1 - \frac{1}{2}\sqrt{(a_i - \alpha_i)^2}; \tag{3.12}$$

$$\Delta_i^2 = -1 + 2 \cdot \exp(-\nu_i \cdot |t_{Ri} - b_i|). \tag{3.13}$$

In (3.10)–(3.13), the parameters of the prototype from rule (3.9) and the parameters of the input FFC are used. The prototype parameters are a_i, b_i, ν_i. Input FFC parameters are α_i, t_{Li}, t_{Ri}. The CF α'_i in (3.11) is an affinity estimate of the input FFC parameters α_i, t_{Ri} and the rule parameters a_i, b_i, respectively. With small differences, estimates (3.12), (3.13) are close to *+1*, with maximum differences, estimates tend to *–1*. If the affinity estimate is obtained on the basis of actual data $t_{Li} = 0$, then it is not adjusted. For aged input data α_i, t_{Ri}, when the parameter $t_{Li} > 0$, the affinity estimate (3.11) is corrected in proportion to the aging rate ν_i so that $\alpha'_i \to 0$.

Aggregation of fuzzy affinity estimates obtained during the matching phase for each input variable of the CFM is carried out as an operation of adding k weighted fuzzy L-R numbers $\{\boldsymbol{X}'_i\}_{i=1,2,\ldots,k}$. The weighting coefficient is the information completeness parameter g_i from the rule (3.9). The operation result is again a fuzzy L-R number.

$$\mathbf{X}'' = (\alpha'', t''_L = \sum_{i=1..k} t'_{Li}, t''_R = \sum_{i=1..k} t'_{Ri}), \tag{3.14}$$

where

$$\alpha'' = \begin{cases} 2(\alpha''' - 0.5), \, if \; \alpha''' \leq +1; \\ +1, \qquad\qquad if \; \alpha''' > +1; \end{cases} \tag{3.15}$$

$$\alpha''' = g_1 \cdot (\alpha'_1 + 1)/2 + \cdots + g_i \cdot (\alpha'_i + 1)/2 + \cdots + g_k \cdot (\alpha'_k + 1)/2. \tag{3.16}$$

In (3.16), the scaling of the components is performed previously so that $0.0 \leq (\alpha'_i + 1)/2 \leq +1.0$, and then the weighted sum of the scaled numbers is found. In the expression (3.15), the inverse scaling operation is performed, so that $-1.0 \leq \alpha \leq +1.0$.

Actualization of FFC values at the CFM output is the final phase of data processing. The result of this operation is the L-R number $\mathbf{X}^{out}$. The operation is as follows. First, on the basis of the found value X'' of the fuzzy L-R number (3.14), the CF cf is calculated by the formula (3.8). Then this value of the CF is used to find the parameters of the L-R number $\mathbf{X}^{out}$, which is the final value of the FFC.

$$\alpha^{out} = cf; \tag{3.17}$$

$$t_R = \begin{cases} 0, \qquad if \; (cf \geq \varepsilon \& {}^{-}q = 0) \; or \; (cf \leq -\varepsilon \, \& {}^{-}q = 1); \\ {}^{-}t_R + 1, \; other \; wise \end{cases} \tag{3.18}$$

$$q = \begin{cases} 1, \;\; if \; (cf \geq \varepsilon \; \& \; {}^{-}q = 0); \\ 0, \;\; if \; (cf \leq -\varepsilon \; \& \; {}^{-}q = 1); \\ {}^{-}q, \; other \; wise; \end{cases} \tag{3.19}$$

$$t_L = MAX(t_{L1}, t_{L2,\ldots,}t_{Lk}), \tag{3.20}$$

where ${}^{-}t_R$, ${}^{-}q$ are the parameter values in the previous calculation step; t_{R1}, t_{R2}, …, t_{Rk} are the FFC parameters of input variables $X_i^{\text{in}}, i = 1, 2, \ldots, k$.

For MSM applications, various kinds of interference are typical, which, ultimately, distort the time of data change. This leads to incorrect calculations of the t_R parameter. To avoid this, a variable $q \in \{0, 1\}$ was introduced in (3.18), which takes the values $+1$ in the range $\varepsilon < cf \leq +1$ and 0 under the condition $-1 \leq cf < -\varepsilon$. The Fig. 3.8 shows an example of changing the value of the parameter α. The Fig. 3.8 shows how, due to interference, instead of a step function (shown by a dotted line) there is a random function of changing values $-1.0 \leq \alpha \leq +1.0$.

In this case, the localization of the moment when the CF changes is impossible. For example, in Fig. 3.8 the values of α at times t_4 and t_6 can be interpreted as data changes and at these times t_R should be reset. To solve this problem, we introduce the variable q, whose behavior is represented by expression (3.19). The variable $q = 1$ in the case when the value of α first falls into the domain $\varepsilon \leq \alpha$ with $q = 0$. In Fig. 3.8 it is time t_5. The variable q in this state will be at the next time instant t_6. A symmetric condition is required to localize the CF change in the range $-1.0 \leq \alpha \leq -\varepsilon$. This is $\alpha \leq -\varepsilon$ with $q = 1$.

According to expression (3.18), the t_R parameter, which characterizes the time interval that has passed since the last CF change, is either reset if the CF has changed,

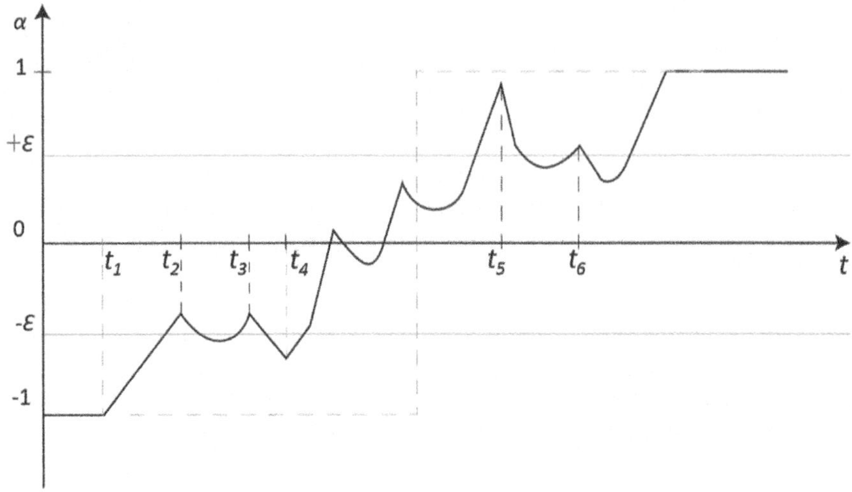

Fig. 3.8 Explanation of the q variable

or is being increased by one-unit time, if CF has not changed. The t_R interval is reset in two cases. When $\alpha \geq \varepsilon$ or $\alpha \leq -\varepsilon$. In applications, can to use one of them. In (3.18) one of the required conditions should be excluded. In the example about the MSM at the road intersection, for processing the input data by the rules (3.4), the EoA is tune to one condition $\alpha \geq \varepsilon$. Interpreting the input data, this allowed to represent dynamic information about events, for example, ***1.2*** only at time intervals since the event occurred (*the object has moved slowly from section 2 to section 1*). This knowledge is required to identify dangerous situations and knowledge of when the situation associated with the disappearance of an event is useless.

3.2.4 Engine of Abstraction

In Sect. 3.1.2, an abstraction model was introduced consisting of three types of abstraction: quantitative abstraction, definitive abstraction and abstraction by generalization. The model of abstraction by generalization was discussed in Sect. 3.2.3. The model of quantitative and definitive abstraction is discussed here. This types of abstraction based on granulation and fuzzification of the input data from sensors.

Granulation and fuzzification data serves as a bridge between data and knowledge [46, 54–58]. At this stage, the input data values are transformed into a semantic representation. The data is transferred to the knowledge portion represented by the zero level KG. The correspondence between the knowledge portion of KG and data is reflected in the parameters of fuzzy characteristics of the zero level facts.

The parameters of the granulation procedure are laid at the stage of perception model designing. Each data source is assigned a subset of informational granules.

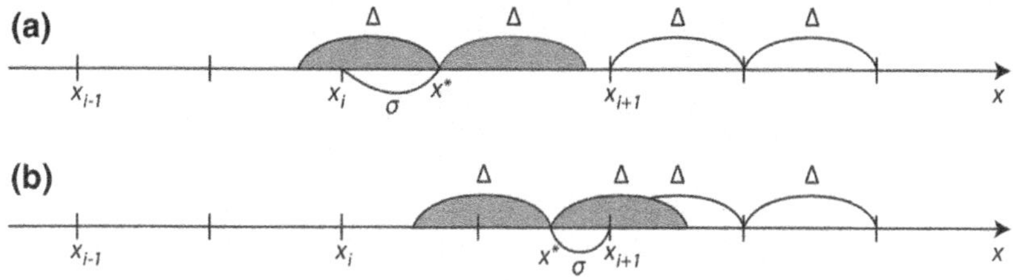

Fig. 3.9 Fuzzification of a granular input numerical variable: **a** $x_i \le x \le x_i + \Delta$; **b** $x_i + \Delta \le x \le x_{i+1}$

For this, the range of possible values of the source data is divided into several informational granules. The number of granules and their sizes depend on the problem to be solved. Quantitative characteristics of each zero level granule is created. Fuzzifier parameters are selected based on the accuracy characteristics of the sensor devices, the data monitoring frequency and the dynamic characteristics of the data source [29]. Thus, the granulation and fuzzification procedure takes into account the specifics caused by the hardware characteristics of the data source and the characteristics of the problem being solved.

Consider the granulation and fuzzification procedure for one numerical value x^* from one data source. Granule size is 2Δ, where Δ is the inaccuracy of the sensor device. The value range of the ith granule is $(x_i, x_i + 2\Delta]$. The amount of granules is d. In Fig. 3.9 two cases are shown when the value of the input numerical variable x refers to the ith granule. The CF for FFC is calculated by the formulas:

For the case when $x_i \le x^* \le x_i + \Delta$ and $\sigma = x^* - x_i$

$$\begin{aligned} \alpha_j &= -1.0,\ j \ne i-1, j \ne i; \\ \alpha_{i-i} &= -\sigma/\Delta; \\ \alpha_i &= \sigma/\Delta. \end{aligned} \tag{3.21}$$

For the case when $x_i + \Delta \le x^* \le x_{i+1}$ and $\sigma = x_{i+1} - x^*$

$$\begin{aligned} \alpha_j &= -1.0,\ j \ne i, j \ne i+1; \\ \alpha_i &= \sigma/\Delta; \\ \alpha_{i+1} &= -\sigma/\Delta. \end{aligned} \tag{3.22}$$

The parameters t_{Rj} and t_{Lj} FFC for all $j = 1, 2, \ldots, i-1, i, i+1, \ldots, d$ are set accordingly:

$$\mathrm{t}_{\mathrm{L_j}} = 0; \tag{3.23}$$

$$t_{Rj} = \begin{cases} 0, & if\ (\alpha_j \ge \varepsilon\ \&\ {}^{-}q_j = 0)\ or\ (\alpha_j \le -\varepsilon\ \&\ {}^{-}q_j = 1); \\ {}^{-}t_{Rj} + 1, & other\ wise; \end{cases} \tag{3.24}$$

$$q_j = \begin{cases} 1, & if\ (\alpha_j \geq \varepsilon\ \&\ {}^{-}q_j = 0); \\ 0, & if\ (\alpha_j \leq -\varepsilon\ \&\ {}^{-}q_j = 1); \\ {}^{-}q_j, & other\ wise; \end{cases} \tag{3.25}$$

Where ${}^{-}t_R' {}^{-}q$ are the parameter values in the previous calculation step.

Consider granulation and fuzzification in dynamics on the example of a data from an US. For example, if accurate distance measurement of the US is $\Delta = \pm 2$ cm than the interval $2\Delta = 4$ cm is considered as an informational granule, which corresponds to the zero level KG. The range of possible values of the distance is divided into $n = 32$ intervals. Figure 3.10 illustrates in dynamics the results of data granulation and fuzzification.

In the Fig. 3.10 the value $x(t - 2T)$ at the time $t - 2T$ is reflected by the straight line segment located above the numbers of the information granules. The Fig. 3.10.a shows the situation at time $t - 2T$, where T is the interval of control. This is the situation when there is no data from the sensor for a long time. The circles below show the zero level facts, which correspond to the granules. In Fig. 3.10b shows the

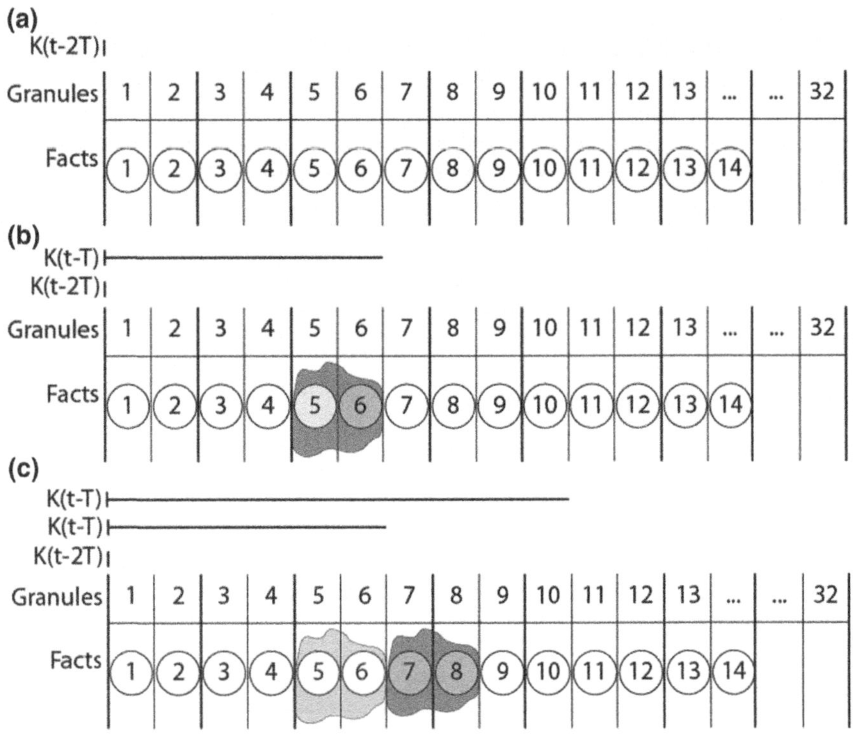

Fig. 3.10 Example of input numeric variable fuzzification: **a** at the time $t - 2T$, there was no data from the sensor for a long time; **b** at the time $t - T$, data corresponding to the 6th granule was obtained; **c** at the time t, the corresponding 7th granule data is obtained

situation at the next moment of control $t - T$. The variable value of the distance falls within the 6th interval and corresponds to the sixth informational granule. At this step, the FFC parameters of the zero level fact are calculated by formulas (3.21) and (3.25). The shading inside the circles of the 5th and 6th facts indicates the value $\alpha = +1.0$, and the light circles correspond to the values $\alpha = -1.0$. The circles in the Fig. 3.10 are surrounded by a background of different levels of shading. The shadow level corresponds to the value of the sum of the parameters $t_L + t_R$. The value $t_L + t_R = 0$ corresponds to the darkest background. The greater value of $t_L + t_R$, the brighter the background surrounding the circles. For large time interval $t_L + t_R$ during which there was no data from the source or the same data was received, there is no darkened background. The Fig. 3.10.c shows the situation when the data corresponding to the 7th granule have been received at the current time t. The α parameters of the FFC of the 7th and 8th facts have the values $\alpha_7 = +\, 0.9$ and $\alpha_8 = -0.3$, and the remaining facts $\alpha_j = -1.0$. Circles depicting the 7th and 8th facts are blacked out; the rest are light. The background surrounding the circles has changed compared to Fig. 3.10.b. The darkest background is around the 7th and 8th facts. A little lighter around the 5th and 6th facts, since the last time tick involved a change in their parameters $\alpha \geq \varepsilon$ (the fuzzification mechanism implements one first condition $\alpha \geq \varepsilon$ in (3.24)). The background around the remaining circles is bright, since there have been no changes in the values of the parameters corresponding to those granules for a long time.

The distribution of the FFC values can serve as a visual model of the data dynamics, represented by the set of zero level facts. In other words, the distribution of the α, t_R and t_L parameters of FFC is a model of data dynamics.

Here was considered the case when the input data are numbers. Due to the granular approach, the features of data processing for different sources are taken into account at the stage of granulation and fuzzification. After the transition from data to knowledge in the form of zero level facts, the knowledge processing procedure using CFM is universal and scalable to arbitrary types of data sources.

The EoA carry out data processing based on the knowledge from Knowledge Base (KB). The KB consists of structured prototypes and has a multi-layered structure. The KGs of prototypes in KB are stratified by levels. The zero level KGs do not refer to any other KGs. Knowledge of these KGs is used to calculate FFCs of corresponded zero level facts. The definitions of the first level KGs have links only to KGs of zero levels. At the second level the KGs have links only to KGs of zero and first levels. The third layer includes KGs, which have links only to KGs of zero, first and second levels. This ensures that there are no KG with cyclic connections between different layers. Such a KB structure greatly simplifies the EoA control algorithm.

In traditional rules-based systems [35, 37, 51], control of inference procedure is organized according to a principle known as "from knowledge to data". The EoA model uses the opposite principle "from data to knowledge". Due to the multi-layer structure of KB and due to the absence the cyclic connections between KGs of different layers the control algorithm "from data to knowledge" significantly increases the efficiency of the EoA.

The EoA consistently selects KGs from the KB for processing. First, the KGs of the first layer are selected in an arbitrary sequence. Then the second and so on layers.

The work of the EoA is completed after processing the last KG of the uppermost layer of the KB. Based on knowledge of each selected KG, the EoA calculates the parameters of the FFC, launching a procedure that implements the FCM, which was set out in Sect. 3.2.3. The result of the EoA operation is the meaning of created fact, represented by the *cf* (3.8). Thus, a fact is created by each KG from the KB. As a result, a set of multilevel facts are created that have the same structure as the prototypes of the KB. The final result is a subset of the facts, the structure mentioned, that received a high value of the certainty factor *cf*.

Interpretation of input data is performed in real time. At the first stage of granulation and fuzzification of input data, asynchronous reception of input data from different sources and calculation of fuzzy characteristics of zero-level facts is carried out. At the second stage, the EoA can operate in one of two modes: regular discrete processing of all the KGs of the KB or event-based processing of a KB subset of the actual KGs.

In the first mode, the granulation mechanisms and the CFM work asynchronously and independently of each other. In the second mode, the granulation and fuzzification mechanism, after processing the current input data from the source, activates the CFM only at those times when CF changes of zero-level facts are detected (conditions (3.24) are satisfied).

Consider these temporal fact-processing models. The difference between them is shown in Fig. 3.11.

The Fig. 3.11 on the vertical axis shows the names of the facts of the zero-level, the meaning of which is "the angle of rotation of the platform *i* degrees" with a resolution of 15 degrees. There is time on the horizontal axis. The crosses lines show events when the CF of the relevant fact has changed. The fact with the name "60" changed its indication at the time $t = 1$ in the ΔT model and at $t_0 = 0$ in the event model. The next in order is the fact "15" has changed its meaning. This happened

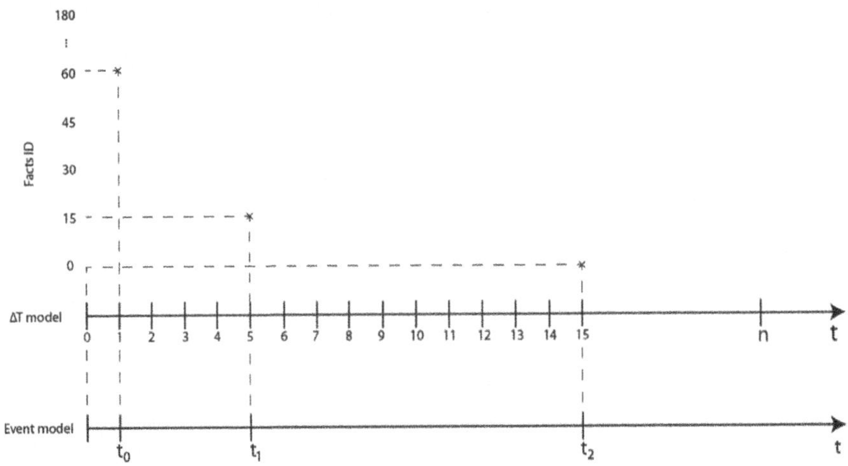

Fig. 3.11 Illustration of the difference between ΔT and event models

at the time $t = 5$ in the model ΔT and at the time $t_1 = 1$ in the event model. The next event in the figure shows the fact "0" changed the value of the CF. This event occurred at the time $t = 15$ in the ΔT model and at the time $t_2 = 3$ in the event model. Despite the intervals between the points t_0, t_1, t_2 on the time axis are different in the event, it is assumed that they are all equal to one-time unit T, that is, $t_1 - t_0 = t_2 - t_1 = T$. In the ΔT model, time is regularly sampled, and the unit is the interval between two adjacent points of the time axis t.

Processing information in real-time systems based on any of these temporary models involves the use of its own "clock". For the ΔT model, this is the traditional regular time sampling with a constant preselected interval T. In the event-based model, "watches" track the flow of events defined on a given set of facts, and for the ith order of events, they form an indication of "watches" in the form $t_i = t_{i-1} + 1$. Thus, the event fact "15" changed its value according to the watch of the ΔT model occurred at time 5, and according to the watch of the event model at time 1.

From the point of view of the required computing resources, an event model prevails. First, the CFM is executed only at the instants of the occurrence of events. This is much less rare than for the ΔT model, since in order not to miss the moment of changing the value of the variable, the sampling interval T is chosen as short as possible. Secondly, for the technology "from data to knowledge" [35, 51], at time of occurrence of an event not all KGs from KB are processed, but only those which correspond to zero level facts, which changed the CF. This subset of different levels KGs includes those KGs that either directly or indirectly have link to the zero level KGs that generated the event. It is assumed that the remaining facts, which are not included in the mentioned subset, have not changed the FFC in the time interval from the previous event. The advantages of this approach are minimization of calculations, a disadvantage is to require additional information about the subsets of facts associated with each generating an event. For a closed KB, this condition is always feasible.

For systems that use heterogeneous data from several sources with different dynamic characteristics, a hybrid model of organization will be more effective [64]. Each data source is associated with a subset of KGs ω_i from the KB. A subset includes all zero level KGs covering the entire range of source data, and upper level KGs that directly or indirectly use information from this source sensors data. The subset ω_i of KGs let's call the ith segment of the KB. The KB is divided into Q segments, where Q is number of sources. The intersection of segments is not necessarily an empty set

$$\Omega = \bigcup_{q=1,\ldots,Q} \omega_q, \quad \omega_i \cap \omega_j \neq \varnothing, \quad i \neq j.$$

For each KB section, the "own" CFM is attributed. It is organized based on a temporary event model. Thus, the processing of KGs in the KB is divided into several CFM, functioning asynchronously and independently from each other. Calculations take place in levels from bottom to top. When processing a fact of some upper level, data of facts of the lower levels can be used, which refer to several subsets for different data sources. Such facts are determined through the *part_of* relationship.

For example, fact ***4.0*** in Fig. 3.5. Its definition is given in (3.4). Data from different sources are processed by the CFM using the ΔT temporal model.

The calculations are organized as parallel and independent processing of the subsets associated with different data sources. Parallelism is synchronized at the levels of knowledge bases, in which there are KGs represented by the *part_of* relationship.

3.3 Expanded Perceptual Model with Tracking Context and Switching Attention Mechanisms

3.3.1 The Problem of Movement Controlling of a Mobile Smart Machine Along a Given Route

MSM solves the problem of interpreting data from sensors about a stochastic dynamic environment with incomplete information. Due to the use of the perception model, the results of interpretation present the situation on abstract level, which allows to make decisions in open-ended environment not designed specifically for MSM [13].

However, the organization of the KB in the form of a set of independent and unordered prototypes does not allow to represent knowledge about the route of MSM movement. To present a route, an ordered sequence of knowledge portions describing individual segments of the route is required. In order not to violate the organization of KB and not to indicate explicitly the connections between prototypes, reflecting set of different routes, contextual links are introduced into the perception model. The context links together the portions of knowledge (KG) at the time of solving the problem. Such a chain of KGs ordered by context is a representation of the route.

However, the problem of the organization of knowledge in accordance with the route description remains unresolved. This limits the ability to control the movement of MSM, taking into account the characteristics of the route.

This section discusses the extended perception model, including context and attention mechanisms [65]. The possibility of using the extended model in the MSM task of interpreting real-time data for make decisions about the continuation of the given route is shown.

Information processing in MSM is divided into three main tasks [25, 37, 65]: perception, route planning and movement control along the route. It is important that all three tasks are based on the same model of knowledge representation and inference on knowledge. Knowledge about situation, obtained by interpreting based on the perception model, is further processed to obtain new knowledge in the form of a plan for achieving the goal, and then this knowledge is used in control the MSM actuators. The following information processing concepts in robotics are known [25, 42, 66–69]: rigid program control; program control with feedback; reflexive or reactive control. Consider these approaches as a basis to implementing the concept of uniform knowledge-oriented model for all stages of information processing and possibilities of their aggregation with the perception model. Let the MSM must to

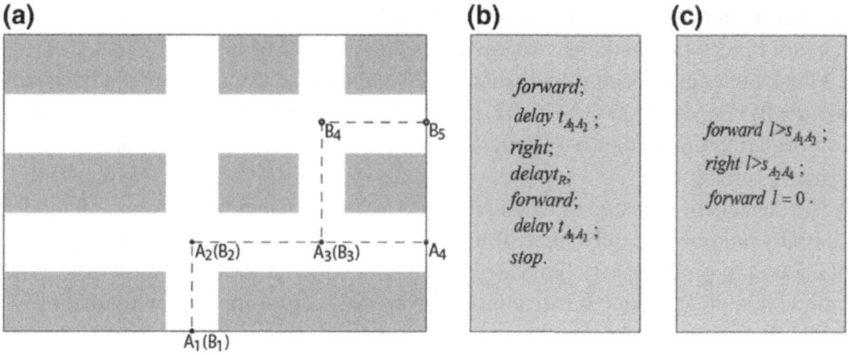

Fig. 3.12 Explanation of MSM control method: **a** route examples; **b** rigid program control for *A* route; **c** program control with feedback for *A* route

follow one of the routes passing through the Route Landmark (RL) as A_1, A_2, A_3, A_4 shown in Fig. 3.12a. An US, which is mounted on a turntable, gives the MSM information about obstacles in any of the four directions around it. An example is being considered with a limited set of control commands: *forward*, *right*, *left*. In the method of rigid program control, the results of the environmental data interpretation are not used. The control commands are implemented strictly sequentially: the *i*th command in order in the control program cannot start execution until the $i - 1$-th command is completed. The control program in Fig. 3.12a contains commands *delay* for delaying the start of the execution of the command following it by *t* units of delay time. Direct program control concept needs a constant MSM speed. Based on this assumption, the required time to move the MSM between adjacent RLs, for example, A_1 and A_2 is calculated. It indicated as $t_{A_1A_2}$, see Fig. 3.12. The MSM moving plan in the form of a rigid control program is shown in Fig. 3.12b, where T_R are the time interval for the execution of turn commands to the right. The control program (Fig. 3.12b) will be considered as a model of knowledge representation about the MSM movement plan.

The rigid program control method is demanding on the preliminary environment ordering: compulsory initial conditions and strict requirements on the speed of movement and acceleration and deceleration modes. If the first requirement can be fulfilled with a certain organization, then the second is almost never satisfied. These limitations are removed in the method of program control with feedback. Here, to identify situations when the MSM is in certain states, data from the sensors are used, and the values of time intervals are not used. Figure 3.12c shows the control program with feedback method similar in functionality to the program shown in Fig. 3.12b. Data from the US about the distance to the obstacle is denoted by *l*. In the control program with feedbacks in commands, for example, $forward\, l > s_{A_1A_2}$ the condition for completing the execution of a command is indicated $l > s_{A_1A_2}$. The MSM control system will form a *forward* command until the condition $l > s_{A_1A_2}$ is met. In this form, the MSM plan is not sensitive to variations in speed and acceleration, however, it is still intended to control in a well-ordered environment: the parameters $s_{A_1A_2}$, $s_{A_2A_4}$ are rigidly tied to the environment in which the MSM moves.

MSM are focused primarily on disordered open-ended environment not designed specifically for MSM [13, 15–17, 42, 65, 66]. In this regard, both the above concepts of information processing in robotics can't be used as a basic for supporting MSM cognitive functions.

The concept of reflexive or reactive control assumes that the environmental model is initially unknown or only some of its general characteristics are known, which are not enough to preliminarily build a plan for achieving the goal and, moreover, how to implement it. Behavior is formed mainly on feedbacks from the environment [37, 40, 41]. For the example considered here, knowledge about route *A* (Fig. 3.12.a) in a simplified form can be represented by an unordered set of rules (3.26).

$$\begin{array}{lll} R_1 & \text{IF Route} = 1,\ P_{OS} = A_1 & \text{THEN } forward, \\ R_2 & \text{IF Route} = 1,\ P_{OS} = A_2 & \text{THEN } right \\ R_3 & \text{IF Route} = 1,\ P_{OS} = A_2, l_L = near, l_R = near, l_F = far & \text{THEN } forward, \\ R_4 & \text{IF Route} = 1,\ P_{OS} = A_3 & \text{THEN } forward, \\ R_5 & \text{IF Route} = 1,\ P_{OS} = A_4, l_L = near, l_R = near, l_F = near & \text{THEN } stop. \end{array} \tag{3.26}$$

Knowledge in embedded rule-based systems are represented by rules (3.26) and facts that describe the current situation based on data from sensors. In rules (3.26) in the IF part, a prototype of a situation is described as a subset of facts: *Route* = *A* is a fact that describing route, *Pos* = A_i is a fact that indicating the location of the MSM, l_L, l_R, l_F are the facts that linguistic estimates the distance to the obstacle on the left, right and forward away from the MSM, respectively. It is assumed that using the set of sensors, the MSM receives information about its location and environment (distance to the obstacle). At the first perception stage, the data from sensors are mapped into knowledge in the form of facts. At following stages, based on this knowledge, a control response is formed according to the rules (3.26). In the KB (3.26) it is easy to introduce new rules. Thus, at the conceptual level, systems based on rules implement mechanism for representing and processing a homogeneous information at all three stages: perception, planning and motion control. However, the use of rule-based systems, including fuzzy ones, in the tasks of situational control encounters two problems. The first problem is related to the task dimension. In classical systems based on rules, it is necessary to fulfill the requirement of pairwise distinguishability of rules. This requires the introduction of new facts and rules with distinguishable combinations of facts in the IF part [29, 38]. In meeting this requirement, there is a duplication of rules that have the same meaning. Thus, the group of rules (3.26) must be supplemented with a similar set with the fact *Route* = *B* to introduce knowledge of the movement along route *B* (Fig. 3.12a). The second problem is low resistance to various kinds of disturbances that violate the movement MSM along the route. For example, in a regular situation, the rules R_1–R_5 in (3.26) must be consistently triggered in time for the MSM to complete its task. However, if an obstacle appears on one of the route sections, forcing the MSM to deviate from the route, then after the obstacle is eliminated, the MSM will not be able to continuation of the given route. Knowledge in the form of rules does not support the mechanism for tracking sequences of any events. Therefore, the MSM cannot restore the section of the route where it was before the obstacle appearance and, therefore, cannot continue its movement.

It is proposed to solve this problem with the help of two mechanisms: context and attention. To do this, perceptual model encapsulates context tracking and attention switching engines.

3.3.2 *The Perception Model Supplemented by Tracking Context and Switching Attention Engines*

The context in cognitive sciences is understood as the general meaning of a fragment of written or oral speech, which makes it possible to clarify the meaning of its individual words and sentences [42–44]. For the formalization of the cognitive phenomenon "the influence of context on the loss of semantic description of spatial-temporal data segments" using paradigm of artificial neural networks and data analysis methods is problematic [70–72]. This chapter discusses the context model, which is implemented as a rules-based system.

The perception model for obtaining the effect of context and attention is proposed to be modified as follows. In addition to the EoA, another Engine of Context Tracking (EoCT) is introduced into the perception model. The EoCT model is rules-based system model. The EoCT KB contains rules that use both the facts created by the EoA and special satellite facts. The EoA produces abstract description of RL as RL Fact (RLF). The EoCT produces and controls special Context Satellite Fact (CSF). Controlling the CSF comes down to changing its FFC. This is done based on the rules of EoCT KB. The CSF is tied with RLF and is satellite fact of RLF. The inference control process has the following features. The CSF processing is carried out only at that time moment when an event occurs (change in the characteristics of the tied RLF). Thus, CSF FFC is not explicitly dependents on data from sensors. Only the above-mentioned events can explicitly changes FFC of CSF.

The rule in which there is a CSF in the THEN part is activated by an event that occurred and described by tied RLF.

And so in rules, EoCT uses knowledge only about CSF and the EoA does not use rules with RLFs. Rules for processing CSFs must in the IF field explicitly contain knowledge about the event. A necessary condition for the activation of the rule is the relevance of the specified event. Sufficient condition of the rule comparability is the comparability of the remaining prototypes of the rule IF field.

In the rule, a reference to an event that is generated by FFC of RLF changes, for example, *Pos* = *A*, will be denoted as *event (A)*. In this case, the rules pairwise distinguishability requirement weakens and reduces to the requirement of dynamic distinguishability of the rules. Only those rules that are activated by the same event should be distinguished. The event processing mechanism significantly reduces the number of rules in KB.

The context tracking is illustrated on Fig. 3.13. In the Fig. 3.13, the RLFs are ordered according to the route. These facts are labeled A_i. The CSFs are beside RLFs. They are labeled A_i^{Csf}, respectively. The context model is represented in dynamics. The actualized fact (value of FFC parameter $\alpha = +1$) is depicted as a darkened circle, and the irrelevant fact (value of FFC parameter $\alpha = -1$) is shown as a light one. The light background surrounding the facts highlights a subset of the facts that belong to attention segment. A darker ellipse covering a couple of facts points out the current position of the context segment: the activation of these facts is facilitated by the activated context. The Fig. 3.13 shows the facts relevance distribution in time for six-time moments.

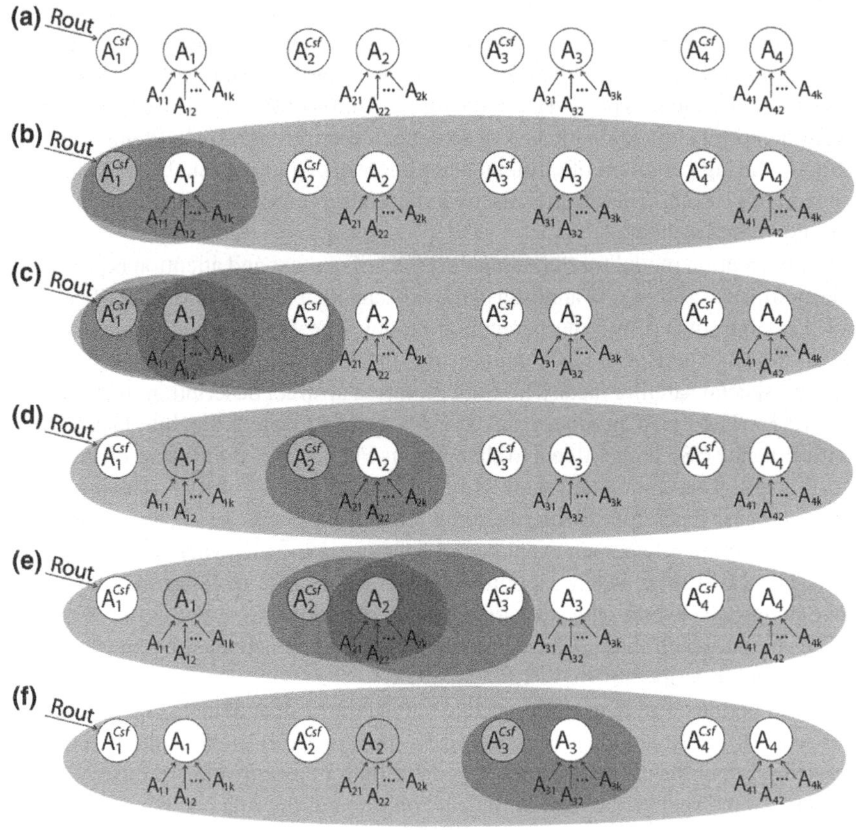

Fig. 3.13 An example of changing the context segment in accordance with the movement of MSM along the route

In the Fig. 3.13 in line (a) shows the distribution at the initial moment before the MSM began moving along the route. In the Fig. 3.13 in line (b) shows at the time when the input data about the route number actualized attention. In the Fig. 3.13 in line (c) describes the situation when the data came from the sensors about the current position (MSM is in RLF A_1 in Fig. 3.12). Line (d) shows the movement of the CS, caused by the update of the CSF A_2^{Csf}. Line (e) corresponds to the point in time when MSM reached the next RL, about which data was obtained from sensors that updated fact A_2. An event caused by the appearance of the value of the parameter $\alpha \approx + 1.0$ FFC A_2 activated the processing of tied CSF. The result of the CSF processing is a change FFC of the CSF (Fig. 3.13, line (e)) and, as a result, the displacement of the CS, as shown in Fig. 3.13, line (f). Activating CSF A_2^{Csf} extends the context segment to CSF A_3^{Csf}. The CS is moved from A_2^{Csf} to A_3^{Csf}. This state is shown in the figure in line (f). From this time point and until reaching the next RL, in the absence of obstacle on the route, two facts will be actualized: RLF A_2, the relevance of which is supported by data from sensors of the A_2 RL, and CSF A_3^{Csf}, the relevance of which is maintained by the context segment until sub-goal will be reached (the RL A_3).

Thus, the essence of the dynamic context model is following. The context segment, for certain events that actualized by data from sensors of an RL, moves in space of facts. The sequence of such events mapped into route of CS motion. The CS is supported in time by the CSF around which it is concentrated. The CS moving is triggered by an event, when data from sensors that describes the RLF features is appear.

3.3.3 *Engine of Tracking Context*

The context tracking model is considered, as before, with using the example of MSM movement control along a given route. Consider the route, for example, *A*, depicted by the dotted line in Fig. 3.12.a. The route passes through the RLs A_1, A_2, A_3, A_4. For each RL, the EoA creates a RLF, possibly structured multi-level, describing the unique features of the RL. For each RLF there are several CSF. The number of CSF corresponds to the number of routes passing through this RLF. CSFs are used to organize the logical processing of information: at current time point it indicates the actual couple of facts (RLF and CSF), reflecting the route segment on which the MSM is located. As MSM moves in environment from one RL of the route to the next, the CS moves as an actual pair of facts in the facts space. At the same time, at the logical level, the current RLF reflects the current location of the MSM, and the current CSF reflects the target RL. The degree of relevance of the CSF, as well as the RLF, is estimated by the certainty factor $-1.0 \leq cf \leq + 1.0$.

The knowledge about the route, for example, *Rout_1* is given by a set of pairs of facts (A_i, A_i^{Csf}) for all the RLs of the route $i = 1, 2,\ldots,n$. RLF A_i are defined in the form (3.2), as facts of different abstraction levels. The previously formulated requirement that the RLF actualization be carried out taking into account the current context is supported by include in (3.2) an additional portion of the knowledge. The definition of RLF A_i in (3.2), additionally included a description of the tied CSF in the form:

$$< A_i, \{< A_{ij}, (a_j, (b_j = t), v_j, g_j) >, j = 1, 2, \ldots, k, \\ < A_i^{csf}, (a = +1.0, b = t, v = 0.0, g_{Csf}) > \} > . \tag{3.27}$$

In (3.27), in addition to the components representing A_i, as a fact, generalizing the segment of data in the form (3.2), one more component < A_i^{Csf}, *(+1.0, t,* $v_{Csf} = 0.0$, g_{Csf}) > is included for the tied CSF. The component represents the knowledge that the fact A_i^{Csf} is actual ($a = +1.0$) for an arbitrary time point ($b = t$) processing of the fact A_i. The value of the parameter $v_{Csf} = 0.0$ indicates that the data about A_i^{Csf} are considered as static. The information completeness parameter g_{Csf} of the fact A_i^{Csf} is selected from the range $0.25 \leq g_{Csf} \leq 0.5$. The completeness parameters for the other facts included in the description (3.27) are corrected, respectively, $g_1 + g_2 + \cdots + g_k + g_{Csf} \geq 1.0$.

As noted in Sect. 3.2.4, EoA uses knowledge (3.27), representing the fact A_i, in the form of rule (3.9).

$$\text{R}_{\text{i1}}\ IF(\{<A_{ij}, (a_j, (b_j = t), v_j, g_j)>, \ j = 1, 2, \ldots, k, \\ <A_i^{csf}, (a = +1.0, b = t, v = 0.0, g_{Csf}) > \})THEN\, \alpha_{Ai}. \tag{3.28}$$

When discussing the conceptual context model, it was noted that the presentation and processing knowledge about CSF A_i^{Csf} have features. The necessary condition for actualization the CSF is an event *event*(A_i), generated by the appearance of a certain input data. This leads to a change in the value of $\alpha = -1.0$ FFC A_i by the value $\alpha = +1.0$. With the occurrence of this event and only at this time moment, the fuzzy characteristics of the CSF A_i^{Csf} changes in accordance with the rules below.

$$\text{R}_{\text{i2}}\ \ IF\, event(A_i)\, and\, (CF\_A_i^{Csf}\ is\ \text{high})\, and\, (CF\_Rout\_1\, is\, \text{high}) \\ THEN\, a_{i+1}^{Csf} = +1, a_i^{Csf} = -1,\ i = 1, 2, \ldots, n. \tag{3.29}$$

Set of rules (3.29) is the route definition. Each rule links a pair of adjacent RLFs. The rule defines the RL A_{i+1} as the next goal after reaching the previous target point A_i. The EoCT solves the task of context tracking by using the knowledge of the route in the form (3.29). The EoCT is activated by the event (*event* (A_i)) and if conditions A_i^{Csf} *is high* and *CF_Rout_1 is high* are met, sets the values of the parameters $\alpha_i^{Csf} = -1$ and $\alpha_{i+1}^{Csf} = +1$ FFC of CSF in accordance with the prototype in *THEN* field of rule.

3.3.4 Engine of Switching Attention

Motion control of MSM based on knowledge in the form of rules (3.28), (3.29) allows for the appearance of obstacles on any part of the route between two adjacent nodes. This event activates the Engine of Switching Attention (EoSA). The attention segment shown in Fig. 3.13, moves from a fragment of a knowledge, including the facts of the active route at that time, to a fragment of a knowledge, the facts of which describe the MSM reaction to this event. Fragment represents the knowledge about obstacle avoidance control based on the strategy of "Feeling using data from sensors". Perhaps this is the avoidance collision trajectory presented in the same way, as the route the form (3.28), (3.29). After the obstacle is eliminated, the attention mechanism should switch the AS back to the KB segment that was relevant before the obstacle appearance. The newly activated AS must restore the conditions for actualization of fact A_i. This restores the conditions for the continuation of the interrupted motion control process to the destination RL of the route.

This is achieved through the mechanism of switching attention. Its formalization is similar to the formalization of the tracking context mechanism.

Let $\{Q_j\}_{j=1,\ldots,N}$ be the set of segments representing the knowledge about possible disturbances or obstacles. The segments Q_j are the union of some segments of facts. Segments Q_j, is a set of multy-level facts. Let D_j be the fact that is at the top level of the segment Q_j. Fact D_j is a generalized semantic representation of the spatial-temporal data segments about jth obstacle.

The obstacle corresponding to D_j is overcome by the trajectory $(B_j, B_j^{Csf})_{j=1,\ldots,M}$, where B_j is a fact representing the RL, B_j^{Csf} is CSF tied with B_j. Knowledge about this trajectory is represented by rules of the type (3.27)–(3.29). In the definition of RLF $\{B_j, B_j^{Csf}\}_{j=1,\ldots,M}$ takes into account not only the context, but attention.

$$\begin{aligned} <B, \big\{< B_w, (a_p, (b_p = t), v_p, g_p)>, \ p = 1, 2, \ldots, P, \\ <B_j^{Csf}, (a = +1.0, b = t, v = 0.0, g_{csf})>, \\ <D_j^{Atten}, (a = +1.0, b = t, v = 0.0, g_{Atten})>\big\} >, j = 1, \ldots, M. \end{aligned} \tag{3.30}$$

In (3.30), in addition to the components indicated in (3.27), an attention component is introduced. The attention component based on Attention Satellite Fact (ASF). The parameter of information completeness g_{Atten} of the D_j^{Atten} must satisfy the condition $0.25 \le g_{Csf} + g_{Atten} \le 0.5$. The completeness parameters for the other facts included in the description (3.30) satisfy the condition $g_1 + g_2 + \cdots + g_P + g_{Csf} + g_{Atten} \ge 1.0$. The RE knowledge (3.30), representing the fact B_j, still, uses in the form of a rule (3.9).

$$\begin{aligned} R_{j1}\ IF\ \ (\{<B_{jp}, (a_p, (b_p = t), v_p, g_p)>, p = 1, 2, \ldots, P, \\ <B_j^{Csf}, (a = +1.0, b = t, v = 0.0, g_{csf})>, \\ < D_j^{Atten}, (a = +1.0, b = t, v = 0.0, g_{Atten}) >\})\ \ THEN\, \alpha_{Bj}. \end{aligned} \tag{3.31}$$

According to the rule (3.31), the fact B_j can be actualized $\alpha_{Bj} \geq 1.0 - \varepsilon$ when both the CSF and the ASF are relevant. The knowledge about context (3.29) in the extended perception model is modified by introducing the attention component into the rule for calculating the CSF certainty factor.

$$\begin{aligned} \mathrm{R_{j2}}\ \ IF\ event(B_j)and\left(CF\_B_j^{Csf}\ is\ \mathrm{high}\right)and\left(CF\_D_j^{Atten}\ is\ \mathrm{high}\right) \\ and(CF\_Rout\_X\ is\ \mathrm{high}) \\ THEN\ \ a_{j+1}^{csf} = +1, a_j^{Csf} = -1, j = 1, 2, \ldots, n. \end{aligned} \tag{3.32}$$

Just as for the CSF, the ASF actualizes the event *event*(D_j), which is similar in origin to the *event (Ai)*. When this event occurs, the FFC of the ASF D_j^{Atten} changes in accordance with the below rule.

$$\mathrm{R_{i3}}\ IF\ event(D_j)\ THEN\ a_j^{Atten} = +1, a^{Rout\_X} = +1, a^{Rout\_1} = -1. \tag{3.33}$$

The EoSA moves the AS back when the last *P*th target point of the obstacle avoidance route is reached.

$$\begin{aligned} \mathrm{R_{i4}}\ \ IF\ event(B_P)and\left(CF\_B_P^{Csf}\ is\ \mathrm{high}\right)and\left(CF\_D_P^{\mathrm{Atten}} is\ \mathrm{high}\right) \\ and(CF\_Rout\_X\ is\ \mathrm{high}) \\ THEN\ a_P^{Csf} = -1, a_j^{Atten} = -1, a^{Rout\_X} = -1, a^{Rout\_1} = +1. \end{aligned} \tag{3.34}$$

$$\mathrm{R_{i5}}\ IF\ event(Rout\_1)THEN\ a_{Rout\ 1}^{Atten} = +1\,. \tag{3.35}$$

The rule (3.35) actualized by the *event (Rout_1)* actualizes the fact *Rout_1* representing the interrupted route.

With the use of the extended perception model, the route knowledge (3.28) and (3.29) also changes. In the description of the facts it is necessary to include a component that reflects the influence of attention on the fact actualization.

$$\begin{aligned} \mathrm{R_{i1}} IF\left(\left\{(< A_{ij}, \left(a_j, (b_j = t), v_j, g_j\right) >, j = 1, 2, \ldots, k)\right\},\right. \\ \left(< A_i^{Csf}, \left(a = +1.0, b = t, v = 0.0, g_{csf}\right) >\right), \\ (< Rout1^{Atten}, (a = +1.0, b = t, v = 0.0, g_{Atten})>)) \\ THEN\ \alpha_{Ai}, i = 1, 2, \ldots, n. \end{aligned} \tag{3.36}$$

$$\begin{aligned} \mathrm{R_{i2}}\ \ IF\ event(A_i)and\left(CF\_A_i^{Csf}\ is\ \mathrm{high}\right)and\left(CF\_D_P^{\mathrm{Atten}} is\ \mathrm{high}\right) \\ and(CF\_Rout\_1\ is\ \mathrm{high}) \\ THEN\ a_{i+1}^{Csf} = +1, a_i^{Csf} = -1. \end{aligned} \tag{3.37}$$

3.3.5 An Example of Using the Expanded Perception Model for the Interpretation of Spatio-Temporal Data

We consider an example of control of MSM motion along route A, passing through the RLs A_1, A_2, A_3, A_4 shown in Fig. 3.12a. The knowledge used in control is divided into three groups: knowledge about the route, knowledge about the environment, knowledge about obstacle avoidance techniques.

Route knowledge is represented by a set of tied facts (A_i, A_i^{Csf}) for the RLs of the route $i = 1, 2, 3, 4$. The orderliness of the facts in accordance with the route is given by unordered set of rules (3.36), (3.37) for all RLs of the route.

The MSM knowledge about environment is knowledge of what control is required to reach the target point A_{i+1} when MSM is in the area where the data about feathers of the RL A_i are available. This portion of knowledge is represented by rules (3.38).

$$\begin{array}{ll} \mathrm{R_{i3}} & IF\ (CF\_A_i\ is\ \text{high})\ and\left(CF\_A_{i+1}^{Cst}\ is\ \text{high}\right) \\ & THEN\ a_{U_{i1}} = +1, a_{U_{i2}} = -1; a_{U_{i3}} = -1, i = 1, 2, \ldots, K. \end{array} \tag{3.38}$$

In rules (3.38), U_{i1}, U_{i2}, U_{i3} are control actions belonging to the set of MSM possible actions. These can be either elementary actions such as *forward* or *left*, discussed earlier, or macro-commands that implement a chain of elementary control actions. The knowledge base stores the rules (3.38) not only for the RLs of the route, but the rules connecting any pair of special points of the workspace. Rules (3.38) are not related to a specific route. They can be activated by data from sensors about the any location of the MSM and form control action, leading to the achievement of the actual goal (A_{i+1}^{Cst}). This mechanism guarantees that the MSM moves along the route in case of deviations when obstacle occurs.

Below we discuss the results of modeling the motion MSM along route A. Motion control based on an extended perception model. To facilitate understanding of the essence of the considered example, the details of the obstacle avoidance technology are not given here. A fragment of the three-block data processing scheme for the second RL of the route is shown in Fig. 3.14a. The names of the blocks correspond to

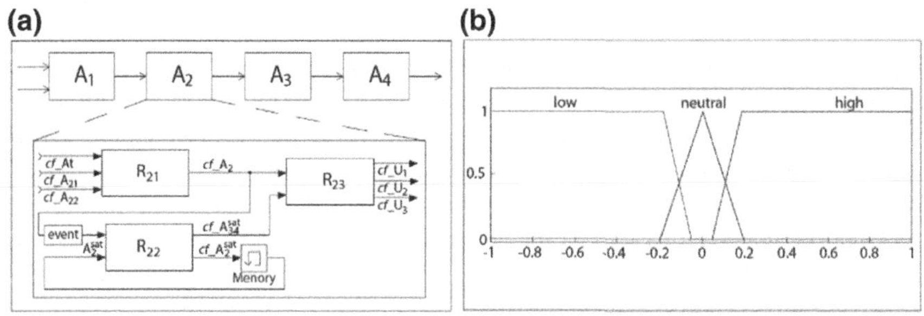

Fig. 3.14 Takagi-Sugeno model: **a** fragment of the scheme; **b** definition of terms of linguistic variables

the names of the rules (3.28)–(3.38). In these blocks, a fuzzy inference is implemented by the Takagi-Sugeno rules-based method. The knowledge base of the R_{21} block contains one $i = 2$ fuzzy rule (3.28); knowledge base of block R_{22} has one $i = 2$ fuzzy rule (3.37); knowledge base of block R_{23} has one for $i = 2$ rule (3.38). The designations of the input and output numerical and linguistic variables in Fig. 3.14a correspond to the notation in the rules. In rules (3.28)–(3.38), *CF_XX* denotes the names of the corresponding linguistic variables. For example, CF_A_i is the name of a linguistic variable that makes sense "FFC A_i certainty factor". All linguistic variables used in the simulation are defined in the same way: three terms on the universe of the certainty factor (Fig. 3.14b). The term set of a linguistic variable is {*high*, *neutral*, *low*}. Terms are defined on the universe of the certainty factor $-1 \leq cf \leq +1$, where *cf* is the numerical value of the certainty. Possible controls set: U_1 is the *forward*; U_2 is the *turn right and straight ahead*; U_3 is the *turn left and straight ahead.*

The simulation results are shown in Tables 3.1 and 3.2. Tables rows correspond to the same points in time. The Table 3.1 reflects the values of CF for the facts, and Table 3.2 shows controls actions. In Table 3.2 the first index *i* in the designation U_{ij} corresponds to the results of modeling obtained for the block A_i (Fig. 3.14a). In the last three columns are the total values of the CF of control actions.

From the analysis of the data it can be seen, that time point 3 corresponds to the appearance of an event related to obtaining information about the RL A_1. This event actualized at the 4th step the purpose of the move in the form of the fact A_2^{Cst}, which, in turn, actualized the control of U_1. At the fifth step, data on reaching the RL A_2 appeared. In the sixth step, this event deactivated the target fact A_2^{Cst} and updated the new target fact A_3^{Cst}. Then, it updated the control action U_2.

3.4 Examples of Interpreting Spatio-Temporal Data Segments

3.4.1 MSM at an Unregulated Crossroads

Data interpretation based on an extended perception model was tested on an application MSM at an unregulated crossroads. The implementation of the model is performed on the resources of the IoT and SM training and research polygon [23]. Wheel robots were used as MSMs, and humanoid robots modeled pedestrians at an intersection. Sources of information: data from ultrasonic sensor on the distance from MSM to the object; type of pedestrian (speeding mode by a humanoid robot). In addition, data from the sound sensor about the noise emitted by the robot, and data from the video camera about the shape of the approaching object are used. In Fig. 3.4 shows the structure of the lower levels' facts for the data from the ultrasonic sensor. In Fig. 3.5 shows an example of the mapping of the semantic representation of a fragment of possible situations at a crossroads into a set of prototypes KGs. The above fragment represents the knowledge about a safe situation in one direction of

Table 3.1 The values of the certainty factor *cf* of facts A_i

t	*Atten*	A_1^{Sat}	A_1	A_2^{Sat}	A_2	A_3^{Sat}	A_3	A_4^{Sat}	A_4
1	−1	−1	−1.0	−1	−1.0	−1	−1.0	−1	−1.0
2	+1	+1	0.0	−1	0.0	−1	0.0	−1	0.0
3	+1	+1	1.0	−1	0.0	−1	0.9	−1	0.0
4	+1	−1	0.9	+1	0.1	−1	0.5	−1	0.0
5	+1	−1	0.9	+1	0.9	−1	0.0	−1	0.0
6	+1	−1	0.1	−1	0.9	+1	0.0	−1	0.0
7	0	−1	−0.5	−1	0.5	+1	0.5	−1	−0.5
8	−1	−1	−0.9	−1	0.5	+1	−0.1	−1	−1.0
9	+1	−1	0.0	−1	0.9	+1	0.9	−1	0.0
10	+1	−1	0.0	−1	0.0	−1	1.0	+1	0.0
11	+1	−1	0.0	−1	0.0	−1	0.0	+1	1.0

Table 3.2 The values of the certainty factor *cf* of control action U_i

t	U_{11}	U_{12}	U_{13}	U_{21}	U_{22}	U_{23}	U_{31}	U_{32}	U_{33}	U_1	U_2	U_3
1	−1.0	−1.0	−1.0	−1.0	−1.0	−1.0	−1.0	−1.0	−1.0	−1.0	−1.0	−1.0
2	−1.0	−1.0	−1.0	−1.0	−1.0	−1.0	−1.0	−1.0	−1.0	−1.0	−1.0	−1.0
3	−1.0	−1.0	−1.0	−1.0	−1.0	−1.0	−1.0	−1.0	−1.0	−1.0	−1.0	−1.0
4	+1.0	−1.0	−1.0	−1.0	−1.0	−1.0	−1.0	−1.0	−1.0	+1.0	−1.0	−1.0
5	+1.0	−1.0	−1.0	−1.0	−1.0	−1.0	−1.0	−1.0	−1.0	+1.0	−1.0	−1.0
6	−1.0	−1.0	−1.0	−1.0	+1.0	−1.0	−1.0	−1.0	−1.0	−1.0	+1.0	−1.0
7	−1.0	−1.0	−1.0	−1.0	+0.5	−1.0	−1.0	−1.0	−1.0	−1.0	0.5	−1.0
8	−1.0	−1.0	−1.0	−1.0	+0.5	−1.0	−1.0	−1.0	−1.0	−1.0	0.5	−1.0
9	−1.0	−1.0	−1.0	−1.0	+1.0	−1.0	−1.0	−1.0	−1.0	−1.0	+1.0	−1.0
10	−1.0	−1.0	−1.0	−1.0	−1.0	−1.0	+1.0	−1.0	−1.0	+1.0	−1.0	−1.0
11	−1.0	−1.0	−1.0	−1–1.0	−1.0	−1.0	+1.0	−1.0	−1.0	+1.0	−1.0	−1.0

the road from the intersection. This is a situation where there are no vehicles near the intersection (the right branch of the semantic network in Fig. 3.5) and the traditional vehicle slowly approaches, and this is accompanied by the traditional noise level (left branch of the semantic network in Fig. 3.5). Section 3.2.2 provides a representation (3.4) of prototypes describing this knowledge. The meaning of the KGs can be seen from the definition of rules (3.4). This is indicated by a portion of knowledge in verbal form.

It should be noted that in the given fragment of possible situations, the KG ***4.0*** at the fourth level of abstraction gives the definition of a situation "at a close distance an object is approaching safely". This situation is determined by five portions of knowledge represented by KGs ***1.0***, ***2.0***, ***3.0***, ***0.8***, ***0.9***. The definition of KG ***4.0*** is given through the *part_of* relation. According to the KG, the situation will be described as safe at a close distance, if at the observation time the object is at a close distance, and approaching, and the speed is not high, and there is sound data about a slowly approaching vehicle and there is an image of a standard vehicle. However, knowledge about the image and sound is not necessarily fixed at the time of processing this rule, in contrast to the first three portions of knowledge. The processing of the fact will take into account the loss of relevance of knowledge about the image and sound associated with the aging of data.

Let us demonstrate the property of openness of the perception model using an example, when, in addition to the two options for a safe situation defined in the (3.4), two more prototypes of situations should be taken into account.

The first prototype is described by a knowledge portion "before the object was far away and was approaching rapidly with deceleration, but now it is getting close, slowly moving and slowing down" and the second one "before the object was close and coming nearer slowly and slowing down, now the object is close and stopped".

In order to correct the concept "at close range the situation is safe", which was determined by KG ***5.0*** in (3.4), you just need to add a number of new KGs (3.39) and rules (3.40) to existing (3.4). A graphic illustration of the definition of new KGs and their aggregation with the already existing shown in Fig. 3.15.

28. <***3.2***, *the object is approaching at medium speed with a slowdown*, {<***2.2***, (1.0, 0, 0.0, 0.5)>, <***2.1***, (1.0, 0, 0.0, 0.5)>}>;
29. <***4.1***, *object close, approaching slowly, slowing*, {<***1.0***, (1.0, 0, 0.0, 0.33)>, <***2.1***, (1.0, 0, 0.0, 0.33)>, <***3.2***, (1.0, 0, 0.0, 0.33)>}>;
30. <***4.2***, *object away, approaching fast, slowing*, {<***1.1***, (1.0, 0, 0.0, 0.33)>, <***2.3***, (1.0, 0, 0.0, 0.33)>, <***3.2***, (1.0, 0, 0.0, 0.33)>}>;
31. <***3.3***, *the object is close stopped*, {<***1.0***, (1.0, 1, 0.0, 0.5)>, <***2.3***, (1.0, 0, 0.0, 0.5)>}>;
32. <***5.1***, *before, the object was far away and was approaching fast with a slowdown, now it is coming closer is approaching slowly and is slowing down*, {<***4.1***, (1.0, 0, 0.15, 0.6)>, <***5.1-1***, (1.0, 0, 0.0, 0.4)>}>;
33. <***5.2***, *before the object was close and approached slowly and slowed down, now the object is close stopped»*, {<***4.1***, (1.0, 1, 0.0, 0.5)>, <***3.3***, (1.0, 0, 0.0, 0.5)>}>;

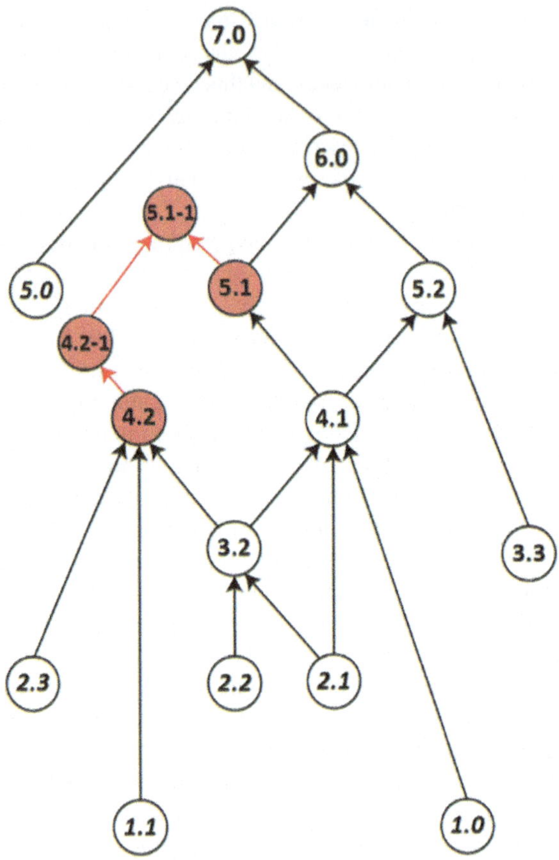

Fig. 3.15 An additional knowledge portions about the dangerous situation

34. <***6.0***, *safely,* {<***5.1***, (1.0, *t*, 0.0, 1.0)>, <**5.2**, (1.0, *t*, 0.0, 1.0)>}>;

35. <**7.0**, *near safe*, {<**5.0**, (1.0, 0, 0.0, 1.0)>, <**6.0**, (1.0, 0, 0.0, 1.0) >}>. (3.39)

$$\begin{array}{c} \mathrm{R}_1 IF\ event\ (\mathbf{4.2}) and (CF\_\mathbf{4.2{-}1}\ is\ \text{high}\ ) THEN a^{Csf}_{\mathbf{5.1}-\mathbf{1}} = +1,\ a^{Csf}_{4.1-1} = -1; \\ \mathrm{R}_2 IF\ event\ (\mathbf{5.1}) and (\text{CF\_}\mathbf{5.1{-}1}\ is\ \text{high}) THEN\ a^{Csf}_{4.1-1} = +1,\ a^{Csf}_{5.1-1} = -1. \end{array} \tag{3.40}$$

In (3.39) the numbering of the definitions of the KGs continues the numbering in (3.4). Comments are required for some KG. The KG ***5.2*** is determined by the temporal relationship *before*. In the formal representation of the relation (3.2), it is necessary to indicate the numerical value *b* of the time interval between the moments of the observed situations. In KG ***5.2***, you need to specify the exact value of the time interval between situation when the object *was* close, approached slowly, slowed down and the situation that are observed *now* when the object is close stopped. Therefore, the definition KG ***5.2*** contains the parameter $b = 1$ in units of event clock. The KG ***5.1*** cannot be determined using the same relationship *before*. This causes difficulties, since the latter situation may appear through different time intervals. It depends on the speed of the approaching object, its slowing down. In determining the sequence of these situations, it is important only that the first situation first appears, and then the second. The time interval between these events does not matter. Such a sequence of changing situations, ordered by the context of the problem being solved, it is advisable to specify using the context model. In Fig. 3.15 are shown beside to facts ***4.2*** and ***5.1*** the tied CSFs ***4.2-1*** and ***5.1-1***, respectively. A CSFs are highlighted in red. At the moment when the situation appears, represented by knowledge of the ***4.2*** fact, with active ***4.2-1*** CSF, the ***5.1-1*** CSF is activated. The ***5.1-1*** CSF will be in this state until the situation appears, represented by ***5.1*** fact. The ***5.1*** fact cannot be actualized with inactive ***5.1-1*** CSF. Its actualization indicates that *before* there was a situation presented by the ***4.2*** fact. The above discussions are formally represented by portions of knowledge in the definitions (3.39), (3.40) of the relevant facts.

Aggregation of the existing knowledge base with new knowledge was accomplished by making references in (3.39), (3.40) to the facts presented in (3.4). The facts defined earlier in (3.4) are shown in italics in (3.39) and in Fig. 3.15. As can be seen, the addition of new knowledge in the KB of the perception model is not related to the adjustment of the definition of facts that were made earlier.

Another important conclusion from this example is that the perception model can solve the problem of interpreting input data on an incomplete knowledge. The results of the interpretation of situations, the prototypes of which are, do not change when new knowledge portions are added. In addition to the already noted advantage, the adaptability property EoA facilitates the development of complex applications by dividing knowledge into fragments and step-by-step debugging of individual fragment.

Below we discuss some data interpretation experiments (Table 3.3) for the knowledge base, in which, apart from (3.4), (3.39), (3.40), there was also knowledge about dangerous situations. Considered dynamic situations such as, for example, the object moves in the direction of the intersection at high speed.

On the section of the road to which fact ***0.6*** was assigned, the object began to slow down and then move on with a constant slowdown to the first section (Fig. 3.4), where it stopped. Data from the US with a given discreteness was submitted to MSM for processing.

The columns of the Table 3.3 correspond to different moments of real time, in which the object is located on a certain section of the road. Column 1 corresponds to the location of the object on the 7th section; column 2 corresponds the 6th and so on, the 6th column corresponds to the location of the MSM at the 1st section. The sections numbers match the fact identifiers in Fig. 3.4. The first column of the table contains facts identifiers. The rows of the table contain the parameters α, t_R, t_L of the FFC for the facts with the specified identifiers. For clarity, the fractional values of the parameter α in the table are rounded to three values *+1*, *–1* and *0*.

Table 3.3 The results of the experiment: the object is approaching the crossroads

ID fact	Values of the α, t_L,t_R					
	1	2	*3*⁻	*4*	*5*⁻	*6*
0.1.	−1, 0, #	−1, 0, #	−1, 0, #	−1, 0, #	−1, 0, #	−1, 0, #
0.2	−1, 0, #	−1, 0, #	−1, 0, #	−1, 0, #	−1, 0, #	+1, 0, 0
0.3	−1, 0, #	−1, 0, #	−1, 0, #	−1, 0, #	+1, 0, 0	−1, 0, 0
0.4	−1, 0, #	−1, 0, #	−1, 0, #	+1, 0, 0	−1, 0, 0	−1, 0, 1
0.5	−1, 0, #	−1, 0, #	+1, 0, #	−1, 0, 1	−1, 0, 2	−1, 0, 3
0.6	−1, 0, #	+1, 0, 0	−1, 0, 0	−1, 0, 2	−1, 0, 3	−1, 0, 4
0.7	+1, 0, #	−1, 0, 0	−1, 0, 1	−1, 0, 3	−1, 0, 4	−1, 0, 5
1.0	−1, 0, #	−1, 0, #	−1, 0, #	0.4, 0, #	0.8, 0, 0	+1, 0, 1
1.1	−1, 0, #	+1, 0, 0	+1, 0, 1	0.6, 0, 3	0.4, 0, 4	−1, 0, 0
1.10	−1, 0, #	−1, 0, #	−1, 0, #	−1, 0, #	+1, 0, 0	−1, 0, 0
1.11	−1, 0, #	−1, 0, #	−1, 0, #	+1, 0, 0	−1, 0, 0	−1, 0, 1
1.12	−1, 0, #	−1, 0, #	+1, 0, 0	−1, 0, 2	−1, 0, 3	−1, 0, 4
1.13	−1, 0, #	+1, 0, 0	−1, 0, 1	−1, 0, 3	−1, 0, 4	−1, 0, 5
1.14	+1, 0, #	+1, 0, #	+1, 0, #	+1, 0, #	−1, 0, 0	−1, 0, 1
2.0	−1, 0, #	+1, 0, 0	+1, 0, 1	+1, 0, 3	+1, 0, 4	+1, 0, #
2.1	−1, 0, #	−1, 0, #	−1, 0, #	−1, 0, #	−1, 0, #	−1, 0, #
2.2	−1, 0, #	−1, 0, #	−1, 0, #	−1, 0, #	−1, 0, #	−1, 0, #
2.3	−1, 0, #	+1, 0, 0	+1, 0, 1	+1, 0, 3	+1, 0, 4	+1, 0, 5
3.0	−1, 0, #	−1, 0, #	−1, 0, #	−1, 0, #	−1, 0, #	−1, 0, #
3.1	−1, 0, #	−1, 0, #	+1, 0, 0	+1, 0, 2	+1, 0, 4	+1, 0, 5
3.2	−1, 0, #	−1, 0, #	−1, 0, #	+1, 0, 0	−1, 0, 0	−1, 0, #
3.3	−1, 0, #	−1, 0, #	−1, 0, #	−1, 0, #	−1, 0, #	−1, 0, #
4.0	−1, 0, #	−1, 0, #	−1, 0, #	−1, 0, #	−1, 0, #	−1, 0, #
4.1	−1, 0, #	−1, 0, #	−1, 0, #	−1, 0, #	−1, 0, #	−1, 0, #
4.2	−1, 0, #	−1, 0, #	−1, 0, #	+1, 0, #	−1, 0, #	−1, 0, #
5.0	+1, 0, #	+1, 0, #	+1, 0, #	+1, 0, #	−1, 0, 0	−1, 0, #
5.1	−1, 0, #	−1, 0, #	−1, 0, #	−1, 0, #	−1, 0, #	−1, 0, #
5.2	−1, 0, #	−1, 0, #	−1, 0, #	−1, 0, #	−1, 0, #	−1, 0, #
6.0	−1, 0, #	−1, 0, #	−1, 0, #	−1, 0, #	−1, 0, #	−1, 0, #
7.0	+1, 0, #	+1, 0, #	+1, 0, #	+1, 0, #	−1, 0, 0	−1, 0, 1

Such rounding does not distort the meaning of the output results. The table uses the # sign to indicate a large unknown value. The Table 3.3 shows the changes over time of the situation description in front of the crossroad related to moving an object from one to another road section. Description of the situation is given at different levels of detail. So at the moment of time, for example, 3 a situation description is obtained: the object is far away (the CF of fact ***1.1*** is α =+*1*) at a distance of 2-3 meters (α =*+1* of the fact ***0.5***), the object just moved in (parameter $t_L = 0$ of fact ***0.5*** and parameter α =*+1* of fact ***1.12***), the object is approaching (α =*+1* of the fact ***2.0***) and the object approaches quickly (α =*+1* of the fact ***2.3***). If more detailed information is needed, this can be obtained from the analysis of the facts of the lower levels. For example, the object has moved from the 6th to the 5th section (*a =+1* of the fact ***1.12***) and so on. The description of the situation at the most generalized level "crossing is recommended", since, α =+1 of the fact ***7.0*** (*near safely*). The CF is obtained due to α =*+1* value of fact ***1.14*** (*no near the object*). From the analysis of the structure of knowledge in Fig. 3.5 it is seen that the results of the interpretation correspond to the knowledge (3.4), (3.39), (3.40).

3.4.2 Interpretation of Spatio-Temporal Data About the Navigable Path Between Landmarks of the MSM Route

In cognitive and intelligent robotics, one of the most significant tasks is robot navigation [73, 74]. The navigational system of a mobile robot can be divided into four types navigation system [75]. The navigation systems described as follows: map-based navigation, behavior-based navigation, learning based navigation, and communication-based navigation. Expanded perception model solves tasks map-based navigation and behavior-based navigation. One of them is discussed in Sect. 3.3.5. The MSM control of movements along a given route is based on the results of environment data interpretation. The route is known in advance and presented by knowledge about the RL features. Previously unknown in this formulation of the problem are the characteristics of the environment associated with the obstacle.

Being able to build a map of the environment and simultaneously one's localized within this map is popular task for mobile robots navigating in unknown environment in absence of external referencing systems such as GPS. This Simultaneous Localization and Mapping (SLAM) problem [73] has been next task of mobile robotics. The SLAM problem can be solved on the base of expanded perception model in the following formulation. The MSM relocation route is not known in advance. According to the results of the interpretation of data on the MSM path, it is required to determine the localization and environment characteristics. Consider some of these applications.

Smart vacuum cleaner realizes their functions by regular or random movement in a limited space of the room. On the base of reflection sensor, a stain is detected that stands out on the main background the smart vacuum cleaner must turn on a special cleaning mode only for this area of coverage. One of the possible approaches is to, after detecting a spot, MSM bypasses this spot along the boundary. For this purpose, data from the matrix of reflection sensors and a control system with feedback are used. When walking around the spot, path data in the form of turning angles and lengths of the path sections are received. Based on these data, it is required to classify the spot

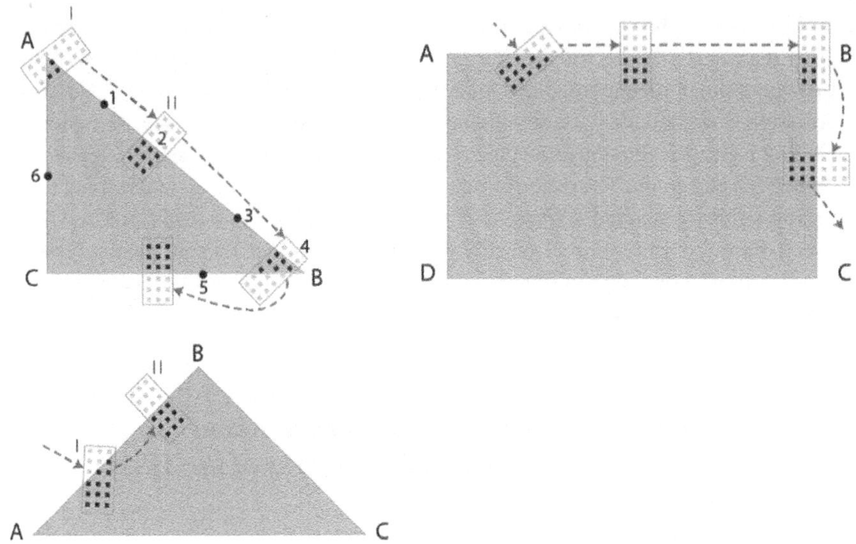

Fig. 3.16 Illustration of the task of traversing a shape along the border

shape. A previously developed optimal cleaning route is associated with the shape class. An illustration of the spot bypass task is shown in Fig. 3.16.

The task of the movements control of a smart lawnmower is similar. If the boundaries of the lawn are not previously unknown, then the lawnmower pre-passes along the contour of the lawn, determines the class of the figure to which the shape of the lawn belongs. For this class of the figure, the route of movement of the lawn mower is chosen optimal in terms of either time or energy costs.

The following application applies to police patrol smart SMs. To perform critical tasks, it is envisaged to preserve the functionality of the system in various extreme conditions. One of these situations is the loss of communication with satellites. In this case, moving along the city streets, the MSM should be able to localize its location on a map. This is SLAM problem.

Patrol MSMs that perform security functions in offices, warehouses and other should be able to autonomously move along specified patrol and be able to localize their location on the map after deviating from the route when a non-standard situation occurs in remote places. This class of tasks includes military applications. To preserve secrecy, MSM must be able to navigate through some areas autonomously in passive mode without external communication. There are SLAM problems, too.

SLAM problem can be considered as the task of interpreting the spatio-temporal date about the path covered by the MSM. Below is an example in which idealized segments are considered, shown in Fig. 3.16. In real problems, the segments are certainly much more complicated. However, this does not exclude the possibility of applying the approach to the interpretation of real data, including segmented video images.

An example is considered when knowledge about three routes are represented in the form of (3.27), (3.29). The route describes a path along the contour of one geometric figure in Fig. 3.16. When detecting the boundary of the contour, as shown in the Fig. 3.16, the movement of the MSM along the contour is controlled by a certain strategy based on data from the reflection sensors. For correction of deviations from the boundary of the figure, proportional-integral differential (PID) control algorithm is used. The control algorithm supports the MSM movement in such a way that the right three columns of the reflection sensors (Fig. 3.16) are located above the darkened surface (inside figure), and the left three columns of sensors are outside the figure. Of the possible strategies, the simplest one was adopted: a clockwise bypass.

The MSM moving is controlled by the commands straight, left, right. After completing the execution of commands, the corresponding signs (*motion_straight* , *turn_left_on_angle_1*,..., *turn_left-side_on_angle_n*, *turn_right_on_angle_1*,..., *turn_right_on_angle_n*) are formed, which are accepted by the corresponding MSM sensors. In the example, for the demonstration purposes, the accuracy of approximation of the boundary is adopted 15 degrees. In Figs. 3.17 and 3.18 graphically shows the KGs, representing the routes for bypassing the figures, which are shown in Fig. 3.16. The identifiers of the KGs of the zero-level are indicated by numbers corresponding to the angle of turn. The maximum angles of turn are 180° to the right and 180° to the left. The right and left turns in the range [−7.5°, +7.5°] correspond the zero-level KG ***0.0***; turns to the right in the range [+7.5°, +22.5°] describes KG ***0.15*** and so on until KG ***0.180***. Correction of deviations from the border of the figure was carried out by the PID algorithm within [−7.5°, +7.5°]. In this regard, corrective control actions did not lead to distortion of data on the path. In Figs. 3.17 and 3.18

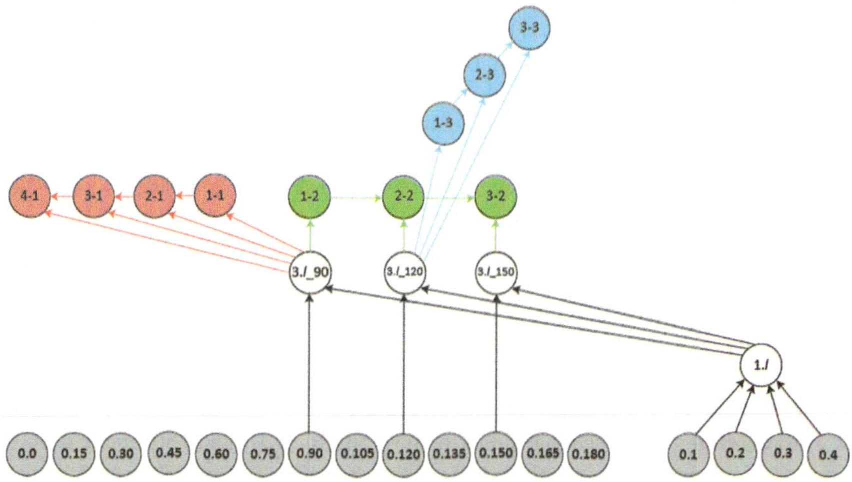

Fig. 3.17 Representation of the prototypes of path of bypassing of the figures: a rectangular triangle (green); equilateral triangle (blue); rectangle (red)

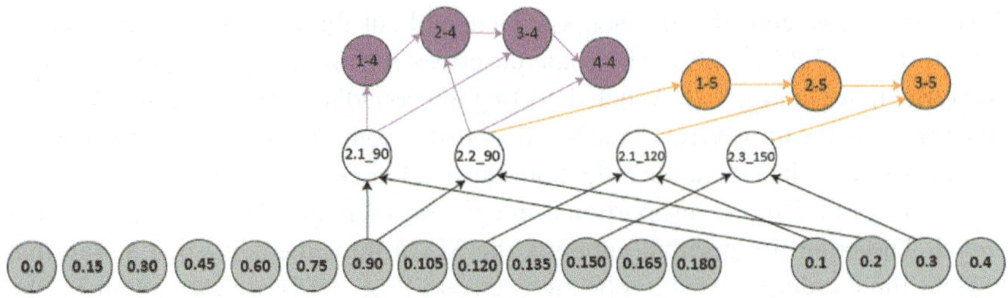

Fig. 3.18 Representation of the prototypes of path of bypassing of the figures: a rectangle (purple) with lengths of sides 1 and 2 unit; rectangular triangle (orange) with lengths of sides 1, 2 and 3 units

the KGs of the zero-level are given only for the right turn, since for the clockwise bypass strategy for the given examples the left turns there was no.

Possible values of the straight section length of the MSM path are divided into four granules corresponding to *1, 2, 3, 4* units of distance. In Figs. 3.17 and 3.18 zero-level KGs ***0.1***, ***0.2***, ***0.3***, ***0.4*** represent knowledge about these granules. The FFC of ***0.1***, ***0.2***, ***0.3***, ***0.4*** are calculated based on data from the sensor “speedometer”, the readings of which are reset each time after completing the turn to any angle that does not belong to the [−7.5°, +7.5°]. Figure 3.17 shows the knowledge presentation about the three routes of MSM movement along the boundaries of the three figures shown in Fig. 3.16. The pass of rectangle shape route is represented by a structure of four CSFs ***1-1***, ***2-1***, ***3-1***, ***4-1***, shown in red. The pass of right triangle shape route with angles of 90°, 30° and 60° is represented by a structure of three CSFs ***1-2***, ***2-2***, ***3-2***, shown in green. The pass of equilateral triangle shape route with angles of 60° is represented by a structure of three CSFs ***1-3***, ***2-3***, ***3-3*** facts shown in blue.

Figure 3.18 shows the knowledge presentation about the two paths of MSM movement along the boundaries of the two figures. This is a rectangle with side lengths of 1 and 2 units and a rectangular triangle with angles of 90°, 30° and 60° and lengths of sides *1*, *2* and *3* units. The structure of the four CSFs ***1–4***, ***2–4***, ***3–4***, ***4–4***, highlighted in purple, describes the pass of rectangle shape route. The structure of the four CSFs ***1–5***, ***2–5***, ***3–5***, highlighted in orange, describes the pass of triangle shape route. The differences in the representation of routes of identical figures in Figs. 3.17 and 3.18 are that in the latter version the lengths of the sides are indicated. The knowledge representation in the form of MSM pass along the boundaries of figure is a prototype of a data segment describing the shape of a figure.

Interpretation of the input data, including a segmented video image, is a segment bypass procedure and forming a zero-level FFC. Based on these values, FFCs of the first, second and third levels are calculated. Shown in Fig. 3.17 KGs ***3.l_90***, ***3.l_120***, ***3.l_150*** represent the knowledge about the turn on 90°, 120° and 150° made during the pass, which corresponds to the angles of the Figs. 90°, 30° and 60°, respectively. The KG of the first level with the identifier ***1.l*** represents the generalized knowledge of the arbitrary side length of the figure. A couple of KGs ***3.l_90***, in Fig. 3.17 and ***2.1_90*** in Fig. 3.18 different in meaning. Fact ***3.l_90*** represents a knowledge portion “MSM passed some distance without changing direction, then turn riht on 90° and started to go straight”. In KG ***2.1_90***, the knowledge about lenth of MSM pass before turning was concretized, namely, “MSM passed a distance equal to one unit of length…”.

Knowledge of the shape of the figure that represents the segment prototype is formally recorded as a route (3.36), (3.37). For the example under consideration, this is knowledge about LRF: ***3.l_90***, ***3.l_120***, ***3.l_150***, ***2.1_90***, ***2.2_90***, ***2.1_120***, ***2.3_150***. Below is their presentation.

1. *<**2.1_90**, MSM has passed the distance equal to the 1 length unit without changing direction, then turned rigt on 90° and began to move straight,* {<**0.1**, (*1.0, 1, 0.15, 0.5*)>, <**0.90**, (*1.0, 0, 0.15, 0.5*)>}>;
2. *<**2.2_90**, MSM has passed the distance equal to the 2 length unit without changing direction, then turned rigt on 90° and began to move straight,* {<**0.2**, (*1.0, 1, 0.15, 0.5*)>, <**0.90**, (*1.0, 0, 0.15, 0.5*)>}>;
3. *<**2.1_120**, MSM has passed the distance equal to the 1 length unit without changing direction, then turned rigt on 120° and began to move straight,* {<**0.1**, (*1.0, 1, 0.15, 0.5*)>, <**0.120**, (*1.0, 0, 0.15, 0.5*)>}>;
4. *<**2.3_150**, MSM has passed the distance equal to the 3 length unit without changing direction, then turned rigt on 150° and began to move straight,* {<**0.3**, (*1.0, 1, 0.15, 0.5*)>, <**0.150**, (*1.0, 0, 0.15, 0.5*)>}>;
5. *<**3.l_90**, MSM has passed an arbitrary length of distance without changing direction, then turned rigt on 90° and began to move straight,* {<***1.l***, (*1.0, 1, 0.15, 0.5*)>, <**0.90**, (*1.0, 0, 0.15, 0.5*)>}>;
6. *<**3.l_120**, MSM has passed an arbitrary length of distance without changing direction, then turned rigt on 120° and began to move straight,* {<***1.l***, (*1.0, 1, 0.15, 0.5*)>, <**0.120**, (*1.0, 0, 0.15, 0.5*)>}>;
7. *<**3.l_150**, MSM has passed an arbitrary length of distance without changing direction, then turned rigt on 150° and began to move straight,* {<***1.l***, (*1.0, 1, 0.15, 0.5*)>, <**0.150**, (*1.0, 0, 0.15, 0.5*)>}>;
8. <**1.*l***, *arbitrary length*, {<**0.1**, (1.0, 0, 0.0, 1.0)>, <**0.2**, (1.0, 0, 0.0, 1.0)>, <**0.3**, (1.0, 0, 0.0, 1.0)>, <**0.4**, (1.0, 0, 0.0, 1.0)>}>;

(3.41)

$$\mathrm{R}_{11}\ IF\ event(\mathbf{3.}l\mathbf{\_90})\ \ and\ CF\_\mathbf{1\text{-}1}\ is\ \text{high}\ THEN\ \alpha_{2-1}^{Csf}=+1,\alpha_{1-1}^{Csf}=-1;$$
$$\mathrm{R}_{12}\ IF\ event(\mathbf{3.}l\mathbf{\_90})\ \ and\ CF\_\mathbf{2\text{-}1}\ is\ \text{high}\ THEN\ \alpha_{3-1}^{Csf}=+1,\alpha_{2-1}^{Csf}=-1;$$
$$\mathrm{R}_{13}\ IF\ event(\mathbf{3.}l\mathbf{\_90})\ \ and\ CF\_\mathbf{3\text{-}1}\ is\ \text{high}\ THEN\ \alpha_{4-1}^{Csf}=+1,\alpha_{3-1}^{Csf}=-1;$$
$$\mathrm{R}_{14}\ IF\ event(\mathbf{3.}l\mathbf{\_90})\ \ and\ CF\_\mathbf{4\text{-}1}\ is\ \text{high}\ THEN\ \alpha^{rect}=+1;$$

$$\mathrm{R}_{21}\ IF\ event(\mathbf{3.}l\mathbf{\_90})\ and\ CF\_\mathbf{1\ \text{-}\ 2}\ is\ \text{high}\ THEN\ \alpha_{2-2}^{Csf}=+1,\alpha_{1-2}^{Csf}=-1;$$
$$\mathrm{R}_{22}\ IF\ event(\mathbf{3.}l\mathbf{\_120})\ and\ CF\_\mathbf{2\text{-}2}\ is\ \text{high}\ THEN\ \alpha_{3-2}^{Csf}=+1,\alpha_{2-2}^{Csf}=-1;$$
$$\mathrm{R}_{23}\ IF\ event(\mathbf{3.}l,\mathbf{150})\ and\ CF\_\mathbf{3\text{-}2}\ is\ \text{high}\ THEN\ \alpha^{triang\ rect}=+1;$$

R_{31} *IF event* (**3.1_120**) *and* *CF*_**1-3** *is* high *THEN* $\alpha_{2-3}^{Csf} = +1, \alpha_{1-3}^{Csf} = -1;$
R_{32} *IF event* (**3.1_120**) *and* *CF*_**2-3** *is* high *THEN* $\alpha_{3-3}^{Csf} = +1, \alpha_{2-3}^{Csf} = -1;$
R_{33} *IF event* (**3.1_120**) *and* *CF*_**3-3** *is* high *THEN* $\alpha^{triang\,equilat} = +1;$

R_{41} *IF event* (**2.1_90**) *and* *CF*_**1-4** *is* high *THEN* $\alpha_{2-4}^{Csf} = +1, \alpha_{1-4}^{Csf} = -1;$
R_{42} *IF event* (**2.2_90**) *and* *CF*_**2-4** *is* high *THEN* $\alpha_{3-4}^{Csf} = +1, \alpha_{2-4}^{Csf} = -1;$
R_{43} *IF event* (**2.1_90**) *and* CF_**3-4** *is* high THEN $\alpha_{4-4}^{Csf} = +1, \alpha_{3-4}^{Csf} = -1;$
R_{44} *IF event* (**2.2_90**) *and* *CF*_**4-4** *is* high *THEN* $\alpha^{rect-1-2} = +1;$

R_{51} *IF event*(**2.2_90**) *and* *CF*_**1-5** *is* high *THEN* $\alpha_{2-2}^{Csf} = +1, \alpha_{1-2}^{Csf} = -1;$
R_{52} *IF event*(**2.1_120**) *and* CF_**2-5** *is* high THEN$\alpha_{3-2}^{Csf} = +1, \alpha_{2-2}^{Csf} = -1;$
R_{53} *IF event*(**2.3_150**) *and* *CF*_**3-5** *is* high *THEN* $\alpha^{triang\,1-2-3} = +1.$
(3.42)

In addition to the CSFs indicated in Figs. 3.17 and 3.18, has entered KGs which are used to illustrate the result of the interpretation. For example, the KG ***rect*** to denote a segment of a rectangle. The conclusion about the compliance of the prototype with the input data segment is derived by the last rule, which actualizes in the verbal form the presentation of the interpretation results.

Knowledge (3.41) and (3.42) about prototype segments does not include attention-switching rules. That is necessary to control the transition from one image segment to another when interpreting a complex image consisting of several, possibly adjacent or even intersecting segments.

The procedure for interpreting one segment of data will be considered in the following example. There are KGs (3.40) and rules (3.42) about five prototypes for five different segments. Rules R_{11}–R_{14} describe a prototype of a rectangular segment with arbitrary lengths of sides, R_{41}–R_{44} with side lengths equal to *1* and *2* units. The prototype of a triangular segment with a right angle with an arbitrary length of the sides is represented by the rules R_{21}–R_{23}, and with sides *1, 2* and *3* units of length is represented by the rules R_{51}–R_{53}. A segment in the form of an equilateral triangle with an arbitrary side length is represented by the rules R_{31}–R_{33}.

The case is considered when input data of the expanded perception model are in the form of, for example, a rectangle with sides of the *2* and *3* units of length. The starting point of the segment bypass is adopted, as shown in Fig. 3.16. The results of the interpretation of the contour tracking data of the rectangle are given in Table 3.4.

The table shows the values of FFC for all the facts shown in Figs. 3.17 and 3.18. In the Table 3.4 lines are the values of FFC for all LR. The columns indicate the FFC values in the form (3.7) of all the facts for one LR. The Table 3.4 shows the values for the five nodal points. The 1-st LR corresponds to the MSM position shown in Fig. 3.16. The 2-nd LR corresponds to the position of the MSM immediately after completing the turn at the angle ABC in Fig. 3.16. The 3-rd LR corresponds to the situation when MSM has rotated at an angle BCD (Fig. 3.16).

The 4th RL is at the CDA angle, and 5th at the DAB angle. At these RLs, data from sensors are received, both about the angle of turn, and the length of the linear path before turning. Based on this data, the FFC of all zero-level facts is updated.

Table 3.4 The results as FFC (α,t_L,t_R) of the interpretation of the contour tracking data of the rectangle

ID Fact	Number RL				
	1	2	3	4	5
0.1	?, #, #	−1, 0, 0	−1, 0, 1	−1, 0, 2	−1, 0, 3
0.2	?, #, #	+1, 0, 0	+1, 0, 1	−1, 0, 0	+1, 0, 0
0.3	?, #, #	−1, 0, 0	−1, 0, 1	+1, 0, 0	−1, 0, 0
0.4	?, #, #	−1, 0, 0	−1, 0, 1	−1, 0, 2	−1, 0, 3
0.0	+1, 0, 0	−1, 0, 1	−1, 0, 2	−1, 0, 3	−1, 0, 4
0.15	−1, 0, 0	−1, 0, 1	−1, 0, 2	−1, 0, 3	−1, 0, 4
0.30	−1, 0, 0	−1, 0, 1	−1, 0, 2	−1, 0, 3	−1, 0, 4
0.45	−1, 0, 0	−1, 0, 1	−1, 0, 2	−1, 0, 3	−1, 0, 4
0.60	−1, 0, 0	−1, 0, 1	−1, 0, 2	−1, 0, 3	−1, 0, 4
0.75	−1, 0, 0	−1, 0, 1	−1, 0, 2	−1, 0, 3	−1, 0, 4
0.90	−1, 0, 0	+1, 0, 0	+1, 0, 1	+1, 0, 2	+1, 0, 3
0.105	−1, 0, 0	−1, 0, 1	−1, 0, 2	−1, 0, 3	−1, 0, 4
0.120	−1, 0, 0	−1, 0, 1	−1, 0, 2	−1, 0, 3	−1, 0, 4
0.135	−1, 0, 0	−1, 0, 1	−1, 0, 2	−1, 0, 3	−1, 0, 4
0.150	−1, 0, 0	−1, 0, 1	−1, 0, 2	−1, 0, 3	−1, 0, 4
0.165	−1, 0, 0	−1, 0, 1	−1, 0, 2	−1, 0, 3	−1, 0, 4
0.180	−1, 0, 0	−1, 0, 1	−1, 0, 2	−1, 0, 3	−1, 0, 4
1.1	?, #, #	+1, 0, 0	+1, 0, 1	+1, 0, 2	+1, 0, 3
2.1_90	?, #, #	−1, 0, 0	−1, 0, 1	−1, 0, 2	−1, 0, 3
2.2_90	?, #, #	+1, 0, 0	+1, 0, 1	−1, 0, 2	+1, 0, 0
2.1_120	?, #, #	−1, 0, 0	−1, 0, 1	−1, 0, 2	−1, 0, 3
2.3_150	?, #, #	−1, 0, 0	−1, 0, 1	−1, 0, 2	−1, 0, 3
3.1_90	?, #, #	+1, 0, 0	+1, 0, 1	+1, 0, 2	+1, 0, 3
3.1_120	?, #, #	−1, 0, 0	−1, 0, 1	−1, 0, 2	−1, 0, 3
3.1_150	?, #, #	−1, 0, 0	−1, 0, 1	−1, 0, 2	−1, 0, 3
1-1	+1, #, 0	−1, #, 0	−1, #, 1	−1, #, 2	−1, #, 3
2-1	−1, #, 0	+1, #, 0	−1, #, 0	−1, #, 1	−1, #, 2
3-1	−1, #, 0	−1, #, 1	+1, #, 0	−1, #, 0	−1, #, 1
4-1	−1, #, 0	−1, #, 1	−1, #, 2	+1, #, 0	−1, #, 0
rect	−1, #, 0	−1, #, 1	−1, #, 2	−1, #, 2	+1, #, 0
1-2	+1, #, 0	+1, #, 1	+1, #, 2	+1, #, 3	+1, #, 4
2-2	−1, #, 0	−1, #, 1	−1, #, 2	−1, #, 3	−1, #, 4
3-2	−1, #, 0	−1, #, 1	−1, #, 2	−1, #, 3	−1, #, 4
Triang_right	−1, #, 0	−1, #, 1	−1, #, 2	−1, #, 3	−1, #, 4

(continued)

Table 3.4 (continued)

ID Fact	Number RL				
	1	2	3	4	5
1-3	+1, #, 0	+1, #, 1	+1, #, 2	+1, #, 3	+1, #, 4
2-3	−1, #, 0	−1, #, 1	−1, #, 2	−1, #, 3	−1, #, 4
3-3	−1, #, 0	−1, #, 1	−1, #, 2	−1, #, 3	−1, #, 4
Triang_equil	−1, #, 0	−1, #, 1	−1, #, 2	−1, #, 3	−1, #, 4
1-4	+1, #, 0	+1, #, 1	+1, #, 2	+1, #, 3	+1, #, 4
2-4	−1, #, 0	−1, #, 1	−1, #, 2	−1, #, 3	−1, #, 4
3-4	−1, #, 0	−1, #, 1	−1, #, 2	−1, #, 3	−1, #, 4
4-4	−1, #, 0	−1, #, 1	−1, #, 2	−1, #, 3	−1, #, 4
rect-1-2	−1, #, 0	−1, #, 1	−1, #, 2	−1, #, 3	−1, #, 4
1-5	+1, #, 0	−1, #, 0	−1, #, 1	−1, #, 2	−1, #, 3
2-5	−1, #, 0	+1, #, 0	+1, #, 1	+1, #, 2	+1, #, 3
3-5	−1, #, 0	−1, #, 1	−1, #, 2	−1, #, 3	−1, #, 4
triang-1-2-3	−1, #, 0	−1, #, 1	−1, #, 2	−1, #, 3	−1, #, 4

Therefore, the second parameter of the FFC is $t_L = 0$ for all facts, starting with the second RL. The values of the third parameter t_R shows the time interval from the time when the FFC changes according to the clock of the event time model.

Lets comment on the results of Table 3.4. For the first RL, when MSM detected the contour of the segment and began to move along the AB side, the length data from sensor are not defined and the zero-level FFC parameters are equal ($\alpha = ?$, $t_L = \#$, $t_R = \#$). The # sign indicates a large unknown number. FFCs that represent knowledge of the turn angles have values ($\alpha =+1.0$, $t_L = 0$, $t_R = 0$) of the fact ***0.0*** and ($\alpha = -1.0$, $t_L = 0$, $t_R = 0$) for the other facts of zero level. Attention switching mechanism set the parameters $\alpha = +1.0$, and $t_L =\#$, $t_R = 0$, for CSF ***1-1***, ***1-2***, ***1-3***, ***1-4***,***1-5***, since the second parameter is not used by the computational model of CSF. For other CSFs ($\alpha = -1.0$, $t_L = \#$, $t_R = 0$). For the second RL, based on the new data from the sensors, the values of the parameters for the facts ***0.2***, ***0.90*** will change to ($\alpha =+1.0$, $t_L = 0$, $t_R = 0$). For the remaining facts of zero level the FFC does not change. Based on the fresh data from this RL calculation, according to the extended perception model, give the values ($\alpha =+1.0$, $t_L = 0$, $t_R = 0$) FFC of ***1.1***, ***2.2_90*** and ***3.1_90***. Events related to changes in the FFC values of these three facts according to the rules (3.41) will change the context segment for two prototypes: a rectangular with arbitrary side lengths and a triangular with a right angle and sides *1*, *2* and *3* units of length. Calculations by the rules (3.41) give values ($\alpha = -1.0$, $t_L = \#$, $t_R = 0$) for ***1-1***, ($\alpha =+1.0$, $t_L = \#$, $t_R = 0$) for ***2-1*** and ($\alpha = -1.0$, $t_L = \#$, $t_R = 0$) for ***1-5***, ($\alpha =+1.0$, $t_L = \#$, $t_R = 0$) for ***2-5***. Similarly, you can analyze the results of calculations for the rest of the RL.

The bypass of the figure represented by the input data is completed at the 5th RL. As a result, the 6th column of Table 3.4 presents the results of interpretation.

In Table 3.4, the values of the fuzzy characteristics of the CSF in the 6th column indicate that the input data represent a segment of a rectangular shape. Since there was no in the KB the prototype of a rectangle with lengths of sides coinciding with the input data, the result of the interpretation is a rectangle with arbitrary side lengths. This is indicated by the fuzzy characteristics of CSFs ($\alpha = +1.0$, $t_L = \#$, $t_R = 0$) for ***rect*** and ($\alpha = -1.0$, $t_L = \#$, $t_R = 5$) for ***triang_rect***, ***triang_equilat***, ***rect-1-2***, ***triang-1-2-3*** facts. Table 3.4 shows the integer values of the CF. This is done to facilitate understanding of the example. For real calculations, values α from the range $[-1.0, +1.0]$ are obtained. In the general case, the result of the interpretation is ranked facts on the value of α.

3.5 Conclusion

The perception model compared to traditional production systems based on rules and fuzzy systems has the following advantages.

1. Abstraction approach to processing data from sensors in rules based engine overcomes problems: dynamic changes in the environment, combining data from different sensors, uncertainty and incompleteness information due to the use of the knowledge representation model in the form of parameterized semantic prototypes and multilevel facts.
2. The scope of rules engine is extended to applications that require consideration of the dynamic properties of a situation. In FLS Type-I, to represent point and interval events, as well as temporal relationships between them, a significant increase in the number of input numeric variables is required, which entails a sharp increase in the number of rules. In expanded perception model, by introducing L-R numbers with the parameters of delay and aging, it became possible to take into account the time history without introducing additional input variables.
3. Application adaptation to changes in input data, facts and rules in the process of debugging and subsequent operation due to the openness of the knowledge base is supported. In FLS Type-I, the introduction of new input variables, terms of linguistic variables entail the need to recycle all previously prepared rules. In expanded perception model, the introduction of new sources of input data and domain knowledge requires the inclusion of a group of new independent rules in the knowledge base. Previously created knowledge remains unchanged.
4. Opened the possibility of using rules engine as expanded perception model for embedded real-time applications thanks to reducing the amount of knowledge base. For expanded perception model applications, the knowledge base size is less than FLS Type-I. For an MSM application at an unregulated crossroads, the FLS Type-I rules base includes 432 complete fuzzy rules. The use of incomplete rules in FLS Type-I distorts the result of fuzzy inference. For this application, expanded perception model knowledge base represents a seven-level structure of 34 knowledge granules.

References

1. Mintchell G (2016) Industry 4.0 Survey: building the digital enterprise. www.themanufacturingconnection.com/2016/09/industry-4-0-survey-building-digital-enterprise. Accessed 25 Jan 2019
2. Ploennigs J, Cohn J, Stanford-Clark (2018) A The future of IoT. IEEE Internet Things Mag 1(1):28–33. https://www.comsoc.org/system/files/2018-10/IOTMAG_2018_Sep.pdf. Accessed 25 Jan 2019
3. Fan X (2015) Real-time embedded systems design principles and engineering practices. Elsevier, 662p
4. Perez J (2017) Artificial intelligence and robotics. https://arxiv.org/ftp/arxiv/papers/1803/1803.10813.pdf. Accessed 28 Jan 2019
5. UK-RAS Conference Proceedings (2018) J Robot Auton Syst 1(1). https://www.ukras.org/wp-content/uploads/2018/10/UK-RAS-Proceedings-2017.pdf. Accessed 28 Jan 2019
6. Buntz B (2017) An executive's guide to Industry 4.0, Smart Factories and beyond. www.ioti.com/industrial-iot-iiot/executive-s-guide-industry-40-smart-factories-and-beyond. Accessed 25 Jan 2019
7. IEEE Computer Society Predicts the Future of Tech: Top 10 Technology Trends for 2019. 19 Dec 2018. https://www.computer.org/web/computingnow/insights/content?g=53319&type=article&urlTitle=ieee-computer-society-predicts-the-future-of-tech-top-10-technology-trends-for-2019. Accessed 25 Jan 2019
8. Rail Technical Strategy Capability Delivery Plan (2017). https://www.rssb.co.uk/rts/Documents/2017-01-27-rail-technical-strategy-capability-delivery-plan-brochure.pdf. Accessed 25 Jan 2019
9. Plinninger T, Hildebrandt A (2017) Comfortable rail travel. Ansys Adv 1:18–23
10. Fraga-Lamas P at al (2017) Towards the Internet of Smart Trains: a review on Industrial IoT-connected railways. Sensors 17(6). https://doi.org/10.3390/s17061457. Accessed 28 Jan 2019
11. Tracy P (2017) Smart trains and the connected railway. https://www.ibm.com/blogs/internet-of-things/smart-trains-connected-railway/. Accessed 25 Jan 2019
12. Duarte F, Ratti C (2018) The impact of autonomous vehicles on cities: a review. J Urban Technol 25(4). https://doi.org/10.1080/10630732.2018.1493883. Accessed 28 Jan 2019
13. Moon M (2019) NVIDIA's new lab aims to develop robotic breakthroughs. https://www.engadget.com/2019/01/12/nvidia-robotics-research-lab/. Accessed 25 Jan 2019
14. Liu D, Wang L, Tan KC (2009) Design and control of intelligent robotic systems. Stud Comput Intel 478
15. Hansen S, Blanke M, Andersen J (2009) Autonomous tractor navigation in orchard—diagnosis and supervision for enhanced availability. In: Proceedings of 7th IFAC symposium on fault detection, supervision and safety of technical processes, vol 42, issue no. 8. Barcelona, Spain, June 30–July 3, pp 360–365
16. Arvanitakis I, Tzes A, Giannousakis K (2018) Synergistic exploration and navigation of mobile robots under pose uncertainty in unknown environments. Int J Adv Robot Syst 15(1). https://doi.org/10.1177/1729881417750785. Accessed 25 Jan 2019
17. Pandey A, Pandey S, Parhi DR (2017) Mobile robot navigation and obstacle avoidance techniques: a review. Int Rob Auto J 2(3):00022. https://doi.org/10.15406/iratj.2017.02.00023. Accessed 25 Jan 2019
18. Bradshaw JM et al (2013) The seven deadly myths of "autonomous systems". IEEE Intel Syst 54–61
19. White paper. Observations and recommendations on connected vehicle security 2017. Cloud Secur Alliance. https://cloudsecurityalliance.org/download/connected-vehicle-security. Accessed 25 Jan 2019
20. Deng Q, Runger G, Tuv E (2012) System monitoring with real-time contrasts. Q J Meth Appl Relat Top 44(1):9–27. https://doi.org/10.1080/00224065.2012.11917878. Accessed 25 Jan 2019

21. Hu H, Brady JM, Du F, Probert PJ (1995) Distributed real-time control of a mobile robot. Intel Autom Soft Comput 1(1):63–83. https://doi.org/10.1080/10798587.1995.10750621. Accessed 25 Jan 2019
22. Munera E et al (2017) Distributed real-time control architecture for ROS-based modular robots. IFAC 50–1:11233–11238. https://doi.org/10.1016/j.ifacol.2017.08.1600. Accessed 25 Jan 2019
23. Kargin A et al (2018) A polygon for smart machine application. In: 2018 IEEE 9th international conference on Dependable Systems, Services and Technologies (DESSERT 2018), Ukraine, Kyiv, 24–27 May 2018, pp 489–494
24. Boano CA et al (2016) Dependability for the Internet of Things—from dependable networking in harsh environments to a holistic view on dependability. e & i Elektrotechnik und Informationstechnik 133(7):304–309. https://link.springer.com/article/10.1007/s00502-016-0436-4. Accessed 25 Jan 2019
25. Liu D et al (2009) Design and control of intelligent robotic system. Stud Comput Intell, 480 p
26. Alippi C (2014) Intelligence for embedded systems: a methodological approach. Springer, 283p
27. Davenport TH, Kirby J (2016) Just how smart are smart machines? MIT Sloan Manag Rev 57(3):21–25. http://ilp.mit.edu/media/news_articles/smr/2016/57306.pdf. Accessed 25 Jan 2019
28. Naujoks F, Kiesel A, Neukum J (2016) Cooperative warning systems: The impact of false and unnecessary alarms on drivers' compliance. Accid Anal Prev 97:162–175. https://doi.org/10.1016/j.aap.2016.09.009. Accessed 28 Jan 2019
29. Kargin A, Petrenko T (2018) Internet of Things smart rules engine. In: 2018 international scientific-practical conference on Problems of Infocommunications. Science and Technology (PIC S&T 2018), Ukraine, Kharkiv, pp 639–644
30. Qanbari S et al (2016) IoT design patterns: computational constructs to design, build and engineer edge applications. In: 2016 IEEE first international conference on Internet-of-Things design and implementation, pp 277–282. http://www.infosys.tuwien.ac.at/Staff/sd/papers/I4T_2016_S_Qanbari.pdf. Accessed 25 Jan 2019
31. Pizurica V (2017) The Waylay engine, Part 1: one rules engine to rule them all. https://blog.waylay.io/waylay-engine-one-rules-engine-to-rule-them-all/. Accessed 25 Jan 2019
32. Sottara D et al (2010) A configurable rete-OO engine for reasoning with different types of imperfect information. IEEE Trans Knowl Data Eng 22(11):1535–1548
33. Koster M (2014) design patterns for an internet of things—a design pattern framework for IoT architecture. http://iot-datamodels.blogspot.in/2014/05/design-patterns-for-internet-of-things.html. Accessed 25 Jan 2019
34. Gorbach G (2016) Five application patterns for the IoT Cloud. https://www.arcweb.com/blog/five-application-patterns-iot-cloud. Accessed 27 Jan 2019
35. Negnevitsky M (2005) Artificial intelligence: a guide to intelligent systems, 2nd edn. Addison-Wesley, 415 p
36. Amazon Web Services (2018) AWS IoT developer guide. Rules for AWS IoT. https://docs.aws.amazon.com/iot/latest/developerguide/iot-dg.pdf#iot-rules. Accessed 25 Jan 2019
37. Russell SJ, Norvig P (2010) Artificial intelligence a modern approach, 3rd edn. Pearson Education, 1151 p
38. Piegat A (2001) Fuzzy modelling and control. Physica-Verlag, 728p
39. Mendel JM (2017) Uncertain rule-based fuzzy systems: introduction and new directions, 2nd edn. Springer, 684p
40. Merrick K (2017) Value systems for developmental cognitive robotics: a survey. Cogn Syst Res 41:38–55
41. Langley P, Laird JE, Rogers S (2009) Cognitive architectures: Research issues and challenges. Cogn Syst Res 10(2):141–160. https://doi.org/10.1016/j.cogsys.2016.08.001. Accessed 27 Jan 2019
42. Asada M (2009) Cognitive developmental robotics: a survey. IEEE Trans Auton Mental Dev 1(1):12–34. http://www.ece.uvic.ca/~bctill/papers/ememcog/Asada_etal_2009.pdf. Accessed 25 Jan 2019

43. Solso RM, MacLin OH, MacLin MK (2004) Cognitive psychology, 7th edn. Allyn & Bacon, 624 p
44. Anderson JR (2009) Cognitive psychology and its implications, 7th edn. Worth Publishers, 469p
45. Schiffman HR (2001) Sensation and perception. An integrated approach, 5th edn. Wiley, 608p
46. Zadeh LA (1997) Toward a theory of fuzzy information granulation and its centrality in human reasoning and fuzzy logic. Fuzzy Sets Syst 90(2):111–127
47. Zadeh LA (2015) Toward a restriction-centered theory of truth and meaning (RCT). In: Magdalena L, Verdegay J, Esteva F (eds) Enric Trillas: a passion for fuzzy sets. A collection of recent works on fuzzy logic. Springer, pp 1–22
48. Mendel JM, Wu D (2010) Perceptual computing: adding people in making subjective judgments. Wiley, 320 p
49. Clancey WJ (1985) Heuristic classification. Artif Intell 27(3):289–350. https://doi.org/10.1016/0004-3702(85)90016-5. Accessed 27 Jan 2019
50. Chandrasekaran B (1986) Generic tasks in knowledge-based reasoning: high-level building blocks for expert systems design. IEEE Expert 1(3):23–30
51. Jackson P (1998) Introduction to expert systems, 3rd edn. Addison-Wesley, 560 p
52. Saba WS (2006) Ontology, types and semantics. In: Proceedings of the 3rd international workshop on natural language understanding and cognitive science, pp 17–26. https://doi.org/10.5220/0002472300170026. Accessed 25 Jan 2019
53. Saba WS (2007) Ontology and formal semantics (integration overdue). https://arxiv.org/ftp/arxiv/papers/0712/0712.1529.pdf. Accessed 25 Jan 2019
54. Yao YY (2008) A unified framework of granular computing. Handbook of granular computing. Wiley, Hoboken, pp 401–410
55. Pedrycz W, Chen S (2011) Granular computing and intelligent systems design with information granules of higher order and higher type. Springer, 305 p
56. Skowron A, Jankowski A, Dutta S (2015) Toward problem solving support based on big data and domain knowledge: Interactive granular computing and adaptive judgement. In: Big data analysis: new algorithms for a new society. Springer, pp 49–90
57. Skowron A (2016) Interactive granular computing. Granular Computing 1(2):95–113
58. Liu A (2016) Rule-based systems: a granular computing. Granular Comput 1:259–274
59. Hirsch P, Nolden S, Declerck M, Koch I (2018) Common cognitive control processes underlying performance in task-switching and dual-task contexts. Adv Cogn Psychol 14(3):62–74. http://www.ac-psych.org/en/issues/volume/14/issue/3. Accessed 25 Jan 2019
60. Luger GF (2009) Artificial intelligence: structures and strategies for complex problem solving, 6th edn. Addison-Wesley, 754 p
61. Tung WL, Quek C (2010) eFSM—a novel online neural-fuzzy semantic memory model. IEEE Trans Neural Netw 21(1):136–157
62. de Barros LC et al (2017) The extension principle of Zadeh and fuzzy numbers. In: A first course in fuzzy logic, fuzzy dynamical systems, and biomathematics, studies in fuzziness and soft computing, vol 347. Springer, Berlin, Heidelberg. https://doi.org/10.1007/978-3-662-53324-6_2. Accessed 28 Jan 2019
63. Zadeh LA (2006) Generalized theory of uncertainty (GTU)—principal concept and ideas. Comput Stat Data Anal 51(1):15–46
64. Doctor F, Hagras H, Callaghan V (2005) A fuzzy embedded agent-based approach for realizing ambient intelligence in intelligent inhabited environments. IEEE Trans Syst Man Cybern Part A Syst Hum 35(1):55–65. https://doi.org/10.1109/TSMCA.2004.838488. Accessed 25 Jan 2019
65. Kargin A, Panchenko S, Vasiljevs A, Petrenko T (2019) Implementation of cognitive perception functions in fuzzy situational control model. In: ICTE in transport and logist. 2018 (ICTE 2018). Procedia Comput Sci Elsevier 149:231–238
66. Abafogi M, Durdu A, Akdemir B (2018) A new approach to mobile robot navigation in unknown environments. In: ECAI 2018 international conference. 10th edition electronics, computers and artificial intelligence. 28–30 June 2018, Iasi, România. https://www.

researchgate.net/publication/329523524_A_New_Approach_to_Mobile_Robot_Navigation_in_Unknown_Environments. Accessed 25 Jan 2019
67. Pilania V, Gupta K (2018) Mobile manipulator planning under uncertainty in unknown environments. Int J Rob Reaches. https://doi.org/10.1177/0278364918754677. Accessed 25 Jan 2019
68. Zaitseva E, Levashenko V (2016) Construction of a reliability structure function based on uncertain data. IEEE Trans Reliab 65(4):1710–1723. https://doi.org/10.1109/TR.2016.2578948
69. Levashenko V, Zaitseva E, Puuronen S (2002) Usage of new information estimations for induction of fuzzy decision trees, lecture notes in computer science (including subseries lecture notes in artificial intelligence and lecture notes in bioinformatics), vol 2412. In: Proceedings of the 3rd International Conference on Intelligent Data Engineering and Automated Learning (IDEAL 2002), pp 493–499
70. Johansson (2018) A machine learning is bad at context. Here's how we fix it. https://www.computer.org/web/computingnow/insights/content?g=53319&type=article&urlTitle=machine-learning-is-bad-at-context-here-s-how-we-fix-it-. Accessed 25 Jan 2019
71. Jirak D, Wermter S (2018) Potentials and limitations of deep neural networks for cognitive robots. https://arxiv.org/abs/1805.00777. Accessed 25 Jan 2019
72. Sünderhauf N et al (2018) The limits and potentials of deep learning for robotics. https://arxiv.org/pdf/1804.06557.pdf. Accessed 25 Jan 2019
73. Murphy RR, Arkin RC (2000) Introduction to AI Robotics (Intelligent Robotics and Autonomous Agents). A Bradford Book, 487 p
74. Morioka K, Lee J, Hashimoto H (2002) Intelligent space for human centered robotics service robots. In: Proceedings 2002 IEEE international conference on robotics and automation. https://doi.org/10.1109/ROBOT.2002.1014836. Accessed 27 Jan 2019
75. Grisetti G at al (2010) A tutorial on graph-based SLAM. IEEE Intell Transp Syst Mag 2(4):31–43. https://doi.org/10.1109/MITS.2010.939925. Accessed 27 Jan 2019

Chapter 4
Spatio-Temporal Video Segmentation

Sergii Mashtalir and Volodymyr Mashtalir

Abstract This Chapter considers approaches to video skimming into semantically-consistent segments of video streams, which are highly redundant and weakly structured data. In such a way, one of the promising ways is spatial-temporal segmentation as frame partitions represent certain spatial image content. Also, properties were formulated and proved which ultimately determine the characteristics of permissible segmentation transformations when searching for a compromise between over and undo segmentation. Temporal segmentation of multidimensional time series has been examined, which enables structuring video streams and significantly reduce the amount of data that will require online processing. For this, multidimensional time series analysis theory was used, since a completely natural video representation is a sequence of frames, followed by their combination into groups of frames (shots) with the same content. It was shown that various approaches can be used to detect shots with homogeneous characteristics, which are based on VAR models, exponential smoothing and predictive models.

Keywords Content-based video retrieval · Video streams · Video sequences · Shot · Multidimensional data analysis · Time-series analysis · VAR models · Exponential smoothing

4.1 Introduction

Original video represents unstructured or poorly structured data. At the same time, there is an exigence to depict and convey high-level stories. Video-structure is usually extracted with an aim to capture temporal properties of the whole video, such as clips, scenarios, events, episodes, which are presented as long sequences of easily understandable visual content. Video parsing relevance plays the key role in efficient

S. Mashtalir · V. Mashtalir (✉)
Kharkiv National University of Radio Electronics, 14 Nauky Ave, Kharkiv 61166, Ukraine
e-mail: volodymyr.mashtalir@nure.ua

S. Mashtalir
e-mail: sergii.mashtalir@nure.ua

V. Mashtalir et al. (eds.), *Advances in Spatio-Temporal Segmentation of Visual Data*, Studies in Computational Intelligence 876,
https://doi.org/10.1007/978-3-030-35480-0_4

and effective temporal segmentation as its reasonableness has a direct impact on reliability and precision of all the other video processing stages. It is especially important, since the generalized semantic level expresses corresponding semantics of video sections [1].

The basic 'building' block of any video stream is a scene (associated with a shot), i.e. a consecutive sequence of frames which have to be similar in some sense within each particular scene [2, 3]. Of course, frame to frame difference occurs inside a scene, but it is not that significant if compared to inter-scene difference. These fundamental units are considered as primitives to explore the content of video sequences for higher level semantic issues. The search of transition from a scene to the next one is a temporal video segmentation step actually. This means partitioning the video stream with adequate sensitivity into a set of meaningful and manageable parts, desirably performed in a real time mode. Figure 4.1 shows an example of structuring video in the form of scenes.

Fundamentally, there exist two types of scene transitions, that is to say, scene boundaries may be abrupt or gradual. Very slow panorama changing, overall brightness offset, abrupt intense disturbance, video editing (lap dissolve, fade, side curtain wipe etc.) may cause significant obstacles to feasible scene boundary detection. Gradual transitions are much more difficult than abrupt ones but anyway, temporal segmentation depends heavily on the selected feature space. Abrupt scene boundaries normally are found using thresholds, for example, approaches based on frame-to-frame feature differences, while gradual transitions are detected by identifying sustained low-level changing in time series produced by video stream. Figure 4.2

Fig. 4.1 Temporal structure of video stream

Fig. 4.2 Examples of scene transitions

Fig. 4.3 Key fames of scenes present in Fig. 4.1

illustrates the 5-th and the 6-th scenes from Fig. 4.1 in more detail: scene frames curling from the bottom left corner to the top right one are taken during two seconds (50 frames), and the next scene frames are a bit shifted and located from left to right as well, with the same duration.

The frame that represents the salient visual content of a scene is stated as a key frame [4, 5]. It is clear that key frame set approach can be accepted, when there arises an attempt to perform video summarization, mainly for shortening the navigation and browsing the original video or aiming to find sufficiently simple mapping to high-level concepts. Assuredly, key frame application decreases redundancy among the frames in the same scene, eliminates matching problems with different sequence lengths, simplifies all video processing models and so on. Figure 4.3 shows key frames for the scene structure demonstrated in Fig. 4.1. However, key frame concept may significantly shrink possibilities of semantic video examination, in particular, upon detection of temporal events suitable for hierarchical organization, e.g. by scene clustering. In any case, primary role in general efficiency belongs to image features reasonability.

Current state-of-the-art segmentation techniques typically use a large number of low-level features such as global and local color or gradient histograms, texture or text on features, edge or point neighborhood features, color and grey-level moments, shape based features, Gabor wavelet features, motion features etc. [4, 6, 7]. The major criticism to the most feature sets lays in conceptual inefficiency and lack of reliability, appeared due to semantic gap between extracted low level attributes and required high level semantic relations.

One of the promising options for improving the video content studies is the development of region based approach, namely it is the involvement of spatial segmentation for each frame in a form of an (additional) attribute possibly. The introduction of such a spatial content into consideration certainly enhances the semantics, but reveals new difficulties. First of all, the semantic segmentation problem is being far from ultimate solution, at least owing to finding a compromise between under- and over segmentation. Appearance of partially nested partitions (see Fig. 4.4) significantly complicates frame processing, but it provides supplementary capabilities of content analysis, e.g. when making a decision about partial objects occlusions. Secondly, it is necessary to be able to compare partitions adequately and it is desirable that a similarity measure should be a metric, since it is often necessary to compare a sufficiently large number of partitions, and the triangle inequality may be useful for computation time reduction [8, 9]. Scrutinize stated above in more detail.

Consider video $V = \{I_1(x,y), I_2(x,y), \ldots, I_N(x,y)\}$ where $(x, y) \in D \subset \mathbb{R}^2$, D stands for the field of view. Any image from the ordered in time frames V can be presented in the following form

Fig. 4.4 Examples of spatial nested segmentations

$$I_i(x, y) = \mu(x, y)[\lambda_i^k(x, y)I_i(x, y) + [1 - \lambda_i^k(x, y)]\Phi(x, y)] + \eta(x, y), \quad (4.1)$$

where $\mu(x, y)$ is a function of multiplicative changes in brightness and contrast, $\lambda_i^k(x, y)$ denotes a carrier characteristic function indicating each of $k = \overline{1, K_i}$ objects, regions of visual attention or key points with some surrounding neighborhoods, i.e. regions of $I_i^k(x, y)$ allocations

$$\lambda_i^k(x, y) = \begin{cases} 1, I_i^k(x, y) \neq 0, \\ 0, I_i^k(x, y) = 0, \end{cases} \quad (4.2)$$

$\Phi(x, y)$ is symbolized for a background, $\eta(x, y)$ designates additive random noise. De facto, estimation of $\lambda_i^k(x, y)$ is nothing else but a spatial image segmentation with partial semantics and an eye to produce content dependent multidimensional time series in image or feature spaces. Generally, $\lambda_i^k(x, y)$ may correspond to sub-frame sets, regular greed, full image partition.

Temporal segmentation usually encompasses online partition exploration, that comprises of consecutive semantically meaningful (according to a given criteria) similar frame sequences declared as scenes. Formally, without loss of generality, temporal segmentation implies that $V = \{S_1, S_2, \ldots, S_l, \ldots, S_L\}$, $S_l = \{I_{l_i}, I_{l_i+1}, \ldots, I_{l_i+\alpha_l}\}$, the scene length is $\alpha_l + 1$ and $S_r \cap S_{r+1} = \emptyset, r \in \{1, 2, \ldots, L-1\}$ as a rule. If gradual transitions exist between the scenes, then either poorly defined sub-scenes are simultaneously included in two consecutive scenes: $S_r \cap S_{r+1} \neq \emptyset$, or such sub-scenes are completely excluded from consideration as having no content meaning, and in this case $\{S_1, S_2, \ldots, S_l, \ldots, S_L\} \subset V$. Found scenes that are the basic units of video analysis can be further used for clustering or stratifying a video stream model.

For valid temporal segmentation, it would be desirable to get feature distinguishing high-level image concepts but we are not able to do this. A reasonable compromise, along with the use of traditional feature spaces, is the presentation of video frames by partitions (segmentations), meaning sui generis spatial content. It is important to understand that special attention should be given to semantically significant time series produced by quotient sets.

Additional facilities of video content analysis are provided by multiple granularities detection in frames, however, it is not known in advance which segmentation approach will produce the most correctly interpreted results. Moreover, many segmentation algorithms have parameters, as a rule, heuristic, i.e. one algorithm can give a large number of partitions or coverings. Thus, there arises a problem of revealing various segmentation elements correspondence, including matching in different video frames. Three situations should also be studied: a positive decision is made, the decision is rejected, or the information needed for an unambiguous decision is not enough and additional analysis is required.

Consider, first, a metric providing partition matching. Suppose for any $\mathrm{A} \subset D$ there exists some number $\mu(\mathrm{A})$ which is a measure (length, area, volume, mass distribution, probability distribution, cardinality, etc.). Let 2^D be a power set in which all subsets are measurable also. Introduce the set $\Pi_D \subset 2^D$ of finite (regarding the number of cosets) partitions of set D, such that $\alpha \in \Pi_D$, $\alpha = \{\mathrm{A}_i\}_{i=1}^n$, $\mathrm{A}_i \in 2^D$, $D = \cup_{i=1}^n \mathrm{A}_i$, $\forall i, j \in \{1, 2, \ldots, n\} : i \neq j \Rightarrow \mathrm{A}_i \cap \mathrm{A}_j = \emptyset$.

Quotient sets defined in (4.2), namely the sets A_i, are actually generated either by different algorithms or by the same algorithm with varied parameters. The appearance of nested partitions corresponding to spatial refinement levels, as noted above, is illustrated in Fig. 4.4. Thus, there arises a necessity to compare partitions and moreover, to compare nested quotient sets. The metric on Cartesian square $\Pi_D \times \Pi_D$ is [7, 10]

$$\rho(\alpha, \beta) = \sum_{i=1}^{n} \sum_{j=1}^{m} \mu(\mathrm{A}_i \Delta \mathrm{B}_j)\mu(\mathrm{A}_i \cap \mathrm{B}_j), \tag{4.3}$$

where $\beta = \{\mathrm{B}_j\}_{j=1}^m$ and $\mathrm{A}_i \Delta \mathrm{B}_j = (\mathrm{A}_i \backslash \mathrm{B}_j) \cup (\mathrm{B}_j \backslash \mathrm{A}_i)$ is a symmetrical difference.

It is easily seen that an equivalent form generalizing Mirkin metric can be obtained for the functional (4.3)

$$\rho(\alpha, \beta) = \sum_{i=1}^{n} [\mu(\mathrm{A}_i)]^2 + \sum_{j=1}^{m} [\mu(\mathrm{B}_j)]^2 - 2 \sum_{i=1}^{n} \sum_{j=1}^{m} [\mu(\mathrm{A}_i \cap \mathrm{B}_j)]^2. \tag{4.4}$$

To apply (4.3), (4.4) efficiently for nested partitions, we have to get expressions realized as one quotient set inclusion into another.

Consider two nested partitions when one is splitting the other. For clarity we assume $\alpha \subseteq \beta$, i.e. some $\mathrm{B}_j \in \beta$ can be found for any $\mathrm{A}_i \in \alpha$, for which $\mathrm{A}_i \subseteq \mathrm{B}_j$. In other words, partition β is divided into m 'sub-partitions', i.e. $\alpha = \{\alpha_1^*, \ldots, \alpha_m^*\}$ and

$$\begin{cases} \alpha_1^* = \{\mathrm{A}_1, \ldots, \mathrm{A}_{k_1}\}, \\ \alpha_2^* = \{\mathrm{A}_{k_1+1}, \ldots, \mathrm{A}_{k_1+k_2}\}, \\ \ldots \\ \alpha_j^* = \{\mathrm{A}_{k_1+k_2+\ldots+k_{j-1}+1}, \ldots, \mathrm{A}_{k_1+k_2+\ldots+k_j}\}, \\ \ldots \\ \alpha_m^* = \{\mathrm{A}_{k_1+k_2+\ldots+k_{m-1}+1}, \ldots, \mathrm{A}_{k_1+k_2+\ldots+k_m}\}, \end{cases}$$

where $\alpha_j (j \in \overline{1, m})$ contains k_j elements of partition α and

$$B_j = \bigcup_{i=k_1+k_2+\ldots+k_{j-1}+1}^{k_1+k_2+\ldots+k_j} A_i,$$

where $k_1 + k_2 + \ldots + k_m = n$.

It is quite easy to show that for any two finite partitions $\alpha, \beta \in \Pi_D$ (relative to number of cosets) of an arbitrary measurable set D, if $\alpha \subseteq \beta$, then we have

$$\rho(\alpha, \beta) = \sum_{j=1}^{m} [\mu(B_j)]^2 - \sum_{i=1}^{n} [\mu(A_i)]^2. \tag{4.5}$$

Now it is necessary to determine the properties of transformations that can be performed with segmentation regions (cuts) or their sets in order to increase the efficiency of subsequent processing and interpreting of images. Suppose we have binary relation $T \subset D \times D$ distinguishing cuts. Introduce indicator function

$$\varphi_T(p, q) = \begin{cases} -1, & \langle A(p), A(q) \rangle \notin T, \\ 0, & decision\,making\,is\,impossible, \\ 1, & \langle A(p), A(q) \rangle \in T, \end{cases} \tag{4.6}$$

where '−1' means that $p = (m', n')$, $q = (m'', n'')$ are points of different cuts, '1' indicates the same cut, 0 implies that information dissatisfies decision-making conditions. Here and subsequently, $M = \{A(p)\}$ and $N = \{A(q)\}$. For the purpose of possible cut transformations choice, it is necessary to study how and under which conditions the maps $F_1{:}D \to \Gamma$, $F_2 : D \to \Gamma$ are modified or remained unchanged (4.6), where Γ is generally an arbitrary set. Suppose that the indicator function $\varphi_T(p, q)$ induces three-valued function with values −1, 0, 1

$$\psi(m, n) = \varphi_T(F_1(p), F_2(q)), \tag{4.7}$$

where $m \in M$, $n \in N$.

Thereby, it is necessary to investigate properties of the binary relation $T \subset D \times D$ ensuring transformation (4.6) into (4.7).

It should be noted that function $\psi(m, n)$ defines two obvious equivalence relations on set D

$$\forall n \in D \Rightarrow m' \sim m'', \text{ if } \psi(m', n) = \psi(m'', n), \tag{4.8}$$

$$\forall m \in D \Rightarrow n' \sim n'' \text{ if } \psi(m, n') = \psi(m, n''). \tag{4.9}$$

In other words, there arise two partitions

$$D = \cup \mathrm{M}_i, \mathrm{M}_i \cap \mathrm{M}_j = \emptyset \quad (i \neq j), \tag{4.10}$$

$$D = \cup \mathrm{N}_i, \mathrm{N}_i \cap \mathrm{N}_j = \emptyset \quad (i \neq j), \tag{4.11}$$

which are induced by expressions (4.8) and (4.9) respectively. We emphasize two properties of indicator function $\psi(m,n)$.

At first,

$$\forall n \in D(\forall m \in D)\exists m \in D(\exists n \in D) \Rightarrow \psi(m, n) = 0. \tag{4.12}$$

Indeed, from F_1, F_2 definitions it follows that for any n there will be an element $m \in D$ such that $F_1(m) = F_2(n)$, but then $\psi(m, n) = \varphi_\mathrm{T}(F_1(p), F_2(q)) = 0$. Clearly that the symmetric statement specified in brackets is correct also.

Secondly, if $\psi(m', n') = \psi(m'', n') = 0$ for some element $n' \in D$ for all $n \in D$, the equality $\psi(m', n) = \psi(m'', n)$ is fulfilled. This property is true with the argument permutation

$$\exists m' \in D : \psi(m', n') = \psi(m', n'') = 0 \Rightarrow \psi(m, n') = \psi(m, n''). \tag{4.13}$$

Let us verify (4.13) (correctness of symmetric property is established in a similar manner). If $\psi(m', n') = \psi(m', n'') = 0$, then

$$\varphi_\mathrm{T}(F_1(m'), F_2(n)) = \varphi_\mathrm{T}(F_1(m'), F_2(n'')) = 0$$

that suggests $F_2(n') = F_1(m') = F_2(n'')$. Then, for any $n \in D$ it follows immediately that $\psi(m, n') = \varphi_\mathrm{T}(F_1(m), F_2(n')) = \varphi_\mathrm{T}(F_1(m), F_2(n'')) = \psi(m, n'')$, as it was required.

From now on, for simplicity of notation, if it is not specifically stipulated, we will consider only the first argument properties, implying correctness for the symmetric ones. It may be concluded from properties (4.12) and (4.13) that in partitions (4.10) and (4.11) there is a one-to-one correspondence between sets M and N, namely each subset M_i is associated with such a subset N_j that for some $m \in \mathrm{M}_i$ and $n \in \mathrm{N}_j$ the equality $\psi(m, n) = 0$ is fulfilled. We will be convinced of the validity of this statement.

From properties (4.12) and (4.13), functions $\psi(m, n)$ equivalence $m' \sim m''$ (accordingly $n' \sim n''$) follows, if and only if $\psi(m', n) = \psi(m'', n) = 0$ (accordingly $\psi(m, n') = \psi(m, n'') = 0$).

Indeed, if at least for one element n' equality $\psi(m', n') = \psi(m'', n') = 0$ is fulfilled, then for all $n \in D$ equality $\psi(m', n) = \psi(m'', n) = 0$ holds by (4.13), i.e. $m' \sim m''$. Equivalence $n' \sim n''$ is established absolutely similarly.

Let us choose now arbitrary set M_i. If $m \in \mathrm{M}_i$, from (4.12) it follows that there exists an element $n \in D$ such that $\psi(m, n) = 0$. Let $n \in \mathrm{N}_j$, then if $q \in \mathrm{N}_k$, $k \neq j$ from equalities $\psi(m, n) = 0$, $\psi(m, q) = 0$ and properties established above, it may be deduced that $n{\sim}q$, i.e. $\mathrm{N}_k = \mathrm{N}_j$ which is impossible. Hence for $\psi(m, n) \neq 0$

all $n \notin \mathrm{N}_j$ takes place. Thereby, it is established that strictly determined subset N_j corresponds to each subset M_i, such that for all $m \in \mathrm{M}_i, n \in \mathrm{N}_j$ equality $\psi(m, n) = 0$ is fulfilled.

If $\mathrm{M}_i \neq \mathrm{M}_j$, then corresponding classes $\mathrm{N}_l, \mathrm{N}_k$ also do not coincide, as otherwise for elements $m_i \in \mathrm{M}_i$ $m_j \in \mathrm{M}_j, n \in \mathrm{N}_l$ it follows that $\psi(m_i, n) = \psi(m_j, n) = 0$, where the contradiction would follow, i.e. $\mathrm{M}_i = \mathrm{M}_j$.

So, if M_i scans set M, then N_i scans N, since for each element $n \in D$ there is an element $m \in D$ according to (4.12), such that $\psi(m, n) = 0$, as it was required.

There is no loss of generality in the assumption that equivalence classes M_i and N_i have such enumerations that corresponding classes have the same numbers. Introduce a binary relation $\tilde{\mathrm{T}}$ for equivalence classes M_i in the following way: $\langle \mathrm{M}_i, \mathrm{M}_j \rangle \in \tilde{\mathrm{T}}$ if $m \in \mathrm{M}_i$ $n \in \mathrm{N}_j$, $\psi(m, n) = 0$ or $\psi(m, n) = 1$. Since classes M_i and N_j correspond each other, the condition $\langle \mathrm{M}_i, \mathrm{N}_j \rangle \in \tilde{\mathrm{T}}$ under $\psi(m, n) = 0$ $(m \in \mathrm{M}_i, n \in \mathrm{N}_j)$ means $i = j$, i.e. a binary relation $\tilde{\mathrm{T}}$ on the set M is reflexive. In this regard, it is possible to introduce the supreme analogue of indicator function (4.6) for equivalence classes, i.e. indicator function of quotient sets interpreting segmentations

$$\varphi_{\tilde{\mathrm{T}}}(\mathrm{M}_i \times \mathrm{M}_j) = \begin{cases} -1, \langle \mathrm{M}_i, \mathrm{M}_j \rangle \notin \tilde{\mathrm{T}}, \\ 0, \mathrm{M}_i = \mathrm{M}_j, \\ 1, \langle \mathrm{M}_i, \mathrm{M}_j \rangle \in \tilde{\mathrm{T}}. \end{cases} \tag{4.14}$$

Now it is possible to specify the functions F_1 and F_2, giving them the following explanation

$$\left. \begin{array}{r} \forall m \in \mathrm{M}_i \, F_1(m) = \mathrm{M}_i, \\ \forall n \in \mathrm{N}_j \, F_2(n) = \mathrm{N}_j. \end{array} \right\} \tag{4.15}$$

The next statement is an answer to a question raised by segmentation transformation necessity.

Proposition 4.1 If some indicator function $\varphi_{\mathrm{T}}(m, n)$ meets the conditions

$$\begin{aligned} &\forall m \in D \exists n \in D \Rightarrow \psi(m, n) = 0, \\ &\exists m', n', n'' \in D : \psi(m', n') = \psi(m', n'') = 0 \\ &\quad \Rightarrow \psi(m, n') = \psi(m, n'')(\forall m \in D), \end{aligned}$$

then $\psi(m, n) = \varphi_{\tilde{\mathrm{T}}}(F_1(m), F_2(n))$ takes place.

Proof Let $\psi(m, n) = 0$, then $m \in \mathrm{M}_i$, and $n \in \mathrm{N}_j$, $F_1(m) = \mathrm{M}_i$, $F_2(n) = \mathrm{M}_i$. Hence $\varphi_{\tilde{\mathrm{T}}}(F_1(m), F_2(n)) = \varphi_{\tilde{\mathrm{T}}}(\mathrm{M}_i, \mathrm{M}_i) = 0$. Now assume that $\psi(m, n) = 0$ and $m \in \mathrm{M}_i$, $n \in \mathrm{N}_j$, then it follows that $\langle \mathrm{M}_i, \mathrm{M}_j \rangle \in \tilde{\mathrm{T}}$ from binary relation $\tilde{\mathrm{T}}$ definition, whence $\varphi_{\tilde{\mathrm{T}}}(\mathrm{M}_i, \mathrm{M}_i) = 1$ or $\varphi_{\tilde{\mathrm{T}}}(F_1(m), F_2(n)) = 1$ holds true according to (4.14). At last, from equality $\psi(m, n) = -1$ under conditions $m \in \mathrm{M}_i$, $n \in \mathrm{N}_j$ we have $\langle \mathrm{M}_i, \mathrm{M}_j \rangle \notin \tilde{\mathrm{T}}$, but then from (4.14) we obtain $\varphi_{\tilde{\mathrm{T}}}(\mathrm{M}_i, \mathrm{M}_i) = -1$ or $\varphi_{\tilde{\mathrm{T}}}(F_1(m), F_2(n)) = -1$, which completes the proof.

Our next goal is to examine indicator function properties depending on the constraints imposed on the relation T.

For correct segmented region transformations, it is necessary to save (at least) belonging points to the same image carrier. In other words, introduction of indicator functions (4.6) and (4.14), binary relations T and $\widetilde{\mathrm{T}}$ was applied without considering of their properties. Let us study various situations when the relation features are given, more precisely, we will establish conditions of indicator function $\psi(m, n)$ representation in the form of (4.7).

Proposition 4.2 Let T be a reflexive and symmetric binary relation between points m and n of the same cut, and let $F_1, F_2 : D \to D$, then the indicator function is representable as $\psi(m, n) = \varphi_{\mathrm{T}}(F_1(m), F_2(n))$, if and only if the next properties are fulfilled

1. $\forall m \in D \exists n \in D \Rightarrow \psi(m, n) = 0$,
2. $\exists m', n', n'' \in D : \psi(m', n') = \psi(m', n'') = 0 \Rightarrow \psi(m, n') = \psi(m, n'')(\forall m \in D)$,
3. $\forall m', m'', n', n'' \in D : \psi(m', n'') = 1, \psi(m', n') = 0, \psi(m'', n'') = 0 \Rightarrow \psi(m'', n') = 1$.

Proof Assume that T is a reflexive and symmetric binary relation. Then equalities $\psi(m', n') = \varphi_{\mathrm{T}}(F_1(m'), F_2(n')) = 0$, $\psi(m'', n'') = \varphi_{\mathrm{T}}(F_1(m''), F_2(n'')) = 0$ hold. Hence $F_1(m') = F_2(n')$, $F_1(m'') = F_2(n'')$, where the chain of equalities come from

$$\psi(m'', n') = \varphi_{\mathrm{T}}(F_1(m''), F_2(n')) = \varphi_{\mathrm{T}}(F_2(n''), F_1(m')) = \psi(m', n'') = 1.$$

Thereby, the necessity is proved. Note, properties (1), (2) repeat the established above properties according to (4.12) and (4.13).

Let us prove that proposition conditions are sufficient.

Suppose that these conditions take place. Then from (1) and (2), and taking into account Proposition 4.1, it follows that $\psi(m, n) = \varphi_{\widetilde{\mathrm{T}}}(F_1(m), F_2(n))$, where the binary relation $\widetilde{\mathrm{T}}$ is defined according to (4.14). We will show that $\widetilde{\mathrm{T}}$ is a symmetric relation (reflexivity has been established by definition). Really, let $\langle \mathrm{M}_i, \mathrm{M}_j \rangle \in \widetilde{\mathrm{T}}\ (i \neq j)$, then for $m \in \mathrm{M}_i$, $n \in \mathrm{N}_j$ the equality

$$\psi(m, n) = 1 \tag{4.16}$$

holds. Let $n' \in \mathrm{N}_j$ and $m'' \in \mathrm{M}_i$ then

$$\psi(m, n') = 1, \quad \psi(m'', n) = 1. \tag{4.17}$$

From (4.16) and (4.17) with property (3) usage, it may be concluded that $\langle \mathrm{M}_j, \mathrm{M}_i \rangle \in \widetilde{\mathrm{T}}$, i.e. the relation is symmetric. This completes the proof.

Proposition 4.3 Let T be a reflexive and antisymmetric binary relation and $F_1, F_2 : D \to D$, then the indicator function is representable as $\psi(m, n) = \varphi_{\mathrm{T}}(F_1(m), F_2(n))$ if and only if the next properties are fulfilled:

1. $\forall m \in D\, \exists n \in D \Rightarrow \psi(m,n) = 0$,
2. $\exists m', n', n'' \in D : \psi(m', n') = \psi(m', n'') = 0 \Rightarrow \psi(m, n') = \psi(m, n'')(\forall m \in D)$,
3. $\forall m, m'', n, n' \in D : \psi(m,n) = 1, \psi(m,n') = 0, \psi(m'', n) = 0 \Rightarrow \psi(m'', n') = -1$.

Proof Suppose that $\psi(m,n) = \varphi_T(F_1(m), F_2(n))$ takes place, where T is a reflexive and antisymmetric relation, then if $\psi(m,n') = 0$, $\psi(m'', n) = 0$ we obtain

$$F_1(m) = F_2(n'), \quad F_1(m'') = F_2(n). \tag{4.18}$$

From (4.18) by anti-symmetry of a binary relation T it follows that

$$\begin{aligned}\psi(m,n) = \varphi_T(F_1(m), F_2(n)) = \varphi_T(F_2(n'),\\ F_1(m'')) = -\varphi_T(F_1(m''), F_2(n')) = \psi(m'', n'),\end{aligned}$$

i.e. $\psi(m'', n') = -1$.

It remains to prove sufficiency. Assume that conditions (1)–(3) hold, then from Proposition 4.1 it may be concluded that $\psi(m,n) = \varphi_{\tilde{T}}(F_1(m), F_2(n))$.

Let us show that the binary relation $\tilde{T}$ is antisymmetric. Indeed, if $\langle M_i, M_j \rangle \in \tilde{T}\,(i \neq j)$ then for $m \in M_i$, $n \in N_j$ according to definition $\tilde{T}$ (see Proposition 4.1), it may be deduced that $\psi(m,n) = 1$. Let $m'' \in M_j$ $n' \in N_i$, then $\psi(m,n') = 0$, $\psi(m'', n) = 0$. Three equalities $\psi(m,n) = 1$, $\psi(m,n') = 0$, $\psi(m'', n) = 0$ from property (3) lead to $\psi(m'', n') = -1$, i.e. $\langle M_j, M_i \rangle \notin \tilde{T}$ that means $\langle M_i, M_j \rangle \in \tilde{T}$, whence it directly follows that $\langle M_j, M_i \rangle \notin \tilde{T}$, i.e. from both direct and inverse relations one is fulfilled only: $\tilde{T} \cap \tilde{T}^{-1} \subseteq E$, i.e. relation $\tilde{T}$ is antisymmetric. The proposition is proved.

Proposition 4.4 Let T be a reflexive and transitive binary relation and $F_1, F_2 : D \to D$, then the indicator function can be presented as $\psi(m,n) = \varphi_T(F_1(m), F_2(n))$ if and only if the next properties are fulfilled:

1. $\forall m \in D \exists n \in D \Rightarrow \psi(m,n) = 0$,
2. $\exists m', n', n'' \in D : \psi(m', n') = \psi(m', n'') = 0 \Rightarrow \psi(m, n') = \psi(m, n'')(\forall m \in D)$,
3. $\forall m, m', n, n' \in D : \psi(m,n) = 1, \psi(m', n) = 0, \psi(m', n') = 1 \Rightarrow \psi(m, n') = 1$.

Proof First, we prove necessity. Assume that $\psi(m,n) = \varphi_T(F_1(m), F_2(n))$ takes place, where T is a reflexive and transitive binary relation. If $\psi(m,n) = 1$, $\psi(m', n) = 0$, $\psi(m', n') = 1$ then $\varphi_T(F_1(m), F_2(n)) = 1$, $F_1(m') = F_2(n)$, $\varphi_T(F_1(m'), F_2(m')) = 1$ whence $\varphi_T(F_2(n), F_2(n')) = 1$ holds, that by transitivity of T it allows to conclude that $\varphi_T(F_1(m), F_2(n')) = 1$ or $\psi(m', n') = 1$, which proves the necessity of (3).

Let us prove now sufficiency. If $\psi(m, n)$ satisfies to the proposition conditions, from (1) and (2), taking into account Proposition 4.1, it may be concluded that $\psi(m, n) = \varphi_{\tilde{T}}(F_1(m), F_2(n))$. Let us show that T is a reflexive and a transitive relation. Indeed, let

$$\langle M_i, M_j \rangle \in \tilde{T}, \quad \langle M_j, M_k \rangle \in \tilde{T}. \tag{4.19}$$

Let further $m' \in M_i, m'' \in M_j, n'' \in N_j, n''' \in N_k$, then we obtain $\psi(m', n'') = 1$, $\psi(m'', n'') = 0$, $\psi(m'', n''') = 1$ according to (4.19). By the assumption of correctness property (3), we get $\psi(m', n''') = 1$ wherefrom it follows immediately that relation $\langle M_i, M_k \rangle \in \tilde{T}$ holds. So, this validity together with (4.19) prove required transitivity. The proof is complete.

Proposition 4.5 Let T be a reflexive and transitive binary relation and $F_1, F_2 : D \to D$, then the indicator function can be presented as $\psi(m, n) = \varphi_T(F_1(m), F_2(n))$ if and only if the property

$$\forall m, m', n, n' \in D : \psi(m, n') = 0, \psi(m', n) = 0, \psi(m, n) \neq 0 \Rightarrow \psi(m', n') \neq 0$$

holds.

Proof Suppose $\psi(m, n) = \varphi_T(F_1(m), F_2(n))$ takes place, where T is reflexive, anti-symmetric and transitive by the conditions of Propositions 4.3 and 4.4, i.e. this binary relation has a partial order of one. Further, from equalities $\psi(m, n') = 0$, $\psi(m', n) = 0$ and from (4.7) we get $F_1(m) = F_2(n')$, $F_1(m') = F_2(n)$. Let $\psi(m, n) = -1$, then it follows $\psi(m, n) = \varphi_T(F_1(m), F_2(n)) = -1$. Further, by virtue of the properties of T it may be concluded that $\varphi_T(F_2(n), F_1(m)) = -\varphi_T(F_1(m), F_2(n)) = 1$ or $\varphi_T(F_1(m'), F_2(n')) = \psi(m', n') = 1$ which proves necessity.

Let proposition conditions be true for $\psi(m, n)$, then the formula $\psi(m, n) = \varphi_{\tilde{T}}(F_1(m), F_2(n))$ takes place. It suffices to prove that $\tilde{T}$ is orderliness relation for set $\{M_i\}$. Let $\langle M_i, M_j \rangle \notin \tilde{T}, m' \in M_i, n' \in N_j, m'' \in M_i, n'' \in N_j$, then $\psi(m', n'') = -1$, $\psi(m'', n'') = 0$, $\psi(m', n') = 0$, hence, according to the proposition conditions we get $\psi(m'', n') = 1$, i.e. $\langle M_j, M_i \rangle \in \tilde{T}$, which was to be shown.

Let us outline the proved propositions. First, we present all the considered conditions in the following form

$$\left.\begin{array}{l} \forall n \in D\, \exists m \in D \Rightarrow \psi(m, n) = 0, \\ \forall m \in D\, \exists n \in D \Rightarrow \psi(m, n) = 0, \end{array}\right\} \tag{4.20}$$

$$\left.\begin{array}{l} \exists m' \in D : \psi(m', n') = \psi(m', n'') = 0 \Rightarrow \psi(m, n') = \psi(m, n''), \\ \exists n' \in D : \psi(m', n') = \psi(m'', n') = 0 \Rightarrow \psi(m', n) = \psi(m'', n), \end{array}\right\} \tag{4.21}$$

$$\begin{array}{l} \forall m', m'', n', n'' \in D : \psi(m', n'') = 1, \psi(m', n') = 0, \\ \psi(m'', n'') = 0 \Rightarrow \psi(m'', n') = 1, \end{array} \tag{4.22}$$

$$\forall m, m'', n, n' \in D : \psi(m, n) = 1, \psi(m, n') = 0,$$
$$\psi(m'', n) = 0 \Rightarrow \psi(m'', n') = -1, \tag{4.23}$$

$$\forall m, m', n, n' \in D : \psi(m, n) = 1, \psi(m', n) = 0,$$
$$\psi(m', n') = 1 \Rightarrow \psi(m, n') = 1, \tag{4.24}$$

$$\forall m, m', n, n' \in D : \psi(m, n') = 0, \psi(m', n) = 0,$$
$$\psi(m, n) \neq 0 \Rightarrow \psi(m', n') \neq 0. \tag{4.25}$$

Under conditions (4.20)–(4.22), relation T represents tolerance, i.e. the usage of T gives possibilities to implement algorithms with 'similarity concept' which is naturally interpreted as a geometrical congruence (presence of key points, boundary parts, shape occlusions and so forth) with reference to image segmentations. Putting it otherwise, we may deal with coverings and make full use of fuzzy logic tools. Under conditions (4.20)–(4.22) and (4.24), the relation T is an equivalence, i.e. implementation of concept of 'the generalized equality' that generates background to analyze partitions from the point of their belonging to one or several image cuts. Performance of conditions (4.20), (4.21), (4.24) gives us a partial order relation, i.e. tools of cut sets analysis from the point of view of nested partitions to seek a reasonable compromise between over and under segmentation. At last, using all collection of conditions (4.20)–(4.25), it is possible to consider the arrangement of cuts with a possibility of comparison of all elements of partitions.

It is necessary to emphasize particularly that the relation T can be represented as superposition set-theoretic (intersection, union, inclusion) and algebraic (product, transitive closure, invertibility) operations over relations. In particular, if T′ and T″ are symmetric relations then (4.22) holds and for

$$\mathrm{T} = \mathrm{T}' \cup \mathrm{T}'', \quad \mathrm{T} = \mathrm{T}' \cap \mathrm{T}'', \quad \mathrm{T} = (\mathrm{T}')^{-1}, \quad \mathrm{T} = (\mathrm{T}'')^{-1}.$$

If relations T′ and T″ are antisymmetric, (4.23) is fulfilled at

$$\mathrm{T} = \mathrm{T}' \cup \mathrm{T}'', \quad \mathrm{T} = (\mathrm{T}')^{-1}, \quad \mathrm{T} = (\mathrm{T}'')^{-1}.$$

Condition (4.24) is correct if T′, T″ are transitive and

$$\mathrm{T} = \mathrm{T}' \cup \mathrm{T}'', \quad \mathrm{T} = (\mathrm{T}')^{-1}, \quad \mathrm{T} = (\mathrm{T}'')^{-1}, \quad \mathrm{T} = \widehat{\mathrm{T}}', \quad \mathrm{T} = \widehat{\mathrm{T}}'',$$

where $\widehat{\mathrm{T}}$ is a transitive closure of T, i.e. such relation that $\forall m, n \in D$, satisfying $\langle m, n\rangle \in \widehat{\mathrm{T}}$, there exists $h_0 = m, h_1, h_2, \ldots, h_n = n$, for which relations $\langle h_0, h_1\rangle \in \mathrm{T}, \ldots, \langle h_{n-1}, h_n\rangle \in \mathrm{T}$ are fulfilled.

In comparison with above two-element interdependence, it is extremely useful to examine relationships between points of several carriers produced by different segmentations. Therefore, generalization of binary results to k-ary relations obtains principal significance.

We will traditionally understand that k-ary relation $\langle m_1, m_2, \ldots, m_k\rangle \in \mathrm{T}$ is some subset Λ of the Cartesian degree $\Omega \times \Omega \times \ldots \times \Omega = \Omega^{(k)}$ of arbitrary set Ω. Remind that k-ary relation T is reflexive if for any $m \in \Omega$ tuple $(m, m, \ldots, m)$ belongs Λ or, that is the same $\langle m, m, \ldots, m\rangle \in \mathrm{T}$. Assume that some subgroup G of full symmetrical group S_k is given, i.e. substitutions $\pi_k = \begin{pmatrix} 1 & 2 & \ldots & k \\ \alpha_1 & \alpha_2 & \ldots & \alpha_k \end{pmatrix}$ are defined.

Relation T with arity k is said to be G-symmetric, if for each tuple $(m_1, m_2, \ldots, m_k) \in \Lambda$ and any substitution $\pi_k \in \mathrm{G}$ expression $\left\langle m_{\alpha_1}, m_{\alpha_2}, \ldots, m_{\alpha_k}\right\rangle \in \mathrm{T}$ holds.

In the sequel, if it is not specifically stipulated, reflexive and G-symmetric relations are considered only. As well as earlier, for each reflexive relation (but now with k-arity) there corresponds a three-valued indicator function

$$\varphi_{\mathrm{T}}(m) = \begin{cases} -1, \langle m_1, m_2, \ldots, m_k\rangle \notin \mathrm{T}, \\ 0, m_1 = m_2 = \ldots = m_k, \exists m_i \neq m_j, \\ 1, \langle m_1, m_2, \ldots, m_k\rangle \in \mathrm{T}. \end{cases} \tag{4.26}$$

Let surjective maps

$$F_i\colon \Xi \to \Omega, i = 1, 2, \ldots, k, \Xi = \{D, D^{(k)}\}, D^{(k)} = \underbrace{D \times D \times \ldots \times D}_{k} \tag{4.27}$$

are given, then indicator function (4.26) and maps (4.27) define the three-valued function identical to (4.7)

$$\psi(m_1, m_2, \ldots, m_n) = \varphi_{\mathrm{T}}(F_1(m_1), F_2(m_2), \ldots, F_n(m_n)). \tag{4.28}$$

Generalizing the above properties, which are in the final form reduced in formulae (4.20) and (4.21), we represent them as

$$\forall m_i, m_i' \in \Xi\ \exists m_1, \ldots, m_{i-1}, m_{i+1}, \ldots, m_n : \psi(m_1, \ldots, m_{i-1}, m_{i+1}, \ldots, m_n) = 0. \tag{4.29}$$

To prove (4.29), assume formula $F_i(m_i) = n_i$ holds, then by maps $F_1(m_1), F_2(m_2), \ldots, F_{i-1}(m_{i-1}), F_{i+1}(m_{i+1}), \ldots, F_k(m_k)$ surjectivity on Ω for some $m_1, m_2, \ldots, m_{i-1}, m_{i+1}, \ldots, m_k \in \Xi$ there are the following equalities

$$F_1(m_1) = F_2(m_2) = \ldots = F_{i-1}(m_{i-1}) = F_{i+1}(m_{i+1}) = \ldots = F_k(m_k) = n_i,$$

where from it is easily seen that $\psi(m_1, m_2, \ldots, m_k) = \varphi_{\mathrm{T}}(n_i, n_i, \ldots, n_i) = 0$, which establishes the required formula.

Suppose $m_i, n_i' \in \Xi$ and the equality

$$\psi(n_1, n_2, \ldots, n_{i-1}, m_i, n_{i+1}, \ldots, n_k) = \psi(u_1, u_2, \ldots, u_{i-1}, m_i', u_{i+1}, \ldots, u_k) = 0 \quad (4.30)$$

are held. If at least for one index $l \neq i$ parity

$$n_l = u_l \quad (4.31)$$

takes place, then $\forall m_1, m_2, \ldots, m_{i-1}, m_{i+1,\ldots,} m_k \in \Xi$ and the expression

$$\psi(m_1, m_2, \ldots, m_{i-1}, m_i, m_{i+1}, \ldots, m_k) = \psi(m_1, m_2, \ldots, m_{i-1}, m_i', m_{i+1}, \ldots, m_k) \quad (4.32)$$

is fulfilled.

Let us verify the property set defined by (4.30)–(4.32). From (4.26), (4.27), (4.30) it is obvious that $F_i(m_i) = F_l(n_l)$, $F_i(m_i') = F_l(u_l)$. Considering $n_l = u_l$, we get $F_i(m_i) = F_i(m_i') = F_l(n_l)$, whence taking into account (4.28), it is definitively discovered

$$\begin{aligned}
&\psi(m_1, m_2, \ldots, m_{i-1}, m_i, m_{i+1}, \ldots, m_k)\\
&= \varphi_{\mathrm{T}}(F_1(m_1), F_2(m_2), \ldots, F_{i-1}(m_{i-1}), F_i(m_i), F_{i+1}(m_{i+1}), \ldots, F_k(\mathfrak{m}_k))\\
&= \varphi_{\mathrm{T}}(F_1(m_1), F_2(m_2), \ldots, F_{i-1}(m_{i-1}), F_i(m_i'), F_{i+1}(m_{i+1}), \ldots, F_k(m_k))\\
&= \psi(m_1, m_2, \ldots, m_{i-1}, m_i', m_{i+1}, \ldots, m_k),
\end{aligned}$$

which was to be shown.

It should be noted that any function $\psi(m_1, m_2, \ldots, m_{i-1}, m_i, m_{i+1}, \ldots, m_k)$ on Cartesian product $\underbrace{\Xi \times \Xi \times \ldots \times \Xi}_{k}$ defines k equivalence relations $\langle m_i, n_i \rangle \in \mathrm{T}_i (i = \overline{1, k})$ if for all the $u_1, u_2, \ldots, u_{i-1}, m_i, u_{i+1}, \ldots, u_k \in \Xi$ equality $\psi(m_1, m_2, \ldots, m_{i-1}, m_i, m_{i+1}, \ldots, m_k) = \psi(m_1, m_2, \ldots, m_{i-1}, n_i, m_{i+1}, \ldots, m_k)$ is fulfilled.

Assume that $\Xi = \cup_j \Xi_{ij}\ \forall j' \neq j''\ \Xi_{ij'} \cap \Xi_{ij''} = \emptyset$ is a partition Ξ induced by relations T_i $(i = \overline{1, k})$. If indicator function $\psi(m_1, m_2, \ldots, m_k)$ satisfies conditions (4.29), (4.30)–(4.32), then sets $L_i = \{\Xi_{ij}\}(i = \overline{1, k})$ of classes composed of T_i-equivalent elements have the same cardinality. One-to-one correspondence between these classes can be established as follows.

For some fixed index i such subset Ξ_{ij} is compared to each subset Ξ_{1l} that for some $m_1 \in \Xi_{1l}$, $m_i \in \Xi_{ij}$, $n_2, n_3, \ldots, n_k \in \Xi$ the equality

$$\psi(m_1, n_2, \ldots, n_{i-1}, m_i, n_{i+1}, \ldots, n_k) = 0 \quad (4.33)$$

holds. Indeed, if $m_1 \in \Xi_{1l}$, then according to (4.29) there will be such elements $m_2, m_3, \ldots, m_k \in \Xi$ that $\psi(m_1, m_2, \ldots, m_i, \ldots, m_k) = 0$. Suppose that $\psi(m_1, m_2', \ldots, m_i', \ldots, m_k') = 0$, then from (4.30)–(4.32) it is evident for any $u_1, u_2, \ldots, u_{i-1}, u_{i+1}, \ldots, u_k \in \Xi$

$$\psi(u_1, u_2, \ldots, u_{i-1}, m_i, u_{i+1}, \ldots, u_k) = \psi(u_1, u_2, \ldots, u_{i-1}, m_i', u_{i+1}, \ldots, u_k),$$

i.e. $\langle m_i, m_i' \rangle \in \mathrm{T}_i$.

Thereby, correspondence $\Xi_{1l} \leftrightarrow \Xi_{ij}$ ($m_1 \in \Xi_{1l}, m_i \in \Xi_{ij}$) is correctly defined, since according to (4.29) the class Ξ_{ij} scans all sets L_i when the class Ξ_{1l} runs set L_1. Thus, various subsets Ξ_{ij} match to various classes $\Xi_{1l}, \Xi_{1l'}$, as from the equality

$$\begin{aligned}&\psi(m_1, m_2, \ldots, m_{i-1}, m_i, m_{i+1}, \ldots, m_k)\\&= \psi(m_1', m_2', \ldots, m_{i-1}', m_i, m_{i+1}', \ldots, m_k') = 0\end{aligned}$$

it follows immediately that $\langle m_1, m_1' \rangle \in \mathrm{T}_1$, as it was required.

From now on, indexing of classes Ξ_{ij} is used such that the index j scans the same index set K and for each fixed value $j \in \mathrm{K}$ classes $\Xi_{1l}, \Xi_{1r}, \ldots, \mathfrak{E}_{kj}$ correspond each other. This numbering is valid because of the aforementioned statement that $\psi(m_1, m_2, \ldots, m_k) = 0$ if and only if $m_1 \in \Xi_{1j}, \ldots, m_k \in \Xi_{kj}$ holds for some $j \in \mathrm{K}$.

Introduce now on the set $W_1 = \{\Xi_{1j}\}$ k-ary relation $\widetilde{\mathrm{T}}$ as follows. Classes $\Xi_{1j_1}, \Xi_{1j_2}, \ldots, \Xi_{1j_k} \in W_1$ are in the relation $\widetilde{\mathrm{T}}$ if for some $m_1 \in \Xi_{1j_1}, m_2 \in \Xi_{1j_2}, \ldots,$ $m_k \in \Xi_{1j_k} \in W_1$ one of equalities $\psi(m_1, m_2, \ldots, m_k) = 0$ or $\psi(m_1, m_2, \ldots, m_k) = 1$ is fulfilled. This definition is valid, which directly follows from the method of finding the class Ξ_{ij} due to value of function $\psi(m_1, m_2, \ldots, m_k)$ that will not vary if an element m_i is replaced by any representative of the class Ξ_{ij} which contains it.

According to (4.26), the introduced k-ary relation $\widetilde{\mathrm{T}}$ on set $W_1 = \{\Xi_{1j}\} = W$ ensures search of the indicator function $\varphi_{\widetilde{\mathrm{T}}}(m)$ defined on $W^{(k)}$. As in the case of binary relations, we define k maps $F_1, F_2, \ldots, F_k$ of set Ξ to set

$$W : F_i(m) = \Xi_{1j}, \text{ if } L_i = \{\Xi_{ij}\}, \;\; i = 1, k. \tag{4.34}$$

Make sure the equality

$$\psi(m_1, m_2, \ldots, m_k) = \varphi_{\widetilde{\mathrm{T}}}(F_1(m_1), F_2(m_2), \ldots, F_k(m_k)) \tag{4.35}$$

is true. Indeed, if $\psi(m_1, m_2, \ldots, m_k) = 0$, then inclusions $m_1 \in \Xi_{1j}, m_2 \in \Xi_{2j}, \ldots, m_k \in \Xi_{kj}$ are true and therefore we obtain

$$\varphi_{\widetilde{\mathrm{T}}}(F_1(m_1), F_2(m_2), \ldots, F_k(m_k)) = \varphi_{\widetilde{\mathrm{T}}}(\Xi_{1j}, \Xi_{1j}, \ldots, \Xi_{1j}) = 0.$$

Suppose that $\psi(m_1, m_2, \ldots, m_k) = 1$ and $m_i \in \Xi_{iji}$, then by $\widetilde{\mathrm{T}}$ definition seeing (4.34), we get

$$\varphi_{\widetilde{\mathrm{T}}}(F_1(m_1), F_2(m_2), \ldots, F_k(m_k)) = \varphi_{\widetilde{\mathrm{T}}}(\Xi_{1j1}, \Xi_{2j2}, \ldots, \Xi_{njk}) = 1.$$

Finally, if $\psi(m_1, m_2, \ldots, m_k) = -1$ and $m_i \in \Xi_{iji}$ according to definition of k-ary relation $\widetilde{\mathrm{T}}$, considering that classes $\Xi_{1j_1}, \Xi_{1j_2}, \ldots, \Xi_{1j_k}$ are not in the relation

$\widetilde{\mathrm{T}}$, we obtain

$$\varphi_{\widetilde{\mathrm{T}}}(F_1(m_1), F_2(m_2), \ldots, F_k(m_k)) = \varphi_{\widetilde{\mathrm{T}}}(\Xi_{1j_1}, \Xi_{2j_2}, \ldots, \Xi_{nj_k}) = -1,$$

which finishes the proof of equality (4.35).

It is actually established that the three-valued indicator function $\psi(m_1, m_2, \ldots, m_k)$, satisfying conditions (4.29)–(4.32), is representable in the form of (4.28), and the relation T is reflexive. Indeed, the equality (4.35) in which $\widetilde{\mathrm{T}}$ is reflexive k-ary relation has been proved, that directly follows from its definition.

Let us proceed to the study of the indicator function, depending on the properties of the relation $\widetilde{\mathrm{T}}$.

Establish now conditions with which three-valued function $\psi(m_1, m_2, \ldots, m_k)$ can be presented in the form of (4.28) where T is G-symmetric k-ary relation.

Proposition 4.6 Assume that T is a reflective and symmetric k-ary relation and $G \in S_n$ is a subgroup of symmetrical group, then indicator function $\psi(m_1, m_2, \ldots, m_k)$ can be represented as (4.28) if and only if conditions (4.29–4.32) are satisfied, and it satisfies such property that if $\psi(m_{11}, m_{22}, \ldots, m_{kk}) = 1$, then

$$\forall \begin{pmatrix} 1 & 2 & \ldots & k \\ \alpha_1 & \alpha_2 & \ldots & \alpha_k \end{pmatrix} \in G, \begin{cases} \psi(m_{11}, m_{12}, \ldots, m_{1k}) = 0, \\ \psi(m_{21}, m_{22}, \ldots, m_{2k}) = 0, \\ \ldots \\ \psi(m_{k1}, m_{k2}, \ldots, m_{kk}) = 0, \end{cases} \Rightarrow \psi(m_{\alpha_1 1}, \ldots, m_{\alpha_k k}) = 1. \tag{4.36}$$

Proof We begin by proving necessity. Suppose that T is reflexive G-symmetric k-ary relation and additionally (4.36) takes place. On the assumption that

$$\begin{aligned} &\psi(m_{11}, m_{22}, \ldots, m_{kk}) = \varphi_{\widetilde{\mathrm{T}}}(F_1(m_{11}), F_2(m_{22}), \ldots, F_k(m_{kk})) = 1, \\ &\psi(m_{11}, m_{12}, \ldots, m_{1k}) = \varphi_{\widetilde{\mathrm{T}}}(F_1(m_{11}), F_2(m_{12}), \ldots, F_k(m_{1k})) = 1, \\ &\ldots \\ &\psi(m_{k1}, m_{k2}, \ldots, m_{kk}) = \varphi_{\widetilde{\mathrm{T}}}(F_1(m_{k1}), F_2(m_{k2}), \ldots, F_k(m_{kk})) = 0 \end{aligned}$$

we get

$$\begin{aligned} &F_1(m_{11}) = F_2(m_{22}) = \ldots = F_k(m_{1k}), \\ &F_1(m_{21}) = F_2(m_{22}) = \ldots = F_k(m_{2k}), \\ &\ldots \\ &F_1(m_{k1}) = F_2(m_{k2}) = \ldots = F_k(m_{kk}). \end{aligned}$$

Let $\alpha_1, \alpha_2, \ldots, \alpha_k$ be arbitrary permutation of numbers $1, 2, \ldots, k$ from G then, taking into account the G-symmetry of the k-ary relation and penultimate equalities, we obtain

$$\begin{aligned}1 &= \psi(m_{11}, m_{22}, \ldots, m_{kk}) = \varphi_{\widetilde{T}}(F_1(m_{11}), F_2(m_{22}), \ldots, F_k(m_{kk}))\\ &= \varphi_{\widetilde{T}}(F_{\alpha_1}(m_{\alpha_1\alpha_1}), F_{\alpha_2}(m_{\alpha_2\alpha_2}), \ldots, F_{\alpha_k}(m_{\alpha_k\alpha_k}))\\ &= \varphi_{\widetilde{T}}(F_{\alpha_1}(m_{\alpha_1\alpha_1}), F_{\alpha_2}(m_{\alpha_2\alpha_2}), \ldots, F_{\alpha_k}(m_{\alpha_k\alpha_k}))\\ &= \varphi_{\widetilde{T}}(F_1(m_{\alpha_1 1}), F_2(m_{\alpha_2 2}), \ldots, F_k(m_{\alpha_k k})) = \psi(m_{\alpha_1 1}, m_{\alpha_2 2}, \ldots, m_{\alpha_k k}),\end{aligned} \tag{4.37}$$

which proves the necessity of condition (4.36).

We turn to the proof of sufficiency. From conditions (4.29)–(4.32) on the basis of (4.35), it may be concluded that

$$\psi(m_1, m_2, \ldots, m_k) = \varphi_{\widetilde{T}}(F_1(m_1), F_2(m_2), \ldots, F_k(m_k)),$$

where the relation $\widetilde{T}$ is reflexive. It remains to prove G-symmetry.

Assume that

$$m_1 \in \Xi_{1j_1}, m_2 \in \Xi_{2j_2}, \ldots, m_k \in \Xi_{kj_k}, \psi(m_1, m_2, \ldots, m_n) = 1, \tag{4.38}$$

then from the definition of $\widetilde{T}$ it follows that classes $\Xi_{1j_1}, \Xi_{2j_2}, \ldots, \Xi_{kj_k}$ are in the relation $\widetilde{T}$. Let $j_{r1}, j_{r2}, \ldots, j_{rk}$ be an arbitrary substitution of numbers $j_1, j_2, \ldots, j_k$. Suppose further that

$$\begin{aligned}&\psi(n_{11}, n_{12}, \ldots, m_{r1}, \ldots, n_{1k}) = 0,\\ &\psi(n_{21}, n_{22}, \ldots, m_{r2}, \ldots, n_{2k}) = 0,\\ &\ldots\\ &\psi(n_{k1}, n_{k2}, \ldots, m_{rk}, \ldots, n_{kk}) = 0.\end{aligned}$$

From these equalities it follows that $n_{11} \in \Xi_{1jr_1}, n_{22} \in \Xi_{2jr_2}, \ldots, n_{k_k} \in \Xi_{2jr_k}$. On the other hand, taking into account (4.38) we get equality $\psi(n_{11}, n_{22}, \ldots, n_{kk}) = 0$, thereby, from $\widetilde{T}$ definition it follows that classes $\Xi_{1jr_1}, \Xi_{2jr_2}, \ldots, \Xi_{2jr_k}$ are in relation $\widetilde{T}$, i.e. it is symmetric. The proof is complete.

Antisymmetric binary relations, i.e. such that $\forall m, n \in D : m \neq n \Rightarrow \langle m, n\rangle \in T$ or $\langle n, m\rangle \in T$, have been investigated above. Transition from a couple (m, n) to a pair (n, m) is fulfilled under the action of unique odd permutation on set $\{1, 2\}$, viz. by transposition $\begin{pmatrix}1\,2\\2\,1\end{pmatrix}$. Based on this, reflexive k-ary relation $\langle m_1, m_2, \ldots, m_n\rangle \in T$ can be called anti-symmetric if conditions

$$\begin{aligned}&\langle m_1, m_2, \ldots, m_k\rangle \in T \Rightarrow \langle m_{\alpha_1}, m_{\alpha_2}, \ldots, m_{\alpha_k}\rangle \in T, g \text{ is an even permutation},\\ &\langle m_1, m_2, \ldots, m_k\rangle \notin T \Rightarrow \langle m_{\alpha_1}, m_{\alpha_2}, \ldots, m_{\alpha_k}\rangle \in T, g \text{ is an odd permutation}\end{aligned}$$

are met for $g = \begin{pmatrix}1 & 2 & \ldots & k\\ \alpha_1 & \alpha_2 & \ldots & \alpha_k\end{pmatrix} \in G$.

Proposition 4.7 Let T be reflexive and symmetric k-ary relation, and $G \in S_k$ is a subgroup of symmetrical group, then indicator function $\psi(m_1, m_2, \ldots, m_n)$ can be presented as (4.28) if and only if conditions (4.29)–(4.32) hold and this function satisfies one of the following properties: if $\psi(m_{11}, m_{22}, \ldots, m_{nn}) = 1$ and $g = \begin{pmatrix} 1 & 2 & \ldots & k \\ \alpha_1 & \alpha_2 & \ldots & \alpha_k \end{pmatrix} \in G \subset S_k$ is an even permutation, or $\psi(m_{11}, m_{22}, \ldots, m_{nn}) = -1$ and $g = \begin{pmatrix} 1 & 2 & \ldots & k \\ \alpha_1 & \alpha_2 & \ldots & \alpha_k \end{pmatrix} \in G \subset S_k$ is an odd permutation, then

$$\left.\begin{array}{r} \psi(m_{\alpha_1 1}, m_{\alpha_1 2}, \ldots, m_{\alpha_1 \alpha_1}, \ldots, m_{\alpha_1 k}) = 0 \\ \psi(m_{\alpha_2 1}, m_{\alpha_2 2}, \ldots, m_{\alpha_2 \alpha_2}, \ldots, m_{\alpha_2 k}) = 0 \\ \ldots \\ \psi(m_{\alpha_k 1}, m_{\alpha_k 2}, \ldots, m_{\alpha_k \alpha_k}, \ldots, m_{\alpha_k k}) = 0 \end{array}\right\} \Rightarrow \psi(m_{\alpha_1 1}, m_{\alpha_2 2}, \ldots, m_{\alpha_k k}) = 1. \tag{4.39}$$

Proof We give the proof only for necessity. The sufficiency proof is carried out absolutely similarly to the sufficiency proof in Proposition 4.6.

Assume that T is reflexive and antisymmetric k-ary relation, and (4.39) takes place. From anti-symmetry definition, in a case when $g \in G \subset S_k$ is an even permutation, k-ary relation T is G-invariant relatively group $G = A_k \subset S_k$ from Proposition 4.6. Hence, the ability to represent an indicator function in the form of (4.28) directly follows.

Suppose that $\psi(m_{11}, m_{22}, \ldots, m_{kk}) = -1$ is an odd permutation and $g = \begin{pmatrix} 1 & 2 & \ldots & k \\ \alpha_1 & \alpha_2 & \ldots & \alpha_k \end{pmatrix} \in G \subset S_k$, as well as (4.39) holds then

$$\begin{aligned} F_1(m_{\alpha_1 1}) &= F_2(m_{\alpha_1 2}) = \ldots = F_k(m_{\alpha_1 k}), \\ F_1(m_{\alpha_2 1}) &= F_2(m_{\alpha_2 2}) = \ldots = F_k(m_{\alpha_2 k}), \\ &\ldots \\ F_1(m_{\alpha_k 1}) &= F_2(m_{\alpha_k 2}) = \ldots = F_k(m_{\alpha_k k}). \end{aligned}$$

Thus, we have

$$\psi(m_{11}, m_{22}, \ldots, m_{kk}) = \varphi_{\tilde{T}}(F_1(m_{11}), F_2(m_{22}), \ldots, F_k(m_{kk})) = -1.$$

Considering last but one relations and an anti-symmetry of T, we obtain

$$\begin{aligned} &\varphi_{\tilde{T}}(F_1(m_{11}), F_2(m_{22}), \ldots, F_k(m_{kk})) \\ &\quad = -\varphi_{\tilde{T}}(F_{\alpha_1}(m_{\alpha_1 1}), F_{\alpha_2}(m_{\alpha_2 2}), \ldots, F_{\alpha_k}(m_{\alpha_k k})) \\ &\quad = -\varphi_{\tilde{T}}(F_1(m_{\alpha_1 1}), F_2(m_{\alpha_2 2}), \ldots, F_k(m_{\alpha_k k})) \\ &\quad = -\psi(m_{\alpha_1 1}, m_{\alpha_2 2}, \ldots, m_{\alpha_k k}). \end{aligned}$$

Therefore $\psi(m_{\alpha_1 1}, m_{\alpha_2 2}, \ldots, m_{\alpha_k k}) = 1$, and proof of necessity is completed.

The obtained results ensure not only the analysis of binary segmentation cuts, but also the number of equivalence classes. These results are far from being conclusive but the conditions of transform segmentations are defined fairly strictly, which provides a rational choice of transformations, that ultimately creates promising prerequisites for producing valid multidimensional time series based on spatial information about each frame.

4.2 Classification of Models for Time Varying Images Semantic Separation

Improved storage technology and data access capabilities speed up the researchers to move from analyzing images exclusively to analyzing visual information in general, which led to the emergence of relevant content-based video retrieval (CBVR) and content-based visual information retrieval (CBVIR).

With the development of CBVR, i.e. with the growth of the video information role, the emphasis in image analysis has shifted, since in addition to spatial analysis, characteristic for the static multimedia data analysis, the time component was added, which in turn added some problems and their possible solutions. Among them in particular, it should be noted: the moving objects detection and recognition in the conditions of interference and disturbances of various kinds [11]; tracking the trajectory of the found objects [12]; motion detection [13]; scene changes detection [5, 6, 14].

To solve these problems, a large number of specialized systems have been developed that carry out the separation of images from moving objects against a complex background, filtering noise, high-speed filtering, separating objects from the background, estimating the speed of each object, identifying and tracking it. These systems are built using different methods and approaches, usually working with data coming in real time.

Conventionally, video processing in such systems can be divided into the following steps: foreground (background) detection; the moving objects detection and classification; moving objects trajectory detection; objects of interest actions recognition and description.

The process of foreground detection is to separate the image moving fragments from the still ones, which are called background or belonging to the background. The complexity of this problem is due to different factors, such as the camera's own wake-ups, a sharp change in scene lighting, falling shadows, movement of tree branches in the wind, etc.

At the second stage of detection and classification, the foreground image is first segmented, that is, there are compact parts moving at the objects same speed that are considered moving. Further, they relate to predefined classes: cars, people, animals, etc.

To detect the trajectory of a moving object (tracking), it is necessary to establish one-to-one correspondence between the objects found on successive frames. At the final stage, the objects actions recognition and description that have been selected.

After objects recognition and classification, it becomes possible to search for information in data stores, when the found object is used as a search criterion. This approach is relevant in tracking objects, editing television programs, when it is necessary to find video data in the existing archive, the context of which corresponds to a given pattern (that is, the key points of a particular video contain a recognized object).

Most content-based image retrieval systems select and calculate characteristics at the data preprocessing stage. Calculated once at the beginning, they act as input data for subsequent tasks of analyzing multimedia information, such as evaluating similarity, annotation and pattern recognition.

Speaking of extracting local characteristics (descriptors of not an entire image, but of some of its area, segment), we come to the image segmentation process. Without reliable segmentation, it is impossible to properly calculate local descriptors, which means that we will encounter a semantic problem when formalizing the original image.

Of course, researchers have made a number of attempts to reduce the dependence on segmentation. For example, include all the obtained segments in the result set and use the so-called flexible metrics or a block of two-dimensional Markov models (MHMMs) to describe the spatial distribution of color and texture [15]. Another possible strategy is grouping perceptual principles using the image structure hierarchical separation [16].

When solving such a problem as searching in video archives and multimedia data bases, an issue arises in what form to store such cumbersome amount of information, because searching for the source data at an acceptable time is not a trivial problem.

One of the approaches to reducing the stored information amount is the key frames detection. Key frames are still images extracted from the original video data that best represent the frames content in an abstract manner. Key frames are often used as a supplement to the video description, although they are often manually selected. Moreover, if they are extracted correctly, they are a very effective visual abstract video content description and are very useful for quick viewing/searching in the video.

The currently developed methods for video key frames analysis can be conditionally combined into several groups: Boundary extraction, Matrix factorization, Clustering, Visual attention model, Genetic algorithms, Histogram difference, Statistics, Neuro network, Ontology.

For the first methods of key frames detection developed back in the 90s of the last century, it was typical to scene boundary elements search. Thus, Nagasaka and Tanaka [17] proposed to extract the first frame of each scene. Despite the simplicity of this approach, the number of key frames is always limited and depends on the scene number; moreover, frames are extracted regardless of the their content importance. Similar approaches fix a specified number of frames, distributing them evenly along the entire length of the video sequence.

Later, methods began to appear that take movement into account as spatio-temporal features [12, 18]. Optical flow methods are usually used to search for a local motion minimum. This solution is very difficult from a computational point of view, good results are achieved only if there is significant movement in video.

Another group of methods relies on matrix factorization, that has been used for a relatively long time to implement a key frame search. This group of methods representatives Gong and Liu [19] proposed to divide similar frames into common clusters. After SVD they obtained reordered feature space, which facilitates the classification. This method allows to eliminate repetitions in the resulting sequence of frames, since each key frame is selected from a cluster with similar frames.

Clustering methods have been used to key frames search since 2000, and they still occupy leading positions in popularity. The only difference between the methods that were used before and now is the computational complexity of the procedure in connection with the increasing power that allows you to maintain at least proximity to real time. The clustering methods main problem is traditionally considered to be that the number of clusters, as a rule, should be set a priori, which requires preliminary information about the video content and user participation. Therefore, approaches are currently being developed for clustering video data in situations where it is necessary to split data of different lengths, which is typical for arbitrary video sequences, and in situations where the number of clusters (segments), into which the source video is partitioned, is not necessarily known [20, 21].

Another group of methods is based on the curve simplification. Such methods allow us to represent a video sequence in the curve form from a multi-dimensional attribute space. Each frame significance can be determined by taking into account the eigenvalues in the curve local maxima and minima [22]. Recently, some other simple methods have become popular, for example, analyzing the difference in color histograms between adjacent frames. When choosing this method, scientists should take into account the fact that frames with similar color histograms may have different content [23]. In addition to the frame difference, the average difference within a group of frames that make up a single scene can be determined to find out a candidate for a key frame. For these purposes, classical methods of clustering or fuzzy clustering are often used [24–26].

To make key frames more focused on semantics, some scientists place emphasis on adding text labels for objects with their subsequent analysis. Such tags are sometimes called annotations. In any case, users, in one way or another, are involved in this process, accepting or denying the correspondence of keywords to objects in a frame. Lexical databases are often used to store meanings and synonyms for keywords. One such frequently used lexical database is Wordnet. Unfortunately, by defining high-level tags, it is impossible to foresee all changes in the content that can occur in a video sequence. Despite the fact that certain success have already been achieved in the semantic concept for the key frames search, the correct machine video data interpretation is still a very difficult problem due to the need to involve users [27].

More recently, methods based on visual attention model have been proposed for key frames searching. They involve the generation of an importance map, or the so-called attention curve built from the data on changes in the background and foreground objects. Genetic algorithms, neural networks and artificial immune networks have also been used to extract key frames, but the authors of these methods warn about extensions and improvements that are needed to increase the key frames extraction quality.

However, it is not always necessary to detect key frames to simplify the video description. In a number of tasks, first of all, it is important to split the source data into segments with similar characteristics, in order to use only the necessary video part (one or some scenes) for solving various tasks, one of which can be the key frames detection. Next, consider video segmentation models and methods designed for the scenes of homogeneous content.

One of the most logical approaches to video analysis from the point of view of video separation/segmentation by the content is to present the initial data as a sequence of frames, which allows analyzing both inner-frame information (spatial segmentation) and inter-frame relationships (temporal segmentation).

Thus, first of all, it is necessary to consider models associated with the time sequence analysis, which allow to find out homogeneous segments in one way or another. One such approach to describe multidimensional data sequence is the so-called econometric models [28, 29]. They can be given in two ways: structural form

$$\sum_{l=0}^{p} B_l x(k-l) + Dz(k) = \eta(k), \tag{4.40}$$

where B_l is a matrix of coefficients for intra-system (endogenous) variables, B_0 is non-singular matrix for current time endogenous variables, D denotes coefficient matrix for extra-system (exogenous) variables, $z(k)$ is a vector of exogenous variables, that includes their delayed values, $\eta(k)$ is a vector disturbing signal with zero mathematical expectation and bounded second moments; and a reduced form

$$x(k) = -B_0^{-1}\left(\sum_{l=1}^{p} B_l x(k-l) + Dz(k) - \eta(k)\right) \tag{4.41}$$

or

$$x(k) = CZ(k) + \xi(k), \tag{4.42}$$

where vector $Z(k)$ includes exogenous variables as well as delayed values of endogenous variables

$$\xi(k) = B_0^{-1}\eta(k).$$

An important problem in econometric research is to identify form parameters (4.40)–(4.42). Indirect, two-step or three-step least square methods are commonly used for this purpose, provided that they are intended for fixed and predefined sets only [28]. Such algorithms as relaxation algorithm, recursive algorithm or fixed-point method [29] are intended for sequential mode. They have insufficient convergence rate for effective signal processing in real time mode.

The following multidimensional time series model is discussed in [30].

$$x(k) = \sum_{l=1}^{p} B_l x(k-l) + \sum_{p=1}^{q} D_p z(k-p) + F\psi(k-1) + \xi(k) \tag{4.43}$$

or

$$B(z^{-1})x(k) = D(z^{-1})z(k-1) + F\psi(k-1) + G(z^{-1})\eta(k), \tag{4.44}$$

where unknown coefficients describing the behavior of controlling sequence are included in matrices B_l, D_p, F of corresponding dimensions or in matrix polynomials $B(z^{-1})$, $D(z^{-1})$, $G(z^{-1})$ of backward shift operator z^{-1}, $\psi(k)$ is some deterministic function describing trend of signal $x(k)$.

The authors [30] also note that the same time series $x(k)$ can be defined using infinite set of multidimensional Eqs. (4.43) or (4.44).

There are three main approaches to parameter estimation of these equations: maximum likelihood method, Bayesian approach and limited information technique. The first two approaches are implemented in a batch form, while the third one—in a form of recursive least square method with ability to process sequentially obtaining observations. Unfortunately, standard recursive least square method, in fact being an identification algorithm with infinite memory, is not appropriate for processing of such non-stationary objects, the signals of which can abruptly change its properties.

So called vector autoregression models (VAR-models) are easy-to-use, effective, and compact way to solve the problem of multidimensional time series property change detection. These models have been presented and researched in [31, 32].

In general case VAR-model associates previous and current observations of vector signal $x(k)$ in the following way:

$$x(k) = B_0 + \sum_{l=1}^{p} B_l x(k-l) + \xi(k), \tag{4.45}$$

where $B_0 = \{b_{0i}\}$ is a $(n \times 1)$ vector of mean values, $B_l = \{b_{l_{ij}}\}$ is $(n \times n)$ parameter matrices, p denotes order of the model.

Aside from formula (4.45), VAR-model can be compactly described in state space the following way

$$\begin{cases} x(k) = \Pi x(k-1) + \Pi_0 + E(k), \\ y(k) = Cx(k), \end{cases} \tag{4.46}$$

where

$$x(k) = \begin{pmatrix} x(k) \\ x(k-1) \\ \vdots \\ x(k-p+1) \end{pmatrix}, \Pi_0 = \begin{pmatrix} B_0 \\ \vec{0} \\ \vdots \\ \vec{0} \end{pmatrix},$$

$$\Pi = \begin{pmatrix} B_1 & \cdots & B_{p-1} & B_p \\ I_n & \cdots & 0 & 0 \\ \vdots & \ddots & \vdots & \vdots \\ 0 & \cdots & I_n & 0 \end{pmatrix}, E(k) = \begin{pmatrix} \xi(k) \\ \vec{0} \\ \vdots \\ \vec{0} \end{pmatrix},$$

$C = (I_n, 0, \ldots, 0)$, $x(k)$ is $(np \times 1)$ vector of states, Π is $(np \times np)$ transfer matrix, $\vec{0}$ and 0 denote $(n \times 1)$ zero-vector and $(n \times n)$ zero-matrix respectively.

The definition proposed in (4.46) enables to use powerful mathematical apparatus of Kalman filtering for multidimensional signal analysis.

Property change detection for multidimensional time series $x(k)$ is related to control of each component $x_i(k), i = 1, 2, \ldots, n$, provided that three following situations may happen:

(a) change of mean values (equations) of components (while $l \leq p$)

$$b_{0i}(k) = \begin{cases} b_{0i}, ifk < k_a, \\ b_{0i}^a, ifk \geq k_a; \end{cases} \tag{4.47}$$

(b) change of characteristics (variations) of disturbances $\xi_i(\sigma_i^2)$ (while $l \leq p$)

$$x_i(k) = \begin{cases} b_{0i} + \sum\limits_{l=1}^{p} \sum\limits_{j=1}^{n} b_{lij} x_j(k-l) + \xi_i(k), ifk < k_a, \\ b_{0i} + \sum\limits_{l=1}^{p} \sum\limits_{j=1}^{n} b_{lij} x_j(k-l) + \xi_i^a(k), ifk \geq k_a; \end{cases} \tag{4.48}$$

(c) change of coefficients b_{lij} that results in autocorrelation property change of non-stationary time series

$$x_i(k) = \begin{cases} b_{0i} + \sum\limits_{l=1}^{p} \sum\limits_{j=1}^{n} b_{lij} x_j(k-l) + \xi_i(k), ifk < k_a, \\ b_{0i} + \sum\limits_{l=1}^{p} \sum\limits_{j=1}^{n} b_{lij}^0 x_j(k-l) + \xi_i(k), ifk \geq k_a, \end{cases} \tag{4.49}$$

where k_a is a moment in time when property change takes place.

Another approach to the video segmentation based on the analysis of multidimensional time series may be their adaptive forecasting, when a noticeable gap between the forecast and the observed data serves as an indicator. Traditional mathematical forecasting methods (statistical, adaptive, neural network, etc.) are based on mathematical models of one kind or another, obtained as a result of solving the structural and parametric identification problem. It is on the basis of mathematical models that the temporal extrapolation problem is solved, and the discrete time argument is explicitly or implicitly present in the model. If there is not enough data to build a mathematical model, the synthesis of a predictive model is simply impossible. In this case, instead of temporal extrapolation, spatial prediction (extrapolation) can be used, which reduces to the vector field estimation from individual observations. Among the methods of spatial extrapolation, one of the most promising to be noted is the multidimensional linear extrapolation [33], which had proved its effectiveness in solving real problems of designing and managing complex multidimensional nonlinear objects.

Let us consider the multidimensional linear extrapolation method as applied to the one-step prediction problem of an n-dimensional nonlinear nonstationary time sequence. Assuming, as before $\cup_{i=1}^{N} x(i)$, then the analyzed series (i-th component) can be described by a nonlinear dependence (NARX model) in the form

$$\begin{aligned}\hat{x}_i(k) &= f_i(x_i(k-1), \ldots, x_i(k-n_{A,i}), x_1'(k-1), \ldots, x_1'(k-n_C),\\ &\quad x_2'(k-1), \ldots, x_2'(k-n_C), \ldots, x_p'(k-l), \ldots, x_q'(k-n_C))\\ &= f_i(R_{i1}(k), \ldots, R_{i,n_{A},i}(k), \ldots, R_{i,n_{A},i+n_{C}q}(k)),\end{aligned} \tag{4.50}$$

where $f_i(\circ)$ is a priori unknown non-linear dependence to be restored on the basis of available observations; $\hat{x}_i(k)$ is evaluation (forecast) of the controlled sequence $x_i(k)$ from the data available up to the $(k-1)$-th instant of time; $i = 1, \ldots, n$; $n_{A,i}$ are depth of the history taken into account in a controlled sequence; $x_p'(k-l)$ is p-th component of the multidimensional exogenous signal $x_i(k)$ affected, $l = 1, \ldots, n_C$; $p = 1, \ldots, q$.

Equation (4.50) can also be represented in a vector-matrix form

$$\widehat{X}(k) = F(X(k-1), \ldots, X(k-n_{A,i}), X'(k-1), \ldots, X'(k-n_C)) = F(R(k)).$$

It should be noted, that in a representative training set, non-linear transformations can be restored in the learning process of one or another artificial neural network, however, if this set is small, the neural network approach turns out to be inoperable, while multidimensional linear extrapolation will provide quite acceptable results.

4.3 Exponential Smoothing and VAR Models for Video Streams Temporal Segmentation

This section covers sequential (real time) detection of property changes in multidimensional time series concurrently with adaptive identification of VAR-process model (4.45). The initial information for identification and property change detection is given by n-dimensional time series $x(k)$ only. Values of $x(k)$ are obtained by adaptive identifier in a real time.

To simplify further computations let $B = (B_0 \vdots B_1 \vdots \ldots \vdots B_p)$ be a composite matrix and $X(k) = (1, x^T(k-1), \ldots, x^T(k-p))^T$ be a composite vector of $((pn+1) \times 1)$ dimension. This enables to present Eq. (4.45) as follows:

$$x(k) = BX(k) + \xi(k), \tag{4.51}$$

where matrix B of a priori unknown parameters contains almost all the necessary information about controlled signal properties.

Identification task consists in matching the following adjustable model to the real signal (4.51)

$$\hat{x}(k) = B(k-1)X(k), \tag{4.52}$$

provided that parameter matrix $B(k)$ of the model is refined at each time tact k. The refinement is made by minimizing identification criterion being used. This criterion is a function of difference between the estimated $\hat{x}(k)$ and experimental data $x(k)$. In addition, the synthesized model (4.52) should be operable in prediction mode. Predictability property violation may confirm the appearance of particular recursive procedures that can be given in a generalized form

$$\begin{cases} B(k) = B(k-1) + \gamma(k)e(k)X^T(k), \\ e(k) = x(k) - \hat{x}(k) = x(k) - B(k-1)X(k), \end{cases} \tag{4.53}$$

where $\gamma(k)$ is a scalar or matrix learning rate parameter, which determines its properties and depends on identification criterion being used, $e(k)$ is a vector error of identification.

The most widespread are algorithms associated with sum of squares minimum criterion of identification errors

$$I(k) = \sum_{u=1}^{k} \beta(u)\|e(u)\|^2 = \sum_{u=1}^{k}\sum_{i=1}^{n} \beta(u)e_i(u)^2 \tag{4.54}$$

and there modifications which are determined by adopted system of weights $\beta(u)$. The most widely used is least square method, all weights of which have equal significance, i.e.

$$I(k) = \sum_{u=1}^{k} \|e(u)\|^2. \tag{4.55}$$

The algorithm corresponding to (4.53), (4.55) is the following

$$\begin{cases} B(k) = B(k-1) + \frac{e(k)X^T(k)P(k-1)}{1+X^T(k)P(k-1)X(k)}, \\ P(k) = P(k-1) - \frac{P(k-1)X(k)X^T(k)P(k-1)}{1+X^T(k)P(k-1)X(k)}. \end{cases} \tag{4.56}$$

When parameters (4.51) are unchangeable, this algorithm provides monotone convergence of estimates $B(k)$ to the true values of parameters B. As stated above, recursive least square method is not suitable for property change detection.

An alternative for (4.55), (4.56) is a one-step procedure

$$B(k) = B(k-1) + \frac{e(k)X^T(k)}{X^T(k)X(k)}. \tag{4.57}$$

This procedure is generated by one-step identification criterion

$$I(k) = \|e(k)\|^2.$$

The procedure is a Kaczmarz algorithm generalization [34] for vector-matrix model (4.52). Having high performance, the procedure (4.57) does not possess filtering properties. That is why it has no ability to distinguish signal changes and the influence of stochastic component $\xi(k)$.

According to the above considerations, it is reasonable to use algorithms with finite memory and smoothing as well as tracking properties. Memory capacity determines the compromise between these properties. Model (4.52) predictability loss and the necessity of algorithm memory rearrangement may serve as a sign for change occurrence.

Turning back to the criterion (4.54), it is important to notice that among criterion-generated group of algorithms the exponentially weighted least squares method became the most distributed with the following criterion

$$I(k) = \sum_{u=1}^{k} \beta^{k-u} \|e(u)\|^2$$

and recursive setup procedure

$$\begin{cases} B(k) = B(k-1) + \frac{e(k)X^T(k)P(k-1)}{\beta+X^T(k)P(k-1)X(k)}, \\ P(k) = \frac{1}{\beta}\left(P(k-1) - \frac{P(k-1)X(k)X^T(k)P(k-1)}{\beta+X^T(k)P(k-1)X(k)}\right), \end{cases} \tag{4.58}$$

where $0 < \beta \leq 1$ is a smoothing parameter.

Here, it is essential to mention that in general case the identifier with exponential smoothing is unstable. This leads to "burst" of covariance matrix parameters, which appears especially often when processing signal $x(k)$ of high dimensions. Thus, application of traditional exponentially weighted recursive least square method becomes more complicated because of ill-conditioning of information matrix

$$\sum_{u=1}^{k} \beta^{k-u} X(u) X^T(u),$$

generated by high level of correlation between $x_i(k)$ components.

This problem is solved using pseudo-inverse operation with Greville theorem instead of weighted information matrix reversal operation based on Sherman-Morrison formula. Thus obtained algorithm is rather cumbersome from computational point of view, that becomes especially evident for high n.

According to this we propose to consider multidimensional modification of exponentially weighted stochastic approximation algorithm

$$\begin{cases} B(k) = B(k-1) + \frac{e(k)X^T(k)}{\beta r(k-1) + \|X(k)\|^2}, \\ r(k) = \beta r(k-1) + \|X(k)\|^2. \end{cases} \tag{4.59}$$

Such modification is somewhat a compromise between procedures (4.57) and (4.58) with required smoothing and tracking properties.

Adaptation mechanism for algorithms (4.58), (4.59) is based on "suppression" of obsolete information. Dynamic properties of these algorithms are memory dependent. The refinement of matrix of current estimates $B(k)$ is performed on the basis of algorithm memory.

As mentioned earlier, algorithms with infinite memory (such as recursive least square method) possess the best filtering properties. At the same time, these procedures have poor tracking properties, if characteristics of controlled signal change in time. This way, low-memory algorithms (like Kaczmarz algorithm) are more preferable to use, though they provide poor results in case of noisy data.

Thus, when identifying systems in which changes may occur, algorithm memory capacity should be chosen based on the compromise between its smoothing and tracking properties. Unfortunately, in practice, disturbance, drift and jump characteristics are a priori unknown, and they can change during operation process of controlled system. In such conditions it is hard to give preference to a particular algorithm with fixed memory. That is why it is reasonable to use variable memory algorithms, capacity of which could be promptly modified depending on presence or absence of changes.

Smoothing parameter β adjustment method is proposed. This method controls statistics that characterizes regular signal prediction error. It is assumed that model parameter setup is made via exponentially weighted Kalman-Mayne algorithm which can be written as follows for i-th component $x_i(k)$

$$\begin{cases} b_i(k) = b_i(k-1) + \frac{(e(k)X^T(k)P_i(k-1))_i}{\beta\sigma_i^2 + X^T(k)P_i(k-1)X(k)}, \\ P_i(k) = \frac{1}{\beta}\left(P_i(k-1) - \frac{P_i(k-1)X(k)X^T(k)P_i(k-1)}{\beta\sigma_i^2 + X^T(k)P_i(k-1)X(k)}\right), \end{cases} \tag{4.60}$$

where $b_i(k)$ is i-th row of matrix $B(k)$, $(\circ)_i$ is i-th row of corresponding matrix product.

If the variances of particular components $\xi_i(k)$ of disturbance vector are unknown, then the following estimate can be used in (4.60):

$$\begin{cases} \sigma_i^2(k) = \sigma_i^2(k-1) + P_i(k-1)(\sigma_i^2(k-1) - l_i^2(k)), \\ P_i(k) = \frac{1}{\beta}\left(P_i(k-1) - \frac{P_i^2(k-1)}{\beta + P_i(k-1)}\right). \end{cases}$$

It is also assumed that the model provides acceptably accurate description of controlled signal on time intervals S of observations, and parameters can change discontinuously at any time k_a. Varying value $\beta(k)$ is adjusted via the following statistics

$$T_i(k) = \sum_{u=k-S}^{k} \frac{l_i^2(u)}{\beta(k-1)\sigma_i^2 + X^T(u)P_i(u-1)X(u)}$$

with χ^2 distribution and S degrees of freedom, provided that $\beta(0) = 1$. Adjustment of $\beta(k)$ is made in discrete moments of time tS by the following rule:

$$\beta(k) = \begin{cases} 1 \text{ when } k < S, k = tS \text{ and } T_i(k) \le \chi_j^2, \\ \beta(k-1) - \Delta\beta \text{ when } k = tS \text{ and } T_i(k) > \chi_j^2, \\ \beta(k-1) \text{ when } tS < k < (t-1)S, \quad t = 1, 2, \ldots \end{cases} \tag{4.61}$$

Here, χ_j^2 is a quantile of χ^2 law with respect to significance level j, $\Delta\beta$ is a step of adjustment.

The rule (4.61) assumes the change of $\beta(k)$ when values k are multiple of S. When values k are intermediate, the value of $\beta(k)$ remains unchangeable, and the fact of change is registered when the second ratio from (4.61) takes place. This lengthy and inertial procedure forces to search for more quick and effective techniques of change detection.

This way, in [35] smoothing parameter β adjustment method is proposed based on Mann-Whitney U-test. In this case the following value is a controllable characteristic:

$$\sum_{u=k-S+1}^{k} sign(x_i(u) - \hat{x}_i(u)) \ge \gamma, \tag{4.62}$$

where γ is a threshold value, S is a value of moving control window,

$$sign(x_i(u) - \hat{x}_i(u)) = \begin{cases} 0 \text{ when } x_i(u) = \hat{x}_i(u), \\ +1 \text{ when } x_i(u) > \hat{x}_i(u), \\ -1 \text{ when } x_i(u) \le \hat{x}_i(u). \end{cases}$$

The process of control begins from the value of $\beta(1) = 0$, corresponding to maximum performance of the algorithm (4.59). In such case, exponentially weighted recursive least square method is naturally invalid. During identification process the following situations may occur:

$$\sum_{u=k-S+1}^{k} sign(x_i(u) - \hat{x}_i(u)) \geq \delta,$$

which means prevalence of stochastic component ξ_i of signal x_i over the "drift" one. In this case it is necessary to improve algorithm smoothing properties, i.e. increase it via the following rule:

$$\beta(k) = \beta(k-1) + \Delta\beta.$$

In (4.62) signal drift component prevails, and the algorithm has no ability to monitor changes at the right time. In this situation, it is necessary to reduce memory

$$\beta(k) = \beta(k-1) - \Delta\beta$$

and to register the fact of change occurrence.

To control multidimensional time series changes, it is proposed to use the following modification of (4.62):

$$\max_i \left(\sum_{u=k-S+1}^{k} sign(x_i(u) - \hat{x}_i(u)) \right) \geq \gamma,$$

i.e. simultaneously control all of the components and register the fact of change occurrence, if the value of at least one component $x_i(k)$ changes significantly.

The most popular procedures used in practical applications are not stochastic, but heuristic ones, such as methods of Chow, Roberts-Reed, Brown, Trigg-Leach, Shown, etc. [36–38]. All these methods are based on one or another heuristic, and that is why they bear some degree of subjectivity. These methods should be preferably used when there is a lack of a priori and current information about characteristics and properties of controlled signal.

The main principle of most heuristic methods is the following: a set of values for smoothing parameter β is specified (for instance, 0; 0.05; 0.1; ... 0.95; 1) with a set of characteristics that determine identification quality. Very often they are the following:

$$e_i(k, \beta) = x_i(k) - \hat{x}_i(k, \beta)$$

– current error of estimation for i-th component,

$$S_i(k, \beta) = e_i(k, \beta) + S_i(k-1, \beta)$$

– cumulative sum of errors,

$$d_i(k, \beta) = (1 - \beta)|e_i(k, \beta)| + \beta d_i(k - 1, \beta)$$

– mean absolute error value,

$$\bar{e}_i(k, \beta) = (1 - \beta)e_i(k, \beta) + \beta\bar{e}_i(k - 1, \beta)$$

– mean error,

$$\tilde{e}_i(k, \beta) = (1 - \beta)\frac{e_i(k, \beta)}{x_i(k)} + \beta\tilde{e}_i(k - 1, \beta)$$

– mean relative error,

$$\bar{e}_i^2(k, \beta) = (1 - \beta)e_i^2(k, \beta) + \beta\bar{e}_i^2(k - 1, \beta)$$

– mean square error.

If the value of chosen controlled characteristic exceeds some threshold γ, then the decision is made about the fact that changes have been occurred and it is necessary to refine smoothing parameter β respectively.

Stationary stochastic signal usually collides with β lying in the interval $0.7 \leq \beta \leq 0.99$ [39]. The simplest form of control is to decrease smoothing parameter by the following rule when the value $\tilde{e}_i(k, \beta)$ with some β exceeds 0.05 threshold:

$$\beta(k) = \beta(k - 1) - \Delta\beta.$$

And then identification process continues with a new $\beta(k)$. If β exceeds the threshold $0.7(\beta(k) \leq 0.7)$, then the decision is made about changes occurrence in signal.

Chow method is more effective, but at the same time it is more complicated. Here, three models with the values of smoothing parameters β, $\beta + \Delta\beta$ and $\beta - \Delta\beta$ are adjusted simultaneously. For instance, if the best result is obtained with the following smoothing parameter value at current moment of time k

$$\beta(k) = \beta + \Delta\beta,$$

then three new smoothing parameters β, $\beta + \Delta\beta$ and $\beta + 2\Delta\beta$ are used at the next tact of time. If the best model is obtained with the following smoothing parameter value

$$\beta(k) = \beta - \Delta\beta,$$

then triple $\beta, \beta - \Delta\beta$ and $\beta - 2\Delta\beta$ is formed, and at last, if the best result is achieved with the following smoothing parameter value

$$\beta(k) = \beta,$$

then the set $\beta, \beta + \Delta\beta$ and $\beta - \Delta\beta$ remains.

More simple and prompt are methods that use control tracking signal which is an indicator of controlled signal changes [38]. There is a variety of tracking signal forms. When tracking signal exceeds some determined threshold, it means the occurrence of changes.

R. Brown proposed to use the following expression as a tracking signal [36]:

$$T_i^B(k) = \frac{\sum_{u=1}^{k} e_i(u)}{\sqrt{\sigma_i^2(k)}}.$$

The physical meaning of this expression is that the sum of errors varies around 0 (because of error randomness), if the adjusted model matches the controlled signal. In addition, this sum of errors does not exceed some threshold which is set a priori for a given probability level with a definite dispersion of prediction error sum, the value of which tends to the following:

$$\lim_{k \to \infty} \sigma_i^2(k) = \frac{1}{1 - (1 - \beta)^{2(pn+1)}} \sigma_i^2.$$

For practical calculations R. Brown proposed to use mean absolute deviation instead of mean square errors σ_i:

$$d = \int_{-\infty}^{\infty} |e - M\{e\}| p(e) de \approx \sum_{u=0}^{n} |e(u) - M\{e\}| P_i.$$

This deviation is proportional to the mean square value of error, because

$$\Delta = (e - M\{e\})/\sigma_i,$$

$$d_i = \sigma_i \int_{-\infty}^{\infty} |\Delta| p(\Delta) d\Delta,$$

while proportionality coefficient changes insignificantly for wide variety of distributions (for normal distribution $d_i/\sigma_i = \sqrt{2\pi} = 0,7979$).

If the value d_i is calculated according to the following expression:

$$d_i(k) = (1 - \beta)|e_i(k)| + \beta d_i(k - 1),$$

and the tracking signal is the following:

$$T_i^{\beta}(k) = \frac{\sum_{u=1}^{k} e_i(u)}{d_i(k)},$$

then $(1 - P_i)\%$ threshold can be presented the following way:

$$\pm\tau_{i,1-P_i} = \pm\frac{\sigma_i}{2}\sqrt{\frac{\pi(1-\beta)}{1-(1-\beta)^{2(pn+1)}}}.$$

D. Trigg and A. Leach proposed to use the following ratio as a tracking signal [37]:

$$T_i^{TL}(k) = \frac{T_i'(k)}{d_i(k)},$$

where $T_i'(k) = (1-\beta)e_i(k) + \beta' T_i'(k-1)$ is not a total sum of deviations, but a smoothed error, provided that the following inequality is observed

$$\beta' \leq \beta.$$

When $\beta' = \beta$, the tracking signal varies between -1 and $+1$. To incorporate automatic feedback, D. Trigg and A. Leach proposed to calculate smoothing parameter according to the ratio:

$$\beta(k) = 1 - \left|T^{TL}(k)\right|$$

and to register signal changes when $\beta(k)$ changes significantly.

Also Shown modification is known with

$$\beta(k) = 1 - \left|T^{TL}(k-1)\right|.$$

Tracking signal increase demonstrates the growth of a gap between the model and controlled sequence. To reduce this gap, more prompt reaction of identifier is needed, which is achieved using lower smoothing parameter value. Thus, negative feedback is obtained.

Brown method is more preferable to use when identifying processes with a drift that is smooth enough. Strong jumps are better distinguished via Trigg-Leach tracking signal. That is why this form of tracking signal can be helpful for analyzing time series changes that are described using the ratios (4.47–4.49).

To conduct an experiment for testing the described approach, a video clip database was created on the grounds of the documentary series "Destroyed in Seconds" by the Discovery Chanel. The choice of such videos is due to the fact that each of them has several different scenes related to instantaneous catastrophes, possibly taken from different angles. Thus, each video sequence contains several segments with homogeneous characteristics, which must be distinguished. We also note that each video contains 550 frames and their spatial segmentation was obtained for each frame. An example of several sequences is shown in Fig. 4.5.

Further, using the feature vector obtained for each segmentation of the frame as input to the considered VAR model and controlling the changes in the signal based on the Trigg-Leach model, we obtain a one-dimensional sequence corresponding to the original video data. An example of the temporal video segmentation results is shown in Fig. 4.6.

It can be seen from the above picture that at the moments when the video characteristics change, there is a surge in the values of the inter-frame difference. In particular, when changing cameras around the 50-th frame. Or at the time of the car explosion in the area near to the 265-th frame.

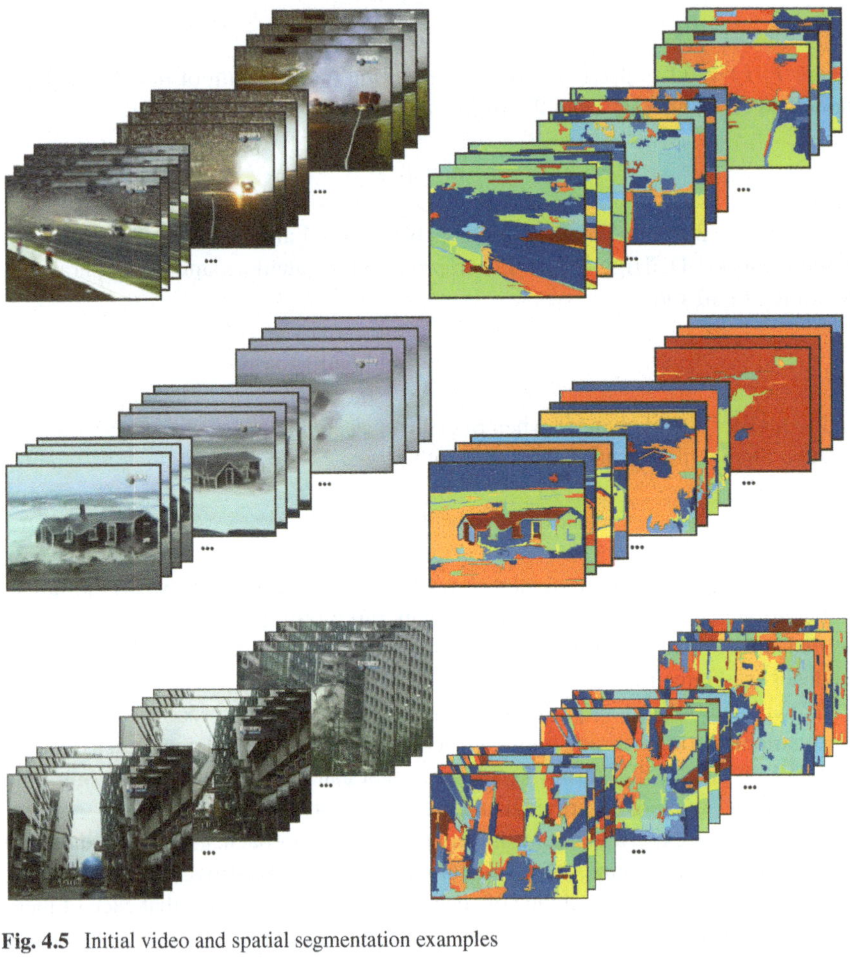

Fig. 4.5 Initial video and spatial segmentation examples

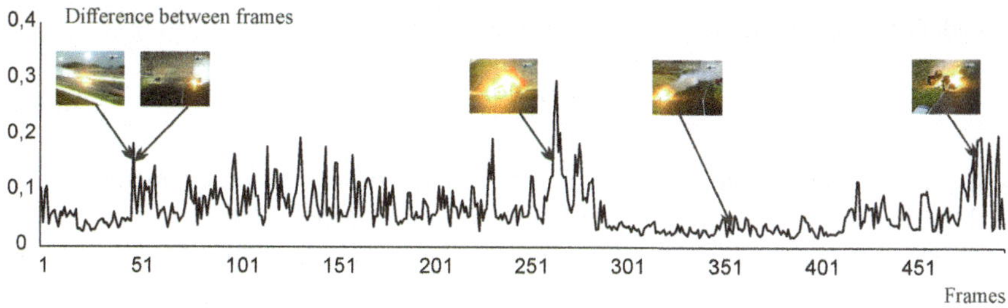

Fig. 4.6 Spatio-temporal segmentation results

At the same time when significant changes in the characteristics do not occur, the values of the characteristics of the resulting sequence also change a little. An example of this is shown in the interval from 295 to 405 frames, during which the camera actually captures the process of a fire in a standing car, i.e. all changes, given the distance to the camera is extremely small. When firefighters appear, a surge of characteristic values is again observed, which is quite consistent with what is happening on the video.

Thus, it can be argued that the considered approaches allow to obtain spatio-temporal video segmentation and highlight both the moments of significant changes in video characteristics and ranges in which changes do not occur. In the next subsection, approaches to fragment-based video segmentation grounded on predictive models are considered.

4.4 Video Segmentation with Forecast Estimation Usage

The multidimensional linear extrapolation issue as applied to the forecasting multidimensional time series problem can be described using the notation (4.50), as follows [33]. Consider the use case matrix (history) in the form

$$\mathrm{P} = \begin{pmatrix} R^T(1), & B^T(1) \\ R^T(2), & B^T(2) \\ \vdots & \\ R^T(N), & B^T(N) \end{pmatrix}$$

of dimensionality $N \times (nn_A + qn_C + n)$. The extrapolation itself is reduced to finding the following evaluation at the time N

$$B'(N+1) = \psi(R(N+1), \mathrm{P}). \tag{4.63}$$

Here, $\psi(\circ)$ is the extrapolation algorithm itself, which should satisfy the requirements, the main of which is that as a result of its use, all the consequences of the precedent matrix should be exactly restored, i.e.

$$\widehat{B}(k) = B(k) = \psi(R(k), \mathrm{P}), k = 1, 2, \ldots N. \tag{4.64}$$

It is important to note that forecasting methods based on certain mathematical models almost never ensure the fulfillment of the condition (4.64).

Among other requirements, it can be noted that the algorithm $\psi(\circ)$ should be vector-based, i.e.

$$\psi = (\psi_1, \ldots, \psi_n), B_i = \psi_i(Z, \mathrm{P}), i = 1, \ldots, n; \tag{4.65}$$

the implementation complexity should grow n and N no faster than linear, the algorithm should work for all N (even if $N = 1$). Note also that for small N, not to mention $N = 1$, a sufficiently accurate mathematical model cannot be constructed at all, although in case of video data analysis, this disadvantage is not so critical.

Following [33], we present an algorithm for multidimensional linear extrapolation with an Euclidean metric in the form of the following sequence of steps:

1. Formation of the predicted process history in the matrix form:

   $R' = (R(1), R(2), \ldots, R(N))$ is $(nn_A + qn_C) \times N$ dimension matrix,
   $B = (B(1), B(2), \ldots, B(N))$ is $n \times N$ dimension matrix.

2. Finding the weight coefficients vector $\lambda = (\lambda_1, \lambda_2 \ldots, \lambda_N)^T$ that are at a minimum rate

$$\left\| R(N+1) - \sum_{k-1}^{N} \lambda(k) R(k) \right\|^2 = \left\| R'(N+1) - R'\lambda \right\|^2.$$

3. Optimal forecast formation in the form of a linear combination

$$\widehat{B}(N+1) = \sum_{k=1}^{N} \lambda_k B(k) = \overline{B}\lambda. \tag{4.66}$$

The minimization of the criteria (4.66) can be carried out using the standard method of least squares, as a result of which we obtain the well-known solution

$$\lambda = (R'^T R')^{-1} R'^T R(N+1) \tag{4.67}$$

which exists only when the matrix $R'^T R'$ is nondegenerate. Since this is not the case for small values $N(N < nn_A + qn_C)$, instead of the matrix inversion operation, it is proposed to use the pseudo-inversion operation, as a result we achieve

$$\lambda = R'^{+} R(N+1), \tag{4.68}$$

whence it follows that, in fact, the task is reduced to finding the orthogonal vector $R(N+1)$ projection onto the linear span formed by the vectors of prehistory R'.

From the computational point of view, there are no difficulties with this procedure implementation, however, the problem solution becomes more complicated if the data for processing is received sequentially in real time. Moreover, all previous relations can be rewritten in the following form:

$$R'_k = (R(1), R(2), \ldots, R(k)), \overline{B}_k = (B(1), B(2), \ldots, B(k)),$$

$$\left\| R(k+1) - \sum_{l=1}^{k} \lambda_l R(l) \right\|^2 = \left\| R(k+1) - R'_k \Lambda_k \right\|^2, \tag{4.69}$$

$$\Lambda_k = (\lambda_1, \lambda_2, \ldots, \lambda_k)^T,$$

$$\widehat{B}(k+1) = \sum_{l=1}^{k} \lambda_l B(l) = \overline{B}_k \Lambda_k, \tag{4.70}$$

$$\Lambda_k = R'^{+}_k R(k+1). \tag{4.71}$$

In [33], it is proposed to use the Grevill formula for calculating the matrix R'^{+}_{k+1} from the available R'^{+}_k and submitted for processing $R(k+1)$ and $B(k+1)$, although it is much easier to use the regularized version (4.67) in the form

$$\Lambda_k = (R'^T_k R'_k + \gamma I_k)^{-1} R'_k R(k+1), \tag{4.72}$$

where γ is regulation parameter, I – $(k \times k)$ is unit matrix.

For processing non-stationary time series, the characteristics of which change unpredictably in time, instead of processing the entire available set, it is advisable to solve the problem on a "sliding window" consisting of the χ latest observations. In this case, relations (4.69–4.71) can be represented as follows:

$$R'_{k,\chi} = (R(k-\chi+1), R(k-\chi+2), \ldots, R(k)),$$

$$\overline{B}_{k,\chi} = (B(k-\chi+1), B(k-\chi+2), \ldots, B(k)),$$

$$\left\| R(k+1) - \sum_{l=k-\chi+1}^{k} \lambda_l R(l) \right\|^2 = \left\| R(k+1) - R'_{k,\chi} \Lambda_{k,\chi} \right\|^2,$$

$$\Lambda_{k,\chi} = (\lambda_{k-\chi+1}, \ldots, \lambda_k)^T,$$

$$\widehat{B}(k+1) = \sum_{l=k-\chi+1}^{k} \lambda_l B(l) = \overline{B}_{k,\chi}\Lambda_{k,\chi}, \tag{4.73}$$

$$\Lambda_{k,\chi} = R'^{+}_{k,\chi} R(k+1).$$

To implement this procedure in real time, a recurrent pseudo-inversion algorithm on a "sliding window" was proposed in [33], which is cumbersome and computationally complex.

In this connection, it is advisable to use modification (4.72) in the "window" version instead of the recurrent pseudo-reversal operation, and its packet form can be written as

$$\Lambda_{k,\chi} = (R'^{T}_{k,\chi} R'_{k,\chi} + \gamma I_\chi)^{-1} R'_{k,\chi} R(k+1), \tag{4.74}$$

and recurrent form [39] as follows

$$\begin{cases} \Lambda_{k+1,\chi} = \Lambda_{k,\chi} + \frac{\mathrm{P}(k)(R(k+1) - \Lambda^{T}_{k,\chi} R(k))}{1 + R^{T}(k)\mathrm{P}(k)R(k)} R(k), \\ \widetilde{\mathrm{P}}(k-1) = \mathrm{P}(k-1) + \frac{\mathrm{P}(k-1)R(k-\chi)R^{T}(k-\chi)\mathrm{P}(k-1)}{1 - R^{T}(k-\chi)\mathrm{P}(k-1)R(k-\chi)}, \\ \mathrm{P}(k) = \widetilde{\mathrm{P}}(k-1) - \frac{\widetilde{\mathrm{P}}(k-1)R(k)R^{T}(k)\widetilde{\mathrm{P}}(k-1)}{1 + R^{T}(k)\widetilde{\mathrm{P}}(k-1)R(k)}, \\ \Lambda_{0,\chi} = 0, \mathrm{P}(k) = \gamma^{-1} I_\chi. \end{cases}$$

Actually, the forecast is calculated according to the relation (4.73). It is clear that the predicting relations (4.73, 4.74) drastically simplifies the method usage, however, a question remains concerning the reasonable choice of the window size χ which is usually set for purely subjective reasons and reduces the effectiveness of the approach as a whole.

The multidimensional extrapolation method proposed below is based on using some proximity (distance) of the last vector of prehistory $R(N+1)$ applied to all previous data $R(1), \ldots, R(N)$ and forming a forecast $\widehat{B}(N+1)$ using the same function.

The method implementation consists of the following sequence of steps:

1. The distance calculation between the vector $R(N+1)$ and all the previous vectors $R(k)$ on the basis of a certain proximity function $d(N+1, k)$ in the simplest case of the Euclidean metric is

$$d(N+1,k) = \left\| B'(N+1) - B'(k) \right\| \forall k.$$

2. The ordering (ranking) of these distances is ascending so that

$$d^1(N+1,k_1) < d^2(N+1,k_2) < \cdots < d^N(N+1,k_N).$$

3. The selection of the first χ vectors for which the following condition is held

$$d^{\chi}(N+1, k_{\chi}) \leq \varepsilon,$$

where ε is some threshold.

4. The set of weights λ_l is composed in the form

$$\lambda_l = \frac{(d^l)^{-1}}{\sum_{l=1}^{\chi} (d^l)^{-1}}, \ 1 \leq l \leq \chi,$$

under condition that $\sum_{l=1}^{\chi} \lambda_l = 1$.

5. Forecast calculation is

$$\widehat{B}(N+1) = \sum_{l=1}^{\chi} \lambda_l R(l).$$

When a new observation $B(N+1)$ of the process is obtained, all iterations are repeated. Thus, at each step in the forecast formation, observations χ are also involved, but this value may change, while it is clear that the smaller χ, the more unsteady is the signal. It is also easy to notice that if $B(k) = const$, then $\chi = N$, $\lambda_l = 1/N$.

This method does not require large amounts of a priori information (small training set), there is no need to solve additional optimization or pseudo-inversion problems, it is extremely simple in computational implementation.

However, for successful use of all traditional forecasting methods, the initial observations set (the time series) must be sufficiently representative. In this case, the more parameters are present in a predictive model, the larger amount of the source information should be available. At the same time, in real-life problems, such situation often arises when the sample is either small in size or the predicted process is non-stationary, so it contains both irregular trends and sudden jumps, so its history cannot be used to find model parameters.

In addition to the proposed approach, it is recommended to use a process characterized by two-dimensional spaces. At the same time, it is necessary to use any hypothesis about either the generation mechanism or the most acceptable mathematical model. For example, you can use the linear transformation operations process b with the same dimensionality matrix z

$$z = Ab = \left(\sum_{j=1}^{v} A_{1j} b_j \vdots \sum_{j=1}^{v} A_{2j} b_j \vdots \ldots \vdots \sum_{j=1}^{v} A_{vj} b_j \right), \tag{4.75}$$

where b_j is the j-th column of the $(n \times v)$-matrix b, A_{ij} is a $(n \times n)$ transformation matrix to be evaluated.

In this case, transformation (4.75) contains $(nv)^2$ unknown parameters.

By introducing further operations of cyclic permutation $[b]^{(j)} = (b_j \vdots b_{j+1} \vdots \ldots \vdots b_v \vdots b_1 \vdots \ldots \vdots b_{j-2} \vdots b_{j-1})$ and $\{b\}_{(j)} = (\vec{0} \vdots \vec{0} \vdots \ldots \vec{0} \vdots b_j \vdots \vec{0} \vdots \ldots \vec{0})$, transformation (4.75) can be rewritten in the form of a matrix convolution

$$z = Ab = \sum_{i=1}^{n} \sum_{j=1}^{v} \left\{ A_{ij}[b]^{(j)} \right\}_{(i)}. \tag{4.76}$$

Equation (4.76) used for the predictive model synthesis is inconvenient because of its complexity and a large number of estimated parameters, which in turn requires large volume of training sets to solve the parametric identification problem. Of course, vectorization and devectorization operations can be applied, while we obtain a $(nv \times 1)$ vector column as a result of the column $b' = (b_1^T, b_2^T, \ldots, b_j^T, \ldots, b_v^T)^T$ vectorization for the $(n \times v)$ matrix, and the result of line vectorization is a $(1 \times nv)$ vector row $b'' = (b_1, b_2, \ldots, b_i, \ldots, b_n)$ where b_j and b_i are the j-th column and the i-th row of the matrix b respectively.

By further introduction of the transformation matrix in the form

$$A = \begin{pmatrix} A_{11} & A_{12} & \ldots & A_{1v} \\ A_{21} & A_{22} & \ldots & A_{2v} \\ \vdots & \vdots & \vdots & \vdots \\ A_{v1} & A_{v2} & \ldots & A_{vv} \end{pmatrix}$$

(here A_{ij} is $(n \times n)$-matrix) and product

$$Ab' = \left(\left(\sum_{j=1}^{v} A_{vj} b_j \right)^T, \left(\sum_{j=1}^{v} A_{vj} b_j \right)^T, \ldots, \left(\sum_{j=1}^{v} A_{vj} b_j \right)^T \right)^T$$

as a result of column devectorization, we get

$$z = (Ab) = \left(\sum_{j=1}^{v} A_{1j} b_j \vdots \sum_{j=1}^{v} A_{2j} b_j \vdots \ldots \vdots \sum_{j=1}^{v} A_{vj} b_j \right).$$

Similarly, a type transformation (4.76) can also be obtained with the row vectorization / devectorization.

Quite often temporal features of video are random, so it is advisable to introduce matrix analog of vector random processes and sequences. So you can use the matrix Markov sequence

$$b(i) = Ab(i-1) + \xi(i), \tag{4.77}$$

where $\xi(i) = \{\xi_{ij}(i)\}$ is a matrix white noise.

In general, for vectorization-devectorization operations, one can rewrite (4.77) in the form $b(i) = (Ab'(i-1)) + \xi(i)$ and use the standard least squares method to estimate the unknown parameters of the matrix A, but the high dimensionality

$((nv)^2)$ greatly complicates its use. In this regard, it was proposed to use the form

$$b(i) = Ab(i-1) + \xi(i) = (\sum_{j=1}^{v} c_{j1} Ab_j(i-1) \vdots \sum_{j=1}^{v} c_{j1} Ab_j(i-1) \vdots$$
$$\ldots \vdots \sum_{j=1}^{v} c_{j1} Ab_j(i-1)) + \xi(i)$$

corresponding to the formula

$$b(i) = Ab(i-1)C + \xi(i), \tag{4.78}$$

where A and C are $(n \times n)$ and $(v \times v)$-transformation matrices. So the description contains $n^2 + v^2 < (nv)^2$ parameters; n is derived from the description (4.78). It is proposed to introduce a VAR matrix model in the form

$$b(i) = \sum_{h=1}^{n_A} A^h b(i-h) + \xi(i) = \sum_{h=1}^{n_A} A^h b(i-h) C^h + \xi(i)$$
$$= \widetilde{A}(b(i-1) \vdots b(i-2) \vdots \ldots \vdots b(i-n_A))\widetilde{C} + \xi(i) = \widetilde{A} b'(i-1)\widetilde{C} + \xi(i),$$

containing $(n^2 + v^2)n_A$ parameters. Here $\widetilde{A} = (A^1 \vdots A^2 \vdots \ldots \vdots A^{n_A})$,

$$\widetilde{C} = \left(C^1 \ldots C^2 \ldots C^{n_A}\right)^T, b'(i-1) = \begin{pmatrix} b(i-1) & & & \\ & 0 & b(i-2) & 0 \\ & & \ddots & \\ & & 0 & b(i-n_A) \end{pmatrix}.$$

It is clear that the indicator values monitored in forecasting problems are significantly influenced by many exogenous factors. In this regard, a predictive matrix model is introduced into consideration along with exogenous inputs in the following form:

$$b(i) = \sum_{h=1}^{n_A} A^h b(i-h) C^h + \sum_{l=1}^{n_K} K^l x(i-l) D^l + \xi(i) \tag{4.79}$$

(here $x(i-l)$ is the $(n \times v)$-matrix of exogenous factors at a discrete point in time $i-l$, $l = 1, 2, \ldots, n_K$) containing $(n^2 + v^2) \times (n_A + n_K)$ parameters to be determined. Introduce further the matrix

$$\widetilde{A} = \left(A^1 \vdots A^2 \vdots \ldots \vdots A^{nA} \vdots K^1 \vdots \ldots \vdots K^{n_K}\right)$$

$n \times n(n_A + n_K)$, matrix

$$\widetilde{C} = \begin{pmatrix} C^1 \\ \vdots \\ C^{nA} \\ D^1 \\ \vdots \\ D^{n_K} \end{pmatrix}$$

with dimension $v(n_A + n_K) \times v$ and

$$b'(i-1) = \begin{pmatrix} b(i-1) & & & & & \\ & \ddots & & & & \\ & & b(i-n_A) & & 0 & \\ & & & x(i-1) & & \\ & 0 & & & \ddots & \\ & & & & & x(i-n_K) \end{pmatrix}$$

with dimension $n(n_A + n_K) \times v(n_A + n_K)$, and thus, the Eq. (4.79) can be rewritten in a compact form

$$b(i) = \widetilde{A}(i)b'(i-1)\widetilde{C} + \xi(i),$$

after that, in accordance with the description (4.44), the following predictable model can be set up

$$b''(i+1) = \widetilde{A}(i)b'(i)\widetilde{C}(i), \tag{4.80}$$

where $\widetilde{A}(i)$, $\widetilde{C}(i)$ are matrices of refined estimates obtained from observational data.

It is clear that at the i-th moment of time, it is not the model (4.80) that is configured, but its "shifted back" analogue

$$b''(i) = \widetilde{A}(i-1)b'(i-1)\widetilde{C}(i-1). \tag{4.81}$$

Since the indicators under control are described by non-stationary stochastic or chaotic sequences for video analysis problems, adaptive identification procedures should be used to estimate the parameters of the model (4.81), however, the known algorithms are not focused on matrix models, therefore their significant modification is required.

An optimal gradient identification algorithm is proposed that minimizes the evaluation criteria in on-line mode

$$\begin{cases} J^A(i) = Sp(b(i) - \widetilde{A}(i-1)b'(i-1)\widetilde{C}(i-1)) \\ \qquad \times (b(i) - \widetilde{A}(i-1)b'(i-1)\widetilde{C}(i-1))^T \\ \quad = SpV^A(i)V^{A^T}(i) = \left\| V^A(i) \right\|^2, \\ J^C(i) = Sp(b(i) - \widetilde{A}(i)b'(i-1)\widetilde{C}(i-1)) \\ \qquad \times (b(i) - \widetilde{A}(i)b'(i-1)\widetilde{C}(i-1))^T \\ \quad = SpV^C(i)V^{C^T}(i) = \left\| V^C(i) \right\|^2 \end{cases}$$

(here $\|\circ\|$ is the Frobenius norm). As a result we have the following form

$$\begin{cases} \widetilde{A}(i) = \widetilde{A}(i-1) + \eta_A(i)V^A(i)\widetilde{C}^T(i-1)b'(i-1), \\ \widetilde{C}(i) = \widetilde{C}(i-1) + \eta_C(i)b'^T(i-1)\widetilde{A}(i)V^C(i), \end{cases} \tag{4.82}$$

where $\eta_A(i)$, $\eta_C(i)$ are scalar parameters of the learning step.

Based on the procedure (4.82), it is possible to introduce a matrix modification of the adaptive additive-multiplicative Kaczmarz algorithm in the following form

$$\begin{cases} \widetilde{A}(i) = \widetilde{A}(i-1) + \gamma(SpV^A(i)\widetilde{C}^T(i-1)b'^T(i-1)) \\ \quad \times b'(i-1)\widetilde{C}(i-1)V^{A^T}(i)(\beta + SpV^A(i)\widetilde{C}^T(i-1)) \\ \quad \times b'^T(i-1)b'(i-1)\widetilde{C}(i-1)\widetilde{C}^T(i-1)b'^T(i-1) \\ \quad \times b'(i-1)\widetilde{C}(i-1)V^{A^T}(i)^{-1}\widetilde{C}^T(i-1)b'^T(i-1), \\ \widetilde{C}(i) = \widetilde{C}(i-1) + \gamma(SpV^{C^T}(i)\widetilde{A}(i-1)b'(i-1)) \\ \quad \times b'^T(i-1)\widetilde{A}^T(i)V^C(i)(\beta + Sp\widetilde{A}(i)b'(i-1)) \\ \quad \times b'^T(i-1)\widetilde{A}^T(i)V^C(i)V^{C^T}(i)\widetilde{A}(i) \\ \quad \times b'(i-1)b'^T(i-1)\widetilde{A}^T(i)^{-1}b'^T(i-1)\widetilde{A}^T(i)V^C(i), \end{cases} \tag{4.83}$$

where $0 < \gamma < 2$, $\beta \geq 0$ are free parameters chosen empirically.

It is also easy to notice that, if $b(i)$ and $b'(i-1)$ are the scalar and the vector, respectively, $\gamma = 1$, $\beta = 0$, then (4.83) automatically turns into usual one-step Kaczmarz algorithm, which is one of the most popular in theory and practice of adaptive identification.

Since in real video monitoring tasks the initial data is "noisy" by intense disturbances, and the controlled sequences themselves are significantly non-stationary, one-step algorithms like (4.83) may not provide the required prediction accuracy, so it makes sense to use windowed procedures with smoothing properties.

To analyze the applicability of the described video segmentation prediction approach, a lot of catastrophes video clips of various kinds were used, similar to the approach described in the previous subsection. Figure 4.7 shows a series of video sequence frames that have certain segment boundary. The following Fig. 4.8 shows the forecast errors for the following frames, which were critical and indicated the presence of a new video segment.

Fig. 4.7 Border-frames from different videos

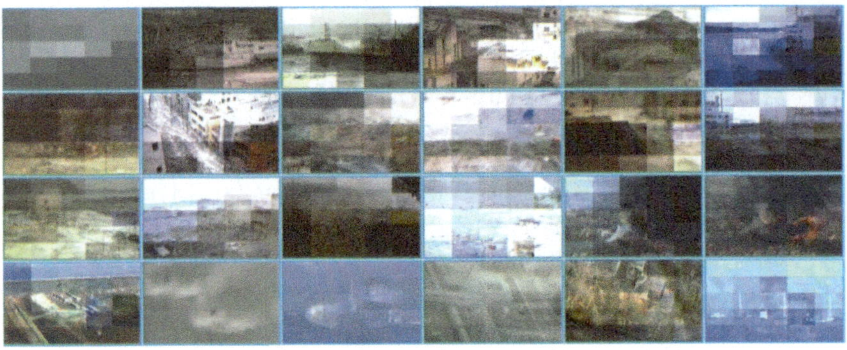

Fig. 4.8 Fragment of prediction errors on the borders of video segments

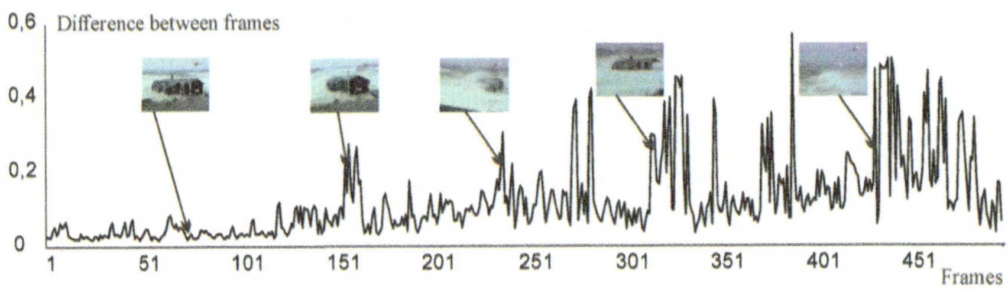

Fig. 4.9 Spatio-temporal video segmentation used for the prediction model

In turn, Fig. 4.9 shows the spatio-temporal segmentation of the video clip as a whole, in the same way as in Fig. 4.6. In this case, there was sometimes a slight discrepancy between the changes in video characteristics in one-dimensional sequence with the original video, and in some cases (for example, in the region of 270 frames) false detections of the scene changes were recorded. This is primarily due to the specifics of a particular video, in which the presence of waves hitting the house is not always a scene change in essence, but it influences on the major changes in spatial segmentation of the video frames, and accordingly on the input characteristic vector, which leads to incorrect results. In order to avoid such situations, it is necessary to change the forecast horizon.

4.5 Dimensionality Reduction and Extension for Temporal Segmentation Adaptation

An important problem in Big Data analysis (both in terms of volume and dimensionality) specified in the time series form is the compression task (dimension reduction). It is designed to identify latent factors determining the internal structure of the monitored signal, which ultimately aims to make the original time series more easily interpreted in terms of detecting property changes.

One of the most effective approaches to solving this problem is the factor analysis, within which the most widely used technique is the main component method, especially in image recognition, image processing, spectral analysis, etc. This method is also known as the Karhunen-Loeve Transform (KLT).

The initial information for the analysis is the $(i \times n)$ matrix of observations

$$B = \begin{pmatrix} b_1(1) & b_2(1) & \dots & b_n(1) \\ b_1(2) & b_2(2) & & b_n(2) \\ \vdots & & & \\ b_1(u) & b_2(u) & \cdots & b_n(u) \\ \vdots & & \ddots & \\ b_1(i) & b_2(i) & \cdots & b_n(i) \end{pmatrix},$$

formed by an array of i n-dimensional observation vectors $B(u) = (b_1(u), b_2(u), \dots, b_n(u))^T$. This is a frame representation for the video data analysis problem, that uses a certain vector of characteristics and its correlation matrix $(n \times n)$ in the form

$$R(i) = \frac{1}{i}\sum_{u=1}^{i}(B(u) - \overline{B})(B(u) - \overline{B})^T = \frac{1}{i}\sum_{u=1}^{i} B^c(u)B^{cT}(u),$$

where $B^c(u) = B(u) - \overline{B}$ are centered to the average source data.

The method of principal components (PCA) consists in projecting the observable source data from the original n-dimensional space into m-dimensional $(n > m \geq 1)$ output. Moreover, it reduces to finding a system of $R(i)$ matrix orthogonal eigenvectors $w^1, w^2, \dots, w^m$, such that $w^1 = (w_1^1, w_2^1, \dots, w_n^1)^T$ corresponds to the largest eigenvalue λ from the matrix $R(i)$, the second $w^2 = (w_1^2, w_2^2, \dots, w_n^2)^T$ corresponds to the second largest proper λ, etc. In other words, we are talking about finding a solution to the matrix equation

$$(R(i) - \lambda_l I)w^l = 0$$

such that $\lambda_1 \geq \lambda_2 \geq \cdots \geq \lambda_n \geq 0$ and $\left\|w^l\right\|^2 = 1$.

Thus, in algebraic terms, this problem solution is closely related to the eigenvalues problem and the correlation matrix rank search; in the geometrical sense, it is the dimensionality reduction problem with minimal information loss; in the statistical sense, it is the orthonormal vectors set search problem in the input space that take the maximum variation of the data, and finally, in the algorithmic sense, it is the sequential determination (extraction) problem of the eigenvectors $w^1, w^2, \ldots, w^m$ set by optimizing each of the local functionals that form the global criterion

$$I_w(i) = \frac{1}{i}\sum_{l=1}^{m}\sum_{u=1}^{i}(B^{cT}(u)w^l)^2$$

under restrictions

$$\begin{cases} w^{lT}w^N = 0, & \text{if } l \neq N, \\ w^{lT}w^N = 1, & \text{else.} \end{cases}$$

The first principal component that carries the maximum information about the monitored signal can be found by maximizing the local criterion

$$I_w^1(i) = \frac{1}{i}\sum_{u=1}^{i}(B^c(u)w^1)^2$$

using the standard method of Lagrange multipliers.

Then, its projection onto the first principal component is subtracted from each vector $B^c(u)$, and the first principal component of the residuals is computed, that is the second principal component of the original data, and the second is orthonormal to the first.

The third principal component is calculated by designing each source vector to the first and second principal components, subtracting this projection from each $B^c(u)$ and finding the first principal component of the residues obtained, which is the third component of the original data. The remaining principal components are calculated recursively according to the described strategy.

Up to date, sufficiently developed mathematical and software parts are available for the KLT implementation, combining one common drawback: the need for a priori assignment of a fixed dimension matrix X. If the data are received consistently in real time, standard factor analysis procedures become inappropriate.

Thus, real-time recurrent procedures should be used for the matrix $R(i)$ eigenvectors search by multidimensional time series $B(1), B(2), \ldots, B(i), B(i+1)\ldots$, implementing sequential processing without calculating the correlation matrix itself.

We introduce an artificial neuron based on an adaptive linear associator to calculate the first main component in a real time. For pre-centered data, the learning algorithm is

$$\begin{cases} w^1(i+1) = w^1(i) + \eta(i+1)(B(i+1) - y(i)w^1(i))y(i+1), \\ y(i+1) = B^T(i+1)w^1(i),\, w^1(0) \neq 0,\, y'(1) = B^T(1)w^1(0), \end{cases} \tag{4.84}$$

where $\eta(i+1)$ is the parameter of the step of adjustment, which is chosen rather small.

Algorithm (4.84) provides vector $w^1(i)$ evaluation

$$\left\| w^1(i) \right\|^2 = 1$$

and the vector $w'(i)$ itself is the matrix $R(i)$ eigenvector corresponding to the maximum eigenvalue, and the output signal $y(i)$.

Paper [40] describes an artificial neuron based on an adaptive neuron characterized by the maximum possible dispersion, i.e. it contains maximum information about the multi-dimensional input signal $b_n(i)$.

Next, the output signal $y(i)$ is subjected to an exponential smoothing, filtering the noise components $\xi(i)$, and the detection of a change in properties occurs using one-dimensional tracking signal $T^{TL}(i)$.

Figure 4.10 illustrates the joint use of the smoothing approach described in Sect. 4.3 and the dimension reduction method for the spatio-temporal video segmentation of the catastrophe clip with a house fall. As it is seen from the result, in this case, the moments of scene change were detected better than in the previous case, and on the basis of this graph, it is easy to extract 3 main segments. The first segment in the range of 275 frames, associated with the first phase of the house fall. The second segment (from 275 to 400 frames) illustrates the damage occurring after

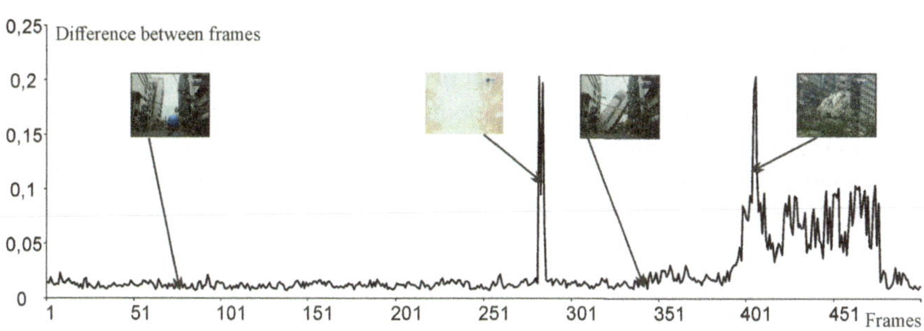

Fig. 4.10 Spatio-temporal video segmentation example

the power line is broken (275 frame), and the last segment shows the results of the accident shot from a different angle.

4.6 Conclusion

There has been a steadily growing challenge to explore summarization and Content Based Video Retrieval techniques. Since video streams are highly redundant and weakly structured data, approaches to video skimming into semantically-consistent segments have been considered. Diminution of semantic gap between detected in each frame low-level features (pixel histograms, local or global features, bag-of-words descriptors, etc.) and high-level interpretation demands is one of the main requirements to temporal video segmentation. One of promising ways is spatio-temporal segmentation as frame partitions represent certain spatial image content. However, different segmentations may correspond to the same image, therefore, three-digit relationships between parts of different partitions have been considered. Properties were formulated and proved which ultimately determine the characteristics of permissible segmentation transformations when searching for a compromise between over and undo segmentation.

Temporal segmentation of multidimensional time series has been examined, which allows to structure video streams and significantly reduce the amount of data that will require online processing. For this, multidimensional time series analysis theory was used, since a completely natural video representation is a sequence of frames, followed by their combination into groups of frames (shots) with the same content. It was shown that various approaches can be used to detect shots with homogeneous characteristics, which are based on VAR models, exponential smoothing and predictive models.

In the case of VAR models and exponential smoothing usage, an approach based on changes detection in a multidimensional time series based on the Trigg-Leach signal to video segments borders extraction was adapted. In this case, the change in the series temporal characteristics allows to select homogeneous video segments (shots, scenes) accurately. When prediction models are applied, the criterion for video segments boundaries detection is the prediction error, if it exceeds the required level. It was experimentally shown in a set of videos that the methods developed for the sequences spatio-temporal segmentation/clustering make it possible to split the data into segments with homogeneous content.

An approach associated with the dimension reduction of the original data based on the KLT to simplify further initial video information analysis was considered also. Its combined use with video segmentation models improves the result quality.

References

1. Thounaojam DM, Trivedi A, Manglem Singh K, Roy S (2014) A survey on video segmentation. Intelligent computing, networking, and informatics. In: Mohapatra D et al (eds) Advances in intelligent systems and computing, vol 243. Springer, New Delhi, pp 903–912
2. Wang W, Shen J, Yang R et al (2018) Saliency-aware video object segmentation. IEEE Trans Pattern Anal Mach Intell 40(1):20–33
3. Xu B, Niu Y (2018) Accurate object segmentation for video sequences via temporal-spatial-frequency saliency model. IEEE Intell Syst 33(1):18–28
4. Haller E, Leordeanu M (2017) Unsupervised object segmentation in video by efficient selection of highly probable positive features. In: Proceedings 2017 IEEE International Conference on Computer Vision (ICCV) IEEE CS CPS, pp 5095–5103
5. Mashtalir S, Mikhnova O (2017) Detecting significant changes in image sequences. Multimedia forensics and security. In: Hassanien A et al (eds) Intelligent systems reference library, vol 115. Springer, Cham, pp 161–191
6. Jiang H, Zhang G, Wang H, Bao H (2015) Spatio-temporal video segmentation of static scenes and its applications. IEEE Trans Multimedia 17(1):3–15
7. Wang W, Shen J, Shao L (2018) Video salient object detection via fully convolutional networks. IEEE Trans Image Process 27(1):38–49
8. Mashtalir S, Mashtalir V (2016) Sequential temporal video segmentation via spatial image partitions. In: Proceedings 2016 IEEE first international conference on Data Stream Mining and Processing (DSMP), pp 239–242
9. Mashtalir S, Mashtalir V, Stolbovyi M (2018) Representative based clustering of long multivariate sequences with different lengths. In: 2018 IEEE second international conference on Data Stream Mining & Processing (DSMP), pp 545–548
10. Mashtalir V, Mikhnova E, Shlyakhov V, Yegorova E (2006) A novel metric on partitions for image segmentation. In: Proceedings IEEE international conference on video and signal based surveillance (AVSS), p 18
11. Sun J (2015) Streaming analysis of track data from video. In: Proceedings of the SPIE 9473, Geospatial informatics, fusion, and motion video analytics V, 947302
12. Liwicki S, Zafeiriou SP, Pantic M (2015) Online Kernel slow feature analysis for temporal video segmentation and tracking. IEEE Trans Image Process 24(10):2955–2970
13. Wang L, Qiao Y, Tang X (2014) Video action detection with relational dynamic-poselets. In: Computer Vision, ECCV2014, pp 565–580
14. Tziakos I, Cavallaro A, Xu L-Q (2009) Video event segmentation and visualisation in non-linear subspace. Pattern Recogn Lett 30:123–131
15. Liu Y, Zhanga D, Lua G, Ma W-Y (2007) A survey of content-based image retrieval with high-level semantics. Pattern Recogn 40(1):262–282
16. Feng L, Bhanu B (2016) Semantic concept co-occurrence patterns for image annotation and retrieval. IEEE Trans Pattern Anal Mach Intell 38(4):785–799
17. Nagasaka A, Tanaka Y (1991) Automatic video indexing and full-video search for object appearances. IFIP Transactions: proceedings of the IFIP TC2/WG 2.6 second working conference on visual database systems, vol A-7. North-Holland Publishing Co., Budapest, Amsterdam, pp 113–127
18. Lin G-S, Chang J-F (2013) Detection of frame duplication forgery in videos based on spatial and temporal analysis. Int J Pattern Recognit Artif Intell 26(7):1–18
19. Gong Y, Liu X (2000) Video summarization using singular value decomposition. In: Proceedings of IEEE conference on computer vision and pattern recognition, Hilton Head Island, vol 2. IEEE CS, Washington, pp 174–180
20. Hu Z, Mashtalir SV, Tyshchenko OK, Stolbovyi MI (2017) Video scenes' matching via various length of multidimensional time sequences. Int J Intell Syst Appl 9(11):10–16
21. Hu Z, Mashtalir SV, Tyshchenko OK, Stolbovyi MI (2018) Clustering matrix sequences based on the iterative dynamic time deformation procedure. Int J Intell Syst Appl 10(7):66–73

22. Billings A, Chen S (1989) Extended model set, global data and threshold model identification of severely non-linear systems. Int J Control 50:1897–1923
23. Rathod GI, Nikam DA (2013) An algorithm for scene boundary detection and key frame extraction using histogram difference. Int J Emerg Technol Adv Eng 3(8):155–163
24. Bezdek JC, Keller J, Krisnapuram R, Pal NR (2005) Fuzzy models and algorithms for pattern recognition and image processing. Springer, NY, p 776
25. Li X-H, Zhan Y-Z, Ke J, Zheng H-W (2011) Scene retrieval based on fuzzy evolutionary AINet and hybrid features. Comput Hum Behav 27(5):1571–1578
26. Vázquez-Martín R, Bandera A (2013) Spatio-temporal feature-based keyframe detection from video scenes using spectral clustering. Pattern Recogn Lett 34(7):770–779
27. Liu B, He X (2015) Multiclass semantic video segmentation with object-level active. In: IEEE conference on Computer Vision and Pattern Recognition (CVPR), pp 4286–4294
28. Isermann R (1984) Process fault detection based modeling and estimating methods—a survey. Automatica 20(4):387–404
29. Nikiforov IV (1991) Sequential detection of changes in stochastic processes. In: Prep. 9-th IFAC/IFORS symposium on identification and system parameter estimation, Budapest, vol 1
30. Badavas PC (1993) Real-time statistical process control. Prentice-Hall, Englewood Cliffs, NJ
31. Pouliezos AD, Stavrakakis GS (1994) Real time fault monitoring of industrial processes. Kluver Academic Publishers, Dordrecht
32. Juselis K (1994) The cointegrated VAR-model. Oxford University Press, NY
33. Rastrigin LA (1974) Systems of extremal control. Nauka, Moscow
34. Kaczmarz S (1993) Approximate solution of system of linear equations. Int J Control 57(5):1269–1271
35. Chow EY, Willsky AS (1980) Issues in the development of a general design algorithm for reliable failure detection. In: Proceedings of the 19-th IEEE Conference on Decis Ant Contr – Albuquerque
36. Brown RG (1963) Smoothing, forecasting, and prediction of discrete time series. Prentice Hall, NY
37. Leach DW, Trigg AG (1967) Exponential smoothing with an adaptive response rate. Oper Res Q 18(1):53–59
38. Montgomery DC, Johnson IA, Gardiner JS (1990) Forecasting and time series analysis. McGraw-Hill, NY
39. Bodyanskiy Y, Rudenko O (2004) Artificial neural networks: architectures, learning, applications. TELETECH, Kharkov (in Russia)
40. Cichocki A, Unbehauen R (1993) Neural networks for optimization and signal processing. Teubner, Stuttgart

Chapter 5
Online Fuzzy Clustering of Data Streams

Yevgeniy Bodyanskiy and Olena Boiko

Abstract Chapter considers new approaches based on computational intelligence methods for solving the tasks of fuzzy clustering-segmentation of data streams sequentially fed for processing. The known methods of probabilistic and potential clustering under conditions of overlapping classes are analyzed and their adaptive recurrent modifications designed for solving problems arising in the framework of Data Stream Mining are introduced. The reasonability of a parallel probabilistic-potential approach is shown, where centroid-prototype clusters are specified using the optimization gradient procedure. For situations where the processed data is "polluted" by abnormal outliers, adaptive fuzzy clustering methods based on robust target functions that suppress outliers are proposed. The close connection of the proposed procedures with self-learning algorithms of T. Kohonen self-organizing cards is shown.

Keywords Data streams · Fuzzy clustering · Fuzzy segmentation · Adaptive online robust fuzzy clustering · Neuro-fuzzy clustering systems · Probabilistic and possibilistic approaches to fuzzy clustering · Online data processing · Self-learning · T. Kohonen's self-organizing map · Kernel fuzzy clustering

5.1 Image Clustering-Segmentation in Data Mining

The task of clustering of data sets of multidimensional observations of different nature, the purpose of which is to find in the samples of these data homogeneous in the accepted sense groups (segments, clusters, classes) of observations, is an integral part of the scientific direction called Data Mining [1–5], and its results can be used in many applications, including the task of segmentation of images of arbitrary nature [6–8].

Y. Bodyanskiy (✉) · O. Boiko
Kharkiv National University of Radio Electronics, 14 Nauky Ave, Kharkiv 61166, Ukraine
e-mail: yevgeniy.bodyanskiy@nure.ua

O. Boiko
e-mail: olena.boiko@nure.ua

V. Mashtalir et al. (eds.), *Advances in Spatio-Temporal Segmentation of Visual Data*, Studies in Computational Intelligence 876,
https://doi.org/10.1007/978-3-030-35480-0_5

Here, the most popular method is the K-means and its subsequent modifications, due to the simplicity of numerical implementation, performance and visibility of the results.

At the same time, there is a large number of tasks involving both the processing of static images and video streams, where traditional methods of clustering-segmentation, based on the self-learning paradigm [9, 10], are ineffective because the amount of information fed for real-time processing is too large, the data itself is "corrupted" with different types of disturbances and noises, including abnormal outliers such as salt and pepper, the image segments themselves tend to overlap to form fuzzy borders.

And if the last problem can be solved efficiently by using fuzzy clustering methods [11–14] and above all the classic Fuzzy C-means (FCM) method and its modifications, the processing of real-time video streams is a rather complicated problem, since well-known fuzzy clustering algorithms are designed to process data in batch mode. In addition, these algorithms do not have robust properties, that is, they are unprotected from the distortion of the input data.

Therefore, the development of adaptive online fuzzy clustering algorithms that have both filtering and tracking properties and capable of sequentially processing of both static images and multi-dimensional time sequences generated by video sequences, is undoubtedly an interesting and useful task.

5.2 Batch Methods of Fuzzy Segmentation-Clustering

The task of clustering of multidimensional observations arriving in online mode is quite common in many tasks related to Data Mining. The traditional approach to solving these problems implies that each observation may belong to only one cluster [15, 16], although it is more natural to have a situation where the input feature vector may belong at the same time to several clusters or classes with different levels of probability or possibility [11, 17, 18]. This situation is the subject of consideration of the fuzzy cluster analysis that is developing intensively nowadays in two main directions: the probabilistic approach and the approach based on possibilities (so-called, possibilistic approach) [11, 19–22].

The problem of fuzzy cluster analysis is widespread and recently the hybrid neuro-fuzzy systems that combine artificial neural networks and clustering methods are widely studied. The results of such studies are presented in [23–25]. Fuzzy clustering methods have been further developed within the adaptive approach [26], which allows monitoring changes in the data structure and adapting the system parameters to these changes. Adaptive online methods are also capable of handling large volumes of data due to the sequential processing of incoming observations.

The input information for both approaches is a sample of observations formed by n-dimensional feature vectors

$$X = \{x(1), x(2), \ldots, x(N), \ldots\}, \quad x(k) \in X, k = 1, 2, \ldots, N, \ldots \tag{5.1}$$

and is limited by N observations in the case of a batch approach to clustering.

The result of the procedure is the split of the input data set into m clusters with some membership levels $w_j(k)$ of the k-th feature vector to the j-th cluster.

The input data that are processed are preliminarily centered and normalized for all features so that all observations belong to the hypercube $[-1, 1]^n$. Centering can be done either relative to the average, calculated according to the expression

$$m_i(k) = m_i(k-1) + \frac{1}{k}(x_i(k) - m_i(k-1)) \tag{5.2}$$

or, in order to add robust properties to this procedure (that is, protection against abnormal observations), relative to the median, calculated according to the recurrence ratio

$$me_i(k) = me_i(k-1) + \eta_m sign(x_i(k) - me_i(k-1)), \quad i = 1, 2, \ldots, n, \tag{5.3}$$

where η_m is the learning rate parameter that is selected in the stationary case according with the conditions of Dvoretzky [27].

Clustering methods based on goal functions [11] and designed to solve clustering tasks by optimizing some pre-set quality criterion are the most correct from the point of mathematics view.

5.2.1 Probabilistic Approach

The target function to be minimized according to the probabilistic approach, has the form

$$E\big(w_j(k), c_j\big) = \sum_{k=1}^{N} \sum_{j=1}^{m} w_j^{\beta}(k) D^2\big(x(k), c_j\big) \tag{5.4}$$

with constraints

$$\sum_{j=1}^{m} w_j(k) = 1, \quad k = 1, \ldots, N, \tag{5.5}$$

$$0 < \sum_{k=1}^{N} w_j(k) < N, \quad j = 1, \ldots, m \tag{5.6}$$

where $w_j(k) \in [0, 1]$ is the membership of the vector $x(k)$ to the j-th cluster; c_j is the prototype (center) of the j-th cluster; β is non-negative parameter—"fuzzyfier", usually $\beta = 2$; $D\big(x(k), c_j\big)$ is the distance between $x(k)$ and c_j according to the chosen metrics.

The result of clustering is a $N \times m$ fuzzy partitioning matrix

$$W = \{w_j(k)\}. \tag{5.7}$$

It should be noted that since the elements of the matrix W can be considered as the probabilities of the hypothesis of the data vectors membership to the clusters, the procedures produced by (5.4) with restrictions (5.5, 5.6), are called probabilistic clustering methods [21].

As a distance function $D(x(k), c_j)$ the Minkowski distance in L^p metrics [28] is usually used:

$$D^p(x(k), c_j) = ||x_i(k) - c_{ji}||_{L_p}^p = \left(\sum_{i=1}^{n} |x_i(k) - c_{ji}|^p\right)^{\frac{1}{p}}, \quad p \geq 1 \tag{5.8}$$

where $x_i(k)$ is the ith component of $(n \times 1)$-vector $x(k)$; c_{ji} is the ith component of $(n \times 1)$–vector c_j.

Let's consider the Lagrange function

$$\begin{aligned} L(w_j(k), c_j, \lambda(k)) &= \sum_{k=1}^{N}\sum_{j=1}^{m} w_j^{\beta}(k)D^2(x(k), c_j) + \sum_{k=1}^{N}\lambda(k)\left(\sum_{j=1}^{m} w_j(k) - 1\right) \\ &= \sum_{k=1}^{N}\left(\sum_{j=1}^{m} w_j^{\beta}(k)D^2(x(k), c_j) + \lambda(k)\left(\sum_{j=1}^{m} w_j(k) - 1\right)\right) \end{aligned} \tag{5.9}$$

where $\lambda(k)$ is undetermined Lagrange multiplier, that ensures the conditions (5.5, 5.6).

Solving the Kuhn–Tucker system of equations

$$\begin{cases} \frac{\partial L(w_j(k), c_j, \lambda(k))}{\partial w_j(k)} = 0, \\ \frac{\partial L(w_j(k), c_j, \lambda(k))}{\partial \lambda(k)} = 0, \\ \nabla_{c_j} L(w_j(k), c_j, \lambda(k)) = \vec{0}, \end{cases} \tag{5.10}$$

we can obtain the result in the form [20]:

$$w_j^{pr}(k) = \frac{(D^2(x(k), c_j))^{\frac{1}{1-\beta}}}{\sum_{l=1}^{m}(D^2(x(k), c_l))^{\frac{1}{1-\beta}}}, \tag{5.11}$$

$$\lambda(k) = -\left(\sum_{l=1}^{m}(\beta D^2(x(k), c_l))^{\frac{1}{1-\beta}}\right)^{1-\beta}, \tag{5.12}$$

$$c_j^{pr} = \frac{\sum_{k=1}^{N} w_j^{\beta}(k)x(k)}{\sum_{k=1}^{N} w_j^{\beta}(k)}. \tag{5.13}$$

Equations (5.11–5.13) generate a broad class of clustering procedures.

With $\beta = p = 2$, that is, in the Euclidean space

$$D^E\big(x(k), c_j\big) = ||x(k) - c_j|| = \sqrt{(x(k) - c_j)^{\mathrm{T}}\big(x(k) - c_j\big)}, \tag{5.14}$$

we obtain a relatively simple and effective procedure for fuzzy C-means clustering by J. Bezdek [11]:

$$w_j^{pr}(k) = \frac{||x(k) - c_j||^{-2}}{\sum_{l=1}^{m} ||x(k) - c_l||^{-2}}, \tag{5.15}$$

$$c_j^{pr} = \frac{\sum_{k=1}^{N} w_j^2(k)x(k)}{\sum_{k=1}^{N} w_j^2(k)}, \tag{5.16}$$

$$\lambda(k) = -\sum_{l=1}^{m}\left(\frac{||x(k) - c_l||^{-2}}{2}\right)^{-1}. \tag{5.17}$$

In spite of the insignificant computational complexity, the procedure (5.15), (5.16) has a disadvantage expressed in the necessity of fulfilling the condition (5.5), common to all probabilistic methods of fuzzy clustering.

In the simplest case of two clusters ($m = 2$), it is easy to see that the observation $x(k)$ that equally belongs to both clusters, and the observation $x(p)$ that does not belong to any of them can have the same membership levels $w_1^{pr}(k) = w_2^{pr}(k) = w_1^{pr}(p) = w_2^{pr}(p) = 0.5$. Obviously, this characteristic can significantly reduce the clustering quality. At the same time, the possibilistic approach to fuzzy clustering [29–31] helps to avoid the above-mentioned situation and thereby improve the quality of the clustering.

5.2.2 *Possibilistic Approach*

For the possibilistic approach to clustering, the minimization criterion has the form

$$E\big(w_j(k), c_j\big) = \sum_{k=1}^{N}\sum_{j=1}^{m} w_j^{\beta}(k)D^2\big(x(k), c_j\big) + \sum_{j=1}^{m}\mu_j\sum_{k=1}^{N}\big(1 - w_j(k)\big)^{\beta} \tag{5.18}$$

where the scalar parameter $\mu_j > 0$ determines the distance at which the membership level takes value 0.5, that is if $D^2\big(x(k), c_j\big) = \mu_j$, then $w_j(k) = 0.5$.

The minimization of the criterion (5.18) by parameters $w_j(k)$, c_j, μ_j leads to the system of equations:

$$\begin{cases} \frac{\partial E(w_j(k),c_j)}{\partial w_j(k)} = 0, \\ \frac{\partial E(w_j(k),c_j)}{\partial \mu(k)} = 0, \\ \nabla_{c_j} E\left(w_j(k), c_j\right) = \vec{0}. \end{cases} \tag{5.19}$$

The solution of the first two equations gives a well-known result:

$$w_j^{pos}(k) = \left(1 + \left(\frac{D^2(x(k), c_j)}{\mu_j}\right)^{\frac{1}{\beta-1}}\right)^{-1}, \tag{5.20}$$

$$\mu_j = \frac{\sum_{k=1}^{N} w_j^{\beta}(k) D^2\left(x(k), c_j\right)}{\sum_{k=1}^{N} w_j^{\beta}(k)}. \tag{5.21}$$

The solution of the third equation of the system (5.19) for the Euclidean norm (5.14) has the form:

$$c_j^{pos} = \frac{\sum_{k=1}^{N} w_j^{\beta}(k) x(k)}{\sum_{k=1}^{N} w_j^{\beta}(k)}. \tag{5.22}$$

It is easy to see that probabilistic and probabilistic methods are quite similar and pass one by one by replacing the expression (5.20) with formula (5.11), and vice versa. A common disadvantage of these methods is the inability to work in online mode when the data are fed, for example, in the form of a video stream.

The procedure (5.11, 5.12) begins with the task of the initial (usually random) partition matrix W^0. Based on its values, the initial set of prototypes c_j^0 is calculated, which are then used to adjust the new matrix W^1. The next step in batch mode is to calculate c_j^1, W^2, ..., W^t, c_j^t, W^{t+1} and so forth, until the difference $||W^{t+1} - W^t||$ becomes less than a predetermined threshold value ε. Thus, the entire sample of data is processed multiple times.

The solution that can be obtained using the probabilistic method is recommended to be used as the initial conditions for the possibilistic method (5.20–5.22) [31, 32], in which the initial values of the distance parameters μ_j^t are selected in accordance with (5.21) by the results of the probabilistic procedure.

5.3 Adaptive Online Methods of Fuzzy Probabilistic and Possibilistic Clustering

5.3.1 Probabilistic Approach

The analysis of Eq. (5.11) shows that for the calculation of membership levels $w_j(k)$ instead of the Lagrange (5.9) function it is possible to use its local modification:

$$L_k\big(w_j(k), c_j, \lambda(k)\big) = \sum_{j=1}^{m} w_j^{\beta}(k) D^2\big(x(k), c_j\big) + \lambda(k)\left(\sum_{j=1}^{m} w_j(k) - 1\right). \quad (5.23)$$

The optimization of the expression (5.23) using the Arrow-Hurwitz-Uzawa procedure [31, 33] leads to the procedure:

$$w_j^{pr}(k) = \frac{\big(D^2(x(k), c_j(k))\big)^{\frac{1}{1-\beta}}}{\sum_{l=1}^{m}\big(D^2(x(k), c_l(k))\big)^{\frac{1}{1-\beta}}}, \quad (5.24)$$

$$\begin{aligned} c_j^{pr}(k+1) &= c_j^{pr}(k) - \eta(k)\nabla_{c_j} L_k\Big(w_j(k), c_j^{pr}(k), \lambda(k)\Big) \\ &= c_j^{pr} - \eta(k) w_j^{\beta}(k) D\Big(x(k+1), c_j^{pr}(k)\Big)\nabla_{c_j} D\Big(x(k+1), c_j^{pr}(k)\Big) \end{aligned} \quad (5.25)$$

where $\eta(k)$ is learning rate parameter; $c_j^{pr}(k)$ are prototypes of the jth cluster that are calculated on the dataset with k observations.

The procedure (5.24, 5.25) is similar to the Chang–Lee learning procedure [34] and with $\beta = p = 2$ coincides with the gradient Park–Dagher clustering procedure [35]:

$$w_j^{pr}(k) = \frac{||x(k) - c_j(k)||^{-2}}{\sum_{l=1}^{m} ||x(k) - c_l(k)||^{-2}}, \quad (5.26)$$

$$c_j^{pr}(k+1) = c_j^{pr}(k) + \eta(k) w_j^2(k)\Big(x(k+1) - c_j^{pr}(k)\Big). \quad (5.27)$$

5.3.2 Possibilistic Approach

In case of the possible approach, the local criterion has the form

$$E_k\big(w_j(k), c_j\big) = \sum_{j=1}^{m} w_j^{\beta}(k) D^2\big(x(k), c_j\big) + \sum_{j=1}^{m} \mu_j\big(1 - w_j(k)\big)^{\beta} \tag{5.28}$$

and the result of its optimization is written as

$$w_j^{pos}(k) = \left(1 + \left(\frac{D^2(x(k), c_j(k))}{\mu_j(k)}\right)^{\frac{1}{\beta-1}}\right)^{-1}, \tag{5.29}$$

$$\begin{aligned} c_j^{pos}(k+1) = c_j^{pos}(k) - \eta(k) w_j^{\beta}(k) \\ \times D\Big(x(k+1), c_j^{pos}(k)\Big) \nabla_{c_j} D\Big(x(k+1), c_j^{pos}(k)\Big), \end{aligned} \tag{5.30}$$

$$\mu_j(k+1) = \frac{\sum_{p=1}^{k} w_j^{\beta}(p) D^2\big(x(p), c_j(k+1)\big)}{\sum_{p=1}^{k} w_j^{\beta}(p)}. \tag{5.31}$$

In the quadratic case (with $\beta = 2$) the procedure (5.29–5.31) becomes a fairly simple construction

$$w_j^{pos}(k) = \frac{\mu_j(k)}{\mu_j(k) + ||x(k) - c_j(k)||^2}, \tag{5.32}$$

$$c_j^{pos}(k+1) = c_j^{pos}(k) + \eta(k) w_j^2(k)\Big(x(k+1) - c_j^{pos}(k)\Big), \tag{5.33}$$

$$\mu_j(k+1) = \frac{\sum_{p=1}^{k} w_j^2(p) ||x(p) - c_j(k+1)||^2}{\sum_{p=1}^{k} w_j^2(p)}. \tag{5.34}$$

The parallel application of adaptive probabilistic and probabilistic procedures leads to the combined procedure of fuzzy clustering [36]

$$\begin{cases} c_j^{pr}(k) = c_j^{pos}(k-1) - \eta(k) w_j^{pos\beta}(k-1) D\Big(x(k), c_j^{pos}(k-1)\Big) \\ \qquad \times \nabla_{c_j} D\Big(x(k), c_j^{pos}(k-1)\Big), \\ w_j^{pr}(k) = \Big(D^2\Big(x(k), c_j^{pr}(k)\Big)\Big)^{\frac{1}{1-\beta}} \left(\sum_{l=1}^{m} \Big(D^2\Big(x(k), c_j^{pr}(k)\Big)\Big)^{\frac{1}{1-\beta}}\right)^{-1}, \\ c_j^{pos}(k) = c_j^{pr}(k-1) - \eta(k) w_j^{pr\beta}(k) D\Big(x(k), c_j^{pr}(k)\Big) \nabla_{c_j} D\Big(x(k), c_j^{pr}(k)\Big), \\ \mu_j(k) = \left(\sum_{p=1}^{k} w_j^{pr\beta}(p) D^2\Big(x(k), c_j^{pos}(k)\Big)\right) \left(\sum_{p=1}^{k} w_j^{pr\beta}\right)^{-1}, \\ w_j^{pos}(k) = \left(1 + \left(\frac{D^2\Big(x(k), c_j^{pos}(k)\Big)}{\mu_j(k)}\right)\right)^{-1}, \quad j = 1, 2, \ldots, m. \end{cases} \tag{5.35}$$

The sign of the correct finding of prototypes (and hence the correct clustering), using the procedure (5.35), is the correctness of the inequality

$$\sum_{l=1}^{m} D^2(c_l^{pr}(k), c_l^{pos}(k)) \leq \varepsilon \tag{5.36}$$

where ε determines the acceptable accuracy of clustering.

For the Euclidean metric, the value of the parameter $\mu_j(k)$ can be calculated according to the recurrent expression directly derived from (5.34):

$$\begin{cases} \beta_q(k) = \beta_q(k-1) + w_j^{pr\,2}(k) s_q(x(k)), \quad q = 0, 1, 2, \\ \mu_j(k) = \frac{\beta_2(k-1) - 2c_j^{pos\,\mathrm{T}}(k)\beta_1(k-1) + \|c_j^{pos}(k)\|^2 \beta_0(k-1)}{\beta_0(k-1)} \end{cases} \tag{5.37}$$

where

$$s_q(x(k)) = \begin{cases} 1, & if\ q = 0, \\ x(k), & if\ q = 1, \\ \|x(k)\|^2, & if\ q = 2. \end{cases} \tag{5.38}$$

The initial values of the parameter $\beta_q(k)$ are selected as

$$\beta_q(N) = \sum_{p=1}^{N} \left(w_j^{pr}(p)\right)^2 s_q(x(p)), \quad q = 0, 1, 2. \tag{5.39}$$

Thus, the adaptive procedure (5.35) can work both in batch mode for iterative processing of a given sample, and in online mode, when the number of observations is determined by the discrete time $k = 1, 2, \ldots, N, N + 1, \ldots$. In the latter case, this procedure sequentially processes the observations that arrive at processing. Consequently, in the case of non-stationary data, the membership levels and clusters prototypes are reconfigured according to new data.

5.4 Adaptive Online Methods of Fuzzy Robust Clustering

Most of the practical problems associated with video processing are characterized by the presence of outliers in the data, which greatly affects the results of clustering using classical methods, which results in the discovery of non significant clusters, the displacement of prototypes and the radii of clusters.

Because of this, the problem of cluster analysis of data generated by distributions with slowly falling (or heavy) "tails" is being paid more and more attention. Various robust modifications of classical clustering procedures for processing data that contains outliers were proposed in [37–39].

At the same time, most of the proposed robust fuzzy clustering methods cannot be used for sequential or real-time operation. In order to overcome this disadvantage, it is expedient to synthesize recurrent procedures for robust fuzzy clustering of time series that have adaptive properties, can be applied to the sequential processing of incoming data, and when the properties of the system generating these data change with time [36, 40].

Estimates related to the quadratic goal function are optimal when the data belong to the bounded variance distribution class. The most important representative of this class is the Gaussian distribution. By changing the value of the parameter p, it is possible to improve the robustness of the clustering procedure. However, it should be remembered that the quality assessment is determined by the distribution of data. For example, the estimates that correspond to $p = 1$ are optimal for the Laplace distribution, but a large number of computations is required to obtain them.

The most important function for approximating the probability density close to the normal distribution is the function [41]

$$p(x_i, c_i) = Se(c_i, s_i) = \frac{1}{2s_i}\sinh^2\left(\frac{x_i - c_i}{s_i}\right) \tag{5.40}$$

where c_i, s_i are parameters that define center and variance of the distribution correspondingly.

This function is close to the Gaussian one in the vicinity of the center, but differs significantly from it in the presence of heavy "tails". Distribution (5.40) is related to the goal function [42, 43]

$$f_i(x_i, c_i) = \beta_i \ln\left[\cosh\left(\frac{x_i - c_i}{\beta_i}\right)\right] \tag{5.41}$$

where β_i is parameter that determines how fast this function is changed.

It should be noted that the function (5.41) is close to the quadratic in the vicinity of the center c_i and tends to the linear with increasing distance from the center. The derivative of this function has the form

$$f_{i'}(x_i) = \phi(x_i) = \tanh\left(\frac{x_i}{\beta_i}\right) \tag{5.42}$$

and coincides with the standard activation function of an artificial neuron [44].

Let's consider the function

$$D^R\big(x(k), c_j\big) = \sum_{i=1}^{n} f_i\big(x_i(k), c_{ji}\big) = \sum_{i=1}^{n} \beta_i \ln\left[\cosh\left(\frac{x_i(k) - c_{ji}}{\beta_i}\right)\right], \tag{5.43}$$

the graph of which for $n = 1$, $\beta_i = 2$ and $c_{ij} = 0$ is shown in Fig. 5.1.

Next, as a distance to the fuzzy clustering goal function, we use the form

$$D\big(x(k), c_j\big) = \big(D^R\big(x(k), c_j\big)\big)^{\frac{1}{2}}. \tag{5.44}$$

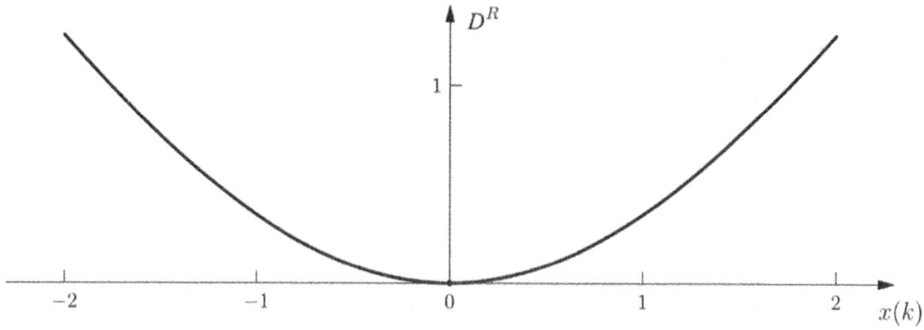

Fig. 5.1 The graph of the function (5.43) for $n = 1$, $\beta_i = 2$ and $c_{ij} = 0$

5.4.1 Probabilistic Approach

Let's consider the goal function for robust probabilistic clustering:

$$E^R\left(w_j(k), c_j\right) = \sum_{k=1}^{N}\sum_{j=1}^{m} w_j^{\beta} D^R(x(k), c_j)$$
$$= \sum_{k=1}^{N}\sum_{j=1}^{m} w_j^{\beta} \sum_{i=1}^{n} \beta_i \ln\left[\cosh\left(\frac{x_i(k) - c_{ji}}{\beta_i}\right)\right]. \tag{5.45}$$

The Lagrange function that corresponds to it has the form:

$$L^R\left(w_j(k), c_j, \lambda(k)\right) = \sum_{k=1}^{N}\sum_{j=1}^{m} w_j^{\beta}(k) \sum_{i=1}^{n} \beta_i \ln\left[\cosh\left(\frac{x_i(k) - c_{ji}}{\beta_i}\right)\right]$$
$$+ \sum_{k=1}^{N} \lambda(k)\left(\sum_{j=1}^{m} w_j(k) - 1\right). \tag{5.46}$$

The saddle point of the Lagrange function (5.46) can be found by solving the Kuhn – Tucker equations system (5.10) in the same way as it is done for the derivation of the procedures (5.11, 5.12). In this case, the solution of the first and second equations of the system (5.10) taking into account the metrics (5.43) will coincide with (5.11) and (5.12), respectively. However, the third equation of the system:

$$\nabla_{c_j} L^R\left(w_j(k), c_j, \lambda(k)\right) = \sum_{k=1}^{N} w_j^{\beta} \nabla_{c_j} D^R(x(k), c_j) = \vec{0} \tag{5.47}$$

obviously, has no analytical solution.

The solution (5.47) can be found numerically based on the local modification of the Lagrange function [45] using the recurrent fuzzy clustering procedure [31]. Finding at the same time the saddle point of the local Lagrange function (5.23) for the metrics (5.43) on the basis of the Arrow–Hurwitz–Uzawa procedure, we obtain the following procedure for fuzzy robust clustering:

$$\begin{cases} w_j^{pr}(k) = \frac{\left(D^R(x(k),c_j)\right)^{\frac{1}{1-\beta}}}{\sum_{l=1}^m \left(D^R(x(k),c_l)\right)^{\frac{1}{1-\beta}}}, \\ c_{ji}(k+1) = c_{ji}(k) - \eta(k)\frac{\partial}{\partial c_{ji}} L_k^R\left(w_j(k), c_j, \lambda(k)\right) \\ \qquad = c_{ji}(k) + \eta(k) w_j^{\beta}(k) \tanh\left(\frac{x_i(k)-c_{ji}(k)}{\beta_i}\right). \end{cases} \tag{5.48}$$

5.4.2 Possibilistic Approach

Within the possibilistic approach, the clustering criterion, taking into account the robust metric (5.43), has the form

$$E^R\left(w_j(k), c_j\right) = \sum_{k=1}^N \sum_{j=1}^m w_j^{\beta}(k) D^R(x(k), c_j) + \sum_{j=1}^m \mu_j \sum_{k=1}^N \left(1 - w_j(k)\right)^{\beta}. \tag{5.49}$$

Solving the system of the Kuhn–Tucker equations which is similar to (5.19) using the metric (5.43) for the first two equations, we can get the solution in the form (5.20, 5.21). However, the third equation of the system (5.19)

$$\nabla_{c_j} E^R(w_j(k), c_j) = \sum_{k=1}^N w_j^{\beta} \nabla_{c_j} D^R\left(x(k), c_j\right) = 0 \tag{5.50}$$

coincides completely with (5.47). It leads to the impossibility of its solution in an analytical form.

Now let's consider the local modification in the criterion (5.49)

$$\begin{aligned} E_k^R\left(w_j(k), c_j\right) &= \sum_{j=1}^m w_j^{\beta}(k) D^R\left(x(k), c_j(k)\right) + \sum_{j=1}^m \mu_j\left(1 - w_j(k)\right)^{\beta} \\ &= \sum_{j=1}^m w_j^{\beta}(k) \sum_{i=1}^n \beta_i \ln\left[\cosh\left(\frac{x_i(k) - c_{ji}}{\beta_i}\right)\right] + \sum_{j=1}^m \mu_i (1 - w_j(k))^{\beta}. \end{aligned} \tag{5.51}$$

Using the Arrow–Hurwitz–Uzawa procedure, we obtain the recurrent procedure of fuzzy possibilistic clustering of the form:

$$\begin{cases} w_j^{pos}(k) = \left(1 + \left(\dfrac{D^R(x(k), c_j(k))}{\mu_j}\right)^{\frac{1}{\beta-1}}\right)^{-1}, \\ c_{ji}(k+1) = c_{ji}(k) - \eta(k)\dfrac{\partial E_k(w_j(k), c_j, \mu_j(k))}{\partial c_{ji}} \\ \qquad = c_{ji}(k) + \eta(k) w_j^{\beta}(k) \tanh\left(\dfrac{x_i(k) - c_{ji}(k)}{\beta_i}\right). \end{cases} \quad (5.52)$$

Here the distance parameter $\mu_j(k)$ is calculated using the formula (5.21) for $k < N$ observations:

$$\mu_j(k+1) = \frac{\sum_{p=1}^{k} w_j^{\beta}(p) D^R\big(x(p), c_j(k+1)\big)}{\sum_{p=1}^{k} w_j^{\beta}(p)}. \quad (5.53)$$

It should be noted that the equations for $c_{ji}(k)$ in systems (5.48), (5.52) completely coincide and are determined by the chosen metric, while the other equations do not depend on the metric, that is, the choice of an arbitrary metric for the clustering procedure will only affect the cluster prototype adjustment procedures, and the equation for weights $w_j^{pr}(k)$ and $w_j^{pos}(k)$ will remain unchanged.

5.4.3 *Robust Recurrent Fuzzy Clustering Method with "Suppression of Outliers"*

As an analog for the metric for the robust recurrent fuzzy clustering method, it is possible to use the function (Fig. 5.2):

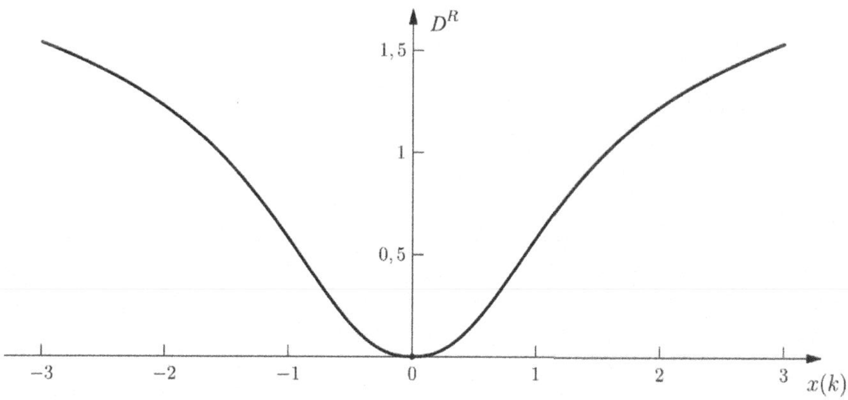

Fig. 5.2 The graph of the function (5.54) for $n = 1$, $c_{ji} = 0$

$$D^R\big(x(k), c_j\big) = \sum_{i=1}^{n}\big(1 - \sinh^2\big(x_i(k) - c_{ji}\big)\big)\big(x_i(k) - c_{ji}\big)^{\frac{2}{5}}, \tag{5.54}$$

that suppresses abnormal outliers in observations.

This function satisfies the axioms of the metric in the vicinity of the center c_j, however, for $|x(k) - c_j| > 0.8762$ it doesn't satisfy the inequalities of the triangle.

Using (5.54), let's write the goal function for robust clustering in the form

$$\begin{aligned} E^R\big(w_j(k), c_j\big) &= \sum_{k=1}^{T}\sum_{j=1}^{m} w_j^{\beta} D^R(x(k), c_j) \\ &= \sum_{k=1}^{T}\sum_{j=1}^{m} w_j^{\beta} \sum_{i=1}^{n}\big(1 - \sinh^2\big(x_i(k) - c_{ji}\big)\big)\big(x_i(k) - c_{ji}\big)^{\frac{2}{5}} \end{aligned} \tag{5.55}$$

and the corresponding Lagrange function

$$\begin{aligned} L^R\big(w_j(k), c_j, \lambda(k)\big) &= \sum_{k=1}^{T}\sum_{j=1}^{m} w_j^{\beta}(k) \sum_{i=1}^{n}\big(1 - \sinh^2\big(x_i(k) - c_{ji}\big)\big)\big(x_i(k) - c_{ji}\big)^{\frac{2}{5}} \\ &\quad + \sum_{k=1}^{T}\lambda(k)\left(\sum_{j=1}^{m} w_j(k) - 1\right). \end{aligned} \tag{5.56}$$

Similarly to the derivation of the procedure (5.48), using the Arrow-Hurwitz-Uzawa method to find the saddle point of the Lagrange function (5.56), we obtain the following robust recurrent fuzzy clustering procedure:

$$\begin{cases} w_j(k) = \dfrac{\big(D^R\big(x(k), c_j\big)\big)^{\frac{1}{1-\beta}}}{\sum_{l=1}^{m}\big(D^R\big(x(k), c_l\big)\big)^{\frac{1}{1-\beta}}}, \\ c_{ji}(k+1) = c_{ji}(k) + \eta(k) w_j^{\beta}(k)\Big(2\sinh^2\big(x_i(k) - c_{ji}(k)\big) \\ \quad \times \tanh\big(x_i(k) - c_{ji}(k)\big)|x_i(k) - c_{ji}(k)|^{\frac{2}{5}} + 0.4\Big(1 - \sinh^2\big(x_i(k) - c_{ji}(k)\big)\Big) \\ \quad \times |x_i(k) - c_{ji}(k)|^{-\frac{3}{5}}\,\mathrm{sign}\big(x_i(k) - c_{ji}(k)\big)\Big). \end{cases} \tag{5.57}$$

The proposed robust recurrent fuzzy clustering method can be used both in batch mode and in one-pass version. The computational complexity of the proposed method is the same as that for other known recurrent clustering procedures [31, 34, 35], and it depends linearly on the number of observations in the data sample.

5.5 Fuzzy Clustering with a Variable Fuzzifier

Let's once more introduce into consideration the Lagrange function

$$L\big(u_j(k), c_j, \lambda(k)\big) = \sum_{k=1}^{N}\sum_{j=1}^{m} u_j^{\beta}(k)\left\|x(k) - c_j\right\|^2 + \sum_{k=1}^{N}\lambda(k)\left(\sum_{j=1}^{m} u_j(k) - 1\right) \tag{5.58}$$

where $\lambda(k)$ denotes an undetermined Lagrange multiplier.

Having solved the system of Karush–Kuhn–Tucker equations, we obtain a solution in the form

$$\begin{cases} u_j(k) = \dfrac{\left(\|x(k)-c_j\|^2\right)^{\frac{1}{1-\beta}}}{\sum_{l=1}^{m}\left(\|x(k)-c_l\|^2\right)^{\frac{1}{1-\beta}}}, \\ c_j = \dfrac{\sum_{k=1}^{N} u_j^{\beta}(k)x(k)}{\sum_{k=1}^{N} u_j^{\beta}(k)}, \\ \lambda(k) = -\left(\left(\sum\limits_{l=1}^{m} \beta\|x(k) - c_l\|^2\right)^{\frac{1}{1-\beta}}\right)^{1-\beta}, \end{cases} \tag{5.59}$$

that with $\beta = 2$ coincides with Fuzzy C-Means (FCM) algorithm by J. Bezdek. With $\beta \to 1$ the result of this solution is close to the results obtained by the traditional K-means crisp clustering algorithm.

As an alternative to the procedure in which the fuzzifier $1 < \beta < \infty$ is used, the objective function for fuzzy probabilistic clustering was proposed in [37]

$$E\big(u_j, c_j\big) = \sum_{k=1}^{N}\sum_{j=1}^{m}\left(\alpha u_j^2(k) + (1-\alpha)u_j(k)\right)\left\|x(k) - c_j\right\|^2 \tag{5.60}$$

where $0 < \alpha \le 1$ is a parameter that defines the nature of the obtained solution.

Introducing the Lagrange function

$$\begin{aligned} L\big(u_j(k), c_j, \lambda(k)\big) &= \sum_{k=1}^{N}\sum_{j=1}^{m}\left(\alpha u_j^2(k) + (1-\alpha)u_j(k)\right)\left\|x(k) - c_j\right\|^2 \\ &\quad + \sum_{k=1}^{N}\lambda(k)\left(\sum_{j=1}^{m} u_j(k) - 1\right) \end{aligned} \tag{5.61}$$

and solving the system of Karush–Kuhn–Tucker equations

$$\begin{cases} \frac{\partial L(u_j(k),c_j,\lambda(k))}{\partial u_j(k)} = \left(2\alpha\, u_j(k) + 1 - \alpha\right)\left\|x(k) - c_j\right\|^2 + \lambda(k) = 0, \\ \nabla_{c_j} L\left(u_j(k), c_j, \lambda(k)\right) = -\sum\limits_{k=1}^{N} 2\left(\alpha\, u_j^2(k) + (1-\alpha)u_j(k)\right)\left(x(k) - c_j\right) = \vec{0}, \\ \frac{\partial L(u_j(k),c_j,\lambda(k))}{\partial \lambda(k)} = \sum\limits_{j=1}^{m} u_j(k) - 1 = 0, \end{cases} \tag{5.62}$$

we obtain the solution in the form

$$\begin{cases} u_j(k) = -\frac{1-\alpha}{2\alpha} + \frac{1+m\frac{1-\alpha}{2\alpha}}{\sum_{l=1}^{m}\frac{\left(\|x(k)-c_j\|^2\right)}{\left(\|x(k)-c_l\|^2\right)}}, \\ c_j = \frac{\sum_{k=1}^{N}\left(\alpha\, u_j^2(k)+(1-\alpha)u_j(k)\right)x(k)}{\sum_{k=1}^{N}\left(\alpha\, u_j^2(k)+(1-\alpha)u_j(k)\right)}. \end{cases} \tag{5.63}$$

With $\alpha = 1$ this procedure coincides with the FCM. It is important to note, that the method (5.63) can't be used to solve Data Stream Mining tasks because it can't process information in online mode. Therefore, the adaptive modification was introduced in [38]

$$\begin{cases} u_j(k+1) = -\frac{1-\alpha}{2\alpha} + \frac{1+m\frac{1-\alpha}{2\alpha}}{\sum_{l=1}^{m}\frac{\left(\|x(k+1)-c_j(k)\|^2\right)}{\left(\|x(k+1)-c_l(k)\|^2\right)}}, \\ c_j(k+1) = c_j(k) + \eta(k)\left(\alpha\, u_j^2(k+1) + (1-\alpha)u_j(k+1)\right)\left(x(k+1) - c_j(k)\right) \end{cases} \tag{5.64}$$

where $\eta(k)$ is the learning rate parameter.

It is easy to see that the second recurrent expression in (5.64) is Kohonen's self-learning rule according to the principle "The Winner Takes More" with the neighborhood function $\alpha\, u_j^2(k+1) + (1-\alpha)u_j(k+1)$.

5.6 Ensemble of Adaptive Neuro-Fuzzy Kohonen Networks

Although the value of the parameter α in formulas (5.63) and (5.64) lies in a much narrower interval than the fuzzifier β, but at present there are no formal rules for choosing and setting this parameter. Therefore, when solving a specific problem in batch mode, this problem is usually solved several times using expression (5.63) with different values of the parameter α (starting from a very small value and increasing to 1).

It is clear that this approach can not effectively solve the problem in online mode. In this case, it is advisable to apply the idea of an ensemble of clustering procedures

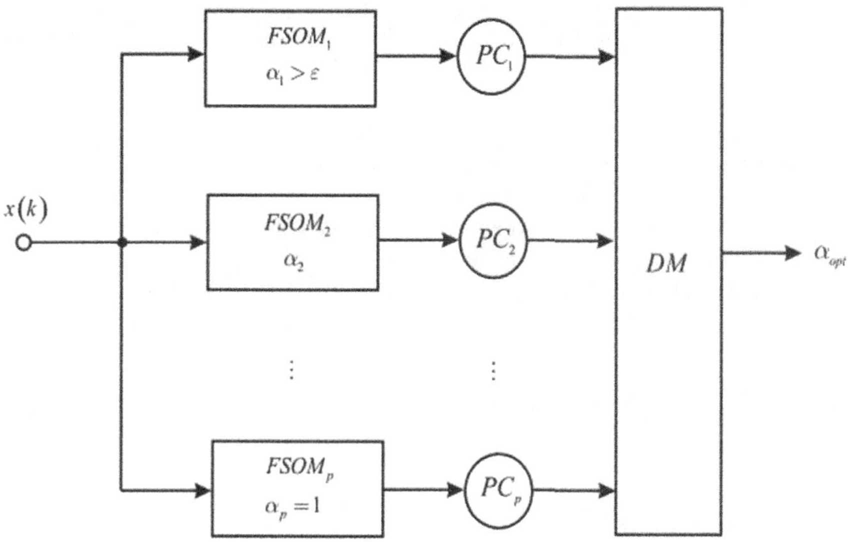

Fig. 5.3 The architecture of the adaptive neuro-fuzzy Kohonen networks ensemble

that work in parallel [39, 40], where each clustering procedure works with a value of parameter α that is different from other procedures.

This ensemble can be easily implemented using adaptive neuro-fuzzy Kohonen networks [41]. They represent a two-layer architecture, where the centers of the clusters are configured in the Kohonen layer (which contains m neurons N_j^K), and the membership levels are calculated in the output layer (which contains m neurons N_j^M).

The ensemble formed by these networks is shown in Fig. 5.3.

The self-learning algorithm for the p th member of the ensemble ($p = 1, 2, \ldots, q$), which contains q neuro-fuzzy networks, can be written as

$$\begin{cases} c_{jp}(k+1) = c_{jp}(k) + \eta(k)\Big(\alpha_p u_{jp}^2(k) + \big(1-\alpha_p\big)u_{jp}(k)\Big)\big(x(k+1) - c_{jp}(k)\big), \\ u_{jp}(k+1) = -\frac{1-\alpha_p}{2\alpha_p} + \frac{1+m\frac{1-\alpha_p}{2\alpha_p}}{\sum_{l=1}^{m}\frac{\left(\left\|x(k+1)-c_{jp}(k)\right\|^2\right)}{\left(\left\|x(k+1)-c_{lp}(k)\right\|^2\right)}}, \end{cases} \tag{5.65}$$

at the same time, the Kohonen layer is adjusted using the first formula in (5.65), and the output layer calculates the membership levels $u_{jp}(k+1)$ for each input observation $x(k+1)$.

The quality of clustering performed by each member of the ensemble can be assessed using any fuzzy clustering index [42]. Among them, one of the simplest and most efficient indices is the so-called partition coefficient (PC), which is calculated by the formula

$$PC_p(k+1) = \frac{1}{k+1} \sum_{\tau=1}^{k+1} \sum_{j=1}^{m} u_{jp}^2(\tau). \tag{5.66}$$

This coefficient has a clear physical sense: the better the clusters are expressed, the higher the value of PC_p (maximum value is $PC_p = 1$). The minimum value $\left(PC_p = m^{-1}\right)$ is reached if the data evenly belong to all clusters. But it is clear that this phenomenon represents the worst solution. This coefficient is suitable within the framework of the proposed system, since it can be calculated in online mode.

It should be noted that expression (5.66) is closely related to the traditional FCM. This coefficient should be modified for the case in question in the form

$$PC_p(k+1) = \frac{1}{k+1} \sum_{\tau=1}^{k+1} \sum_{j=1}^{m} \left(\alpha_p u_{jp}^2(\tau) + \left(1-\alpha_p\right) u_{jp}(\tau)\right). \tag{5.67}$$

5.7 Evolving Adaptive Neuro-Fuzzy Kohonen Network and Its Online Self-learning Algorithm in Fuzzy Clustering Tasks

In [43], an original online evolving fuzzy clustering method (EFCM) was proposed. It is based on the probabilistic approach [44] to solving the problem, while the main parameter that ultimately determines the final result is the radius of the formed clusters, chosen from empirical considerations and ultimately determining the number of possible classes. Despite the effectiveness of the probabilistic fuzzy clustering algorithms, their "weak" place is the "hard" constraint

$$\sum_{j=1}^{m} u_j(k) = 1 \;\forall\, k = 1, 2, \ldots, N, \tag{5.68}$$

that leads to the fact that an observation that belongs equally to all classes, that is, equidistant from all centroids, can have the same membership levels as an observation that is also equidistant from centroids but does not belong to any of the clusters.

The possibilistic approach (PCM) to fuzzy clustering [45] doesn't have this drawback. This approach is associated with the optimization of the objective function

$$E\left(u_j, c_j\right) = \sum_{k=1}^{N} \sum_{j=1}^{m} u_j^{\beta}(k) \left\| x(k) - c_j \right\|^2 + \sum_{j=1}^{m} \mu_j \sum_{k=1}^{N} \left(1 - u_j(k)\right)^{\beta} \tag{5.69}$$

where $c_j = \left(c_{j1}, c_{j2}, \ldots, c_{jn}\right)^{\mathrm{T}}$ is the center of the j th cluster computed during the data processing, $\beta > 1$ is fuzzification parameter that defined the "blurriness" of the borders between clusters and that usually has the value $\beta = 2$, $\mu_j > 0$ is a scalar parameter that defines the distance at which the membership level is equal to 0.5, i.e. if

$$\left\|x(k) - c_j\right\|^2 = \mu_j, \tag{5.70}$$

then $u_j(k) = 0, 5$.

The using of the possibilistic approach leads to the evolving possibilistic clustering method (EPCM), which is conveniently written in the form of a sequence of steps [46]:

Step 1: when the observation $x(1)$ arrives, the first cluster with a centroid c_1 is formed.

Step 2: when the observation $x(2)$ arrives, the condition

$$\|x(2) - c_1\| \leq \Delta \tag{5.71}$$

is checked (here Δ is some threshold that is set a priori). If this condition is met, then the observation $x(1)$ does not form a new centroid, i.e. it is assumed that it belongs to the first cluster with the membership level

$$u_1(2) = \left(1 + \left(\frac{\|x(2) - c_1\|^2}{\mu_1}\right)^{\frac{1}{\beta-1}}\right)^{-1}. \tag{5.72}$$

If the condition

$$\Delta < \|x(2) - c_1\| \leq 2\Delta \tag{5.73}$$

is satisfied, then the centroid correction is performed according to the WTA Kohonen self-learning rule [47]:

$$c_1(2) = c_1(1) + \eta(2)(x(2) - c_1(1)) \tag{5.74}$$

where $\eta(2)$ is the learning rate parameter. At the same time, the centroid c_1 "pulls up" to the vector of observations $x(1)$; if for $x(1)$ the inequality

$$2\Delta < \|x(2) - c_1\| \tag{5.75}$$

holds, then the second cluster is formed with a centroid

$$c_2 = x(2). \tag{5.76}$$

In this case, the membership levels $u_2(1)$ and $u_1(2)$ should be calculated according to the formulas given below.

Step N: so, if there are N observations and m clusters with centroids c_j, the calculations of all membership levels and of the adjusted coordinates of the centroids are estimated according to the relations [48]:

$$\begin{cases} u_j(k) = \left(1 + \left(\frac{\|x(k)-c_j\|^2}{\mu_j}\right)^{\frac{1}{\beta-1}}\right)^{-1}, \\ c_j = \frac{\sum_{k=1}^{N} u_j^{\beta}(k)x(k)}{\sum_{k=1}^{N} u_j^{\beta}(k)}, \\ \mu_j(k) = \frac{\sum_{k=1}^{N} u_j^{\beta}(k)\|x(k)-c_j\|^2}{\sum_{k=1}^{N} u_j^{\beta}(k)} \end{cases} \tag{5.77}$$

obtained by minimizing (5.69) over all estimated parameters. The system of Eq. (5.77) is essentially a batch information processing algorithm, so that when an observation $x(N+1)$ arrives, all calculations must be carried out again. It is clear that with a sufficiently high frequency of data stream the considered approach may be ineffective.

In this regard, it seems appropriate to develop an adaptive neuro-fuzzy system that allows to process the incoming information using recurrent procedures that do not require the storage of previously processed data.

It is possible to organize truly online information processing at the rate of arrival of new data, using instead of traditional packet FCM and PCM their recurrent counterparts [49, 50], which are essentially gradient optimization procedures for the adopted self-learning (clustering) criterion. In [51] it was shown that these procedures are essentially WTM-rules for Kohonen self-learning with a narrowing neighborhood function and that they can be successfully used to train self-organizing maps with a fixed architecture.

It is possible to implement the evolving neuro-fuzzy clustering system based on a two-layer adaptive neuro-fuzzy network [41] with the architecture shown in Fig. 5.4.

The first hidden layer of the network is formed by usual Kohonen neurons N_j^K, connected by lateral connections, through which the competition process is realized. The output layer of the network formed by the nodes N_j^u is designed to calculate the membership levels of each observation $x(k)$ to each j th cluster, $j = 1, 2, \ldots, m$.

To tune the centroids of the clusters, a self-learning recurrent procedure is used, which has the form [51]:

$$\begin{cases} c_j(k+1) = c_j(k) + \frac{u_j^{\beta}(k)}{k+1}\left(x(k+1) - c_j(k)\right), \\ u_j(k+1) = \frac{1}{1+\left(\frac{\|x(k+1)-c_j(k+1)\|^2}{\mu_j(k)}\right)^{\frac{1}{1-\beta}}}, \\ \mu_j(k+1) = \frac{\sum_{p=1}^{k+1} u_j^{\beta}(p)\|x(p)-c_j(k+1)\|^2}{\sum_{p=1}^{k+1} u_j^{\beta}(p)}. \end{cases} \tag{5.78}$$

It is easy to see that the first relation in (5.78) is the WTM self-learning rule with a narrowing neighborhood function $(k+1)^{-1}u_j^{\beta}(k)$.

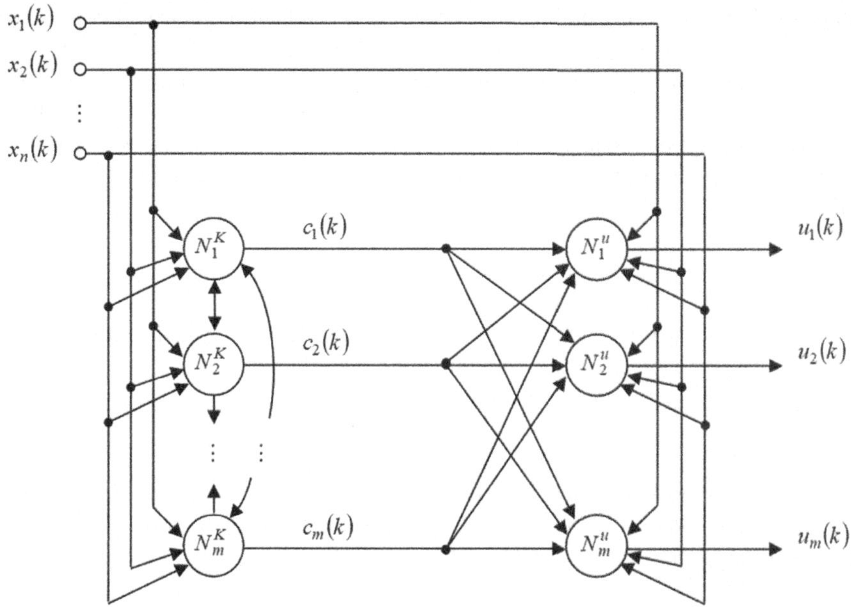

Fig. 5.4 The architecture of the adaptive clustering neuro-fuzzy network

The process of the evolution of the system, like the previous one, begins with a single Kohonen neuron that adjusts the coordinates of the first centroid c_1 . The next neuron is added to the network when the condition (5.75) is satisfied, which in this case takes the form

$$2\Delta < \|x(k) - c_1(k-1)\|. \tag{5.79}$$

At this moment, a neuron with a centroid $c_2(k) = x(k)$ is formed. It should be noted here that since in the Kohonen neural networks data are pre-normalized to a hypersphere, so that

$$\|x(k)\|^2 = \left\|c_j(k)\right\|^2 = 1, \tag{5.80}$$

the inequality (5.75), that determines the necessity of introducing new neurons into the network, takes the form

$$-1 \leq 1 - 2\Delta^2 < c_j^{\mathrm{T}}(k-1)x(k) \leq 1,\ \forall j = 1, 2, \ldots, m, \tag{5.81}$$

or

$$-1 \leq 1 - 2\Delta^2 < \cos\bigl(c_j(k-1), x(k)\bigr) \leq 1. \tag{5.82}$$

Thus, the build-up of the architecture occurs as a result of the constant control of inequalities (5.81) or (5.82) and is done in case when these inequalities are violated. It should be also noted that here the threshold Δ has the same meaning as in (5.75).

Due to the usage of the possibilistic approach, it is advisable to implement another "branch of evolution", namely, if at some point in time it becomes clear that the membership levels of the observation $x(k)$ don't exceed some additional threshold value

$$u_j(k) < \varepsilon \; \forall j = 1, 2, \ldots, m \tag{5.83}$$

i.e. the observation $x(k)$ is far enough away from all already formed centroids, this can also serve as a signal for the formation of a new cluster

$$c_{m+1}(k) = x(k). \tag{5.84}$$

To assess the quality of fuzzy clustering, the popular Xie–Beni index [42] can be used in its extended form [52]. For a fixed dataset containing N observations, this index has the form

$$EXB(N) = \frac{\left(\sum_{k=1}^{N} \sum_{j=1}^{m} u_j^{\beta}(k) \left\| x(k) - c_j(N) \right\|^2\right)/N}{\min\limits_{j \neq l} \left\| c_j(N) - c_l(N) \right\|^2} = \frac{NEXB(N)}{DEXB(N)}. \tag{5.85}$$

For online processing, this index, like the centroids of the clusters, can also be calculated recursively:

$$\begin{aligned} EXB(k+1) &= \frac{NEXB(k+1)}{DEXB(k+1)} \\ &= \frac{NEXB(k) + \frac{1}{k+1}\left(\sum_{j=1}^{m} u_j^{\beta}(k+1) \left\| x(k+1) - c_j(k+1) \right\|^2 - NEXB(k)\right)}{\min\limits_{j \neq l} \left\| c_j(k+1) - c_l(k+1) \right\|^2}. \end{aligned} \tag{5.86}$$

The inclusion of the expression (5.86) in the learning procedure (5.78) allows to organize additional control over the number of clusters formed by the system. So, introducing the third threshold δ and checking the condition

$$EXB(k+1) > \delta \tag{5.87}$$

at each step, it is possible to stop the process of building up neurons in case of violation of the inequality (5.87).

5.8 Neuro-Fuzzy Self-organizing Maps

As already noted, in the processing of real data, often there is a situation where one observation belongs to different classes (Fig. 5.5), and these classes are mutually intersected [53]. In the framework of traditional self-organizing maps formally this circumstance is not taken into account. However, it can be considered from the position of fuzzy cluster analysis.

It is expedient to combine the simplicity and speed of self-organizing Kohonen maps with the possibility of data processing under conditions of mutually intersected classes.

Thus, in [54, 55], a modification of the SOM was proposed, in which the neurons of the original architecture, that are essentially adaptive linear neurons, are replaced by fuzzy sets and fuzzy rules. This neural network has proved its effectiveness in image recognition tasks, but its learning is associated with a number of significant issues. In [56], a modification of a self-organizing map with a fuzzy inference and a combined self-learning algorithm based on the Kohonen and Grossberg rules was proposed. The disadvantage of this network is the presence of undetermined algorithm parameters. If these parameters are not well chosen, it can lead to unsatisfactory quality of clustering. In [46], the so-called fuzzy Kohonen clustering network (FKCN), that

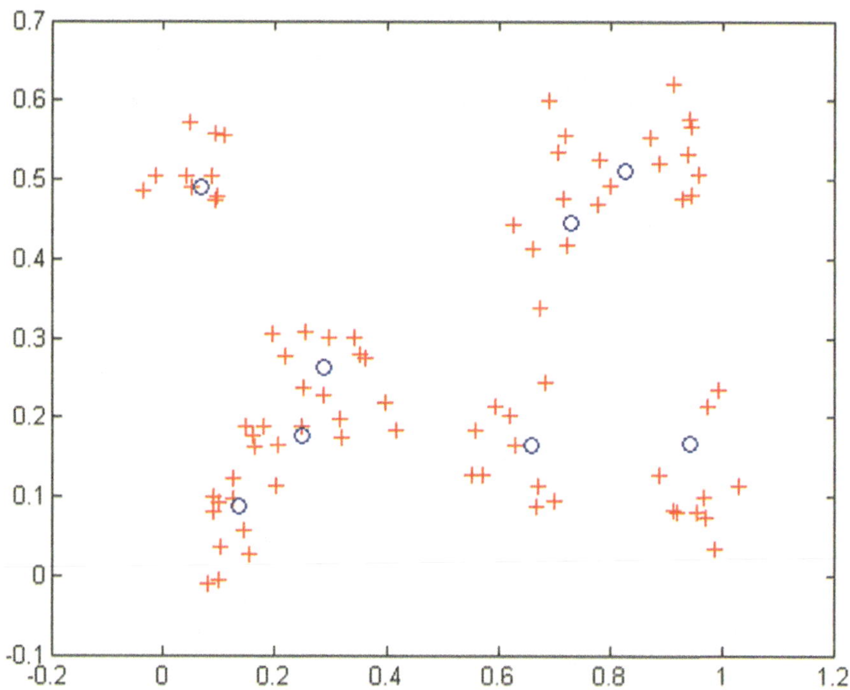

Fig. 5.5 Overlapping clusters

is based on the fuzzy C-means (FCM) algorithm by Bezdek [57], was introduced and it was developed in [58]. The main feature of this neuro-fuzzy system is the batch learning mode, in which the whole data set to be processed must be predefined. In this way, the FKCN can not work in online mode time, analyzing information as it arrives.

In this connection, as an alternative to the SOM and FKCN, an adaptive fuzzy-self-organizing neural network is proposed, that is tuned using a recurrent self-learning algorithm that is a generalization of Kohonen's training rule. It allows not only to find in online mode the prototypes (centroids) of formed clusters, but also to estimate the membership levels of each incoming observation to each cluster.

The architecture of the neuro-fuzzy network is shown in Fig. 5.6. It has a single layer of adaptive linear neurons N_i, $i = 1, 2, \ldots, m$ that forms the Kohonen map.

To the receptive layer of the network observations to be processed are fed sequentially. These observations are $(n \times 1)$ feature vectors $x(k) = (x_1(k), x_2(k), \ldots, x_n(k))^{\mathrm{T}}$.

Adjustable synaptic weights c_{ji}, $j = 1, 2, \ldots, m$; $i = 1, 2, \ldots, n$ determine the coordinates of the centroides m of mutually overlapping clusters $c_j(k) = (c_{j1}(k), c_{j2}(k), \ldots, c_{jn}(k))^{\mathrm{T}}$. The output of the network (in contrast to the SOM, whose output signal is determined only by the neuron-winner) is the $(m \times 1)$-vector $w(k) = (w_1(k), w_2(k), \ldots, w_m(k))^{\mathrm{T}}$ that determines the membership level of the

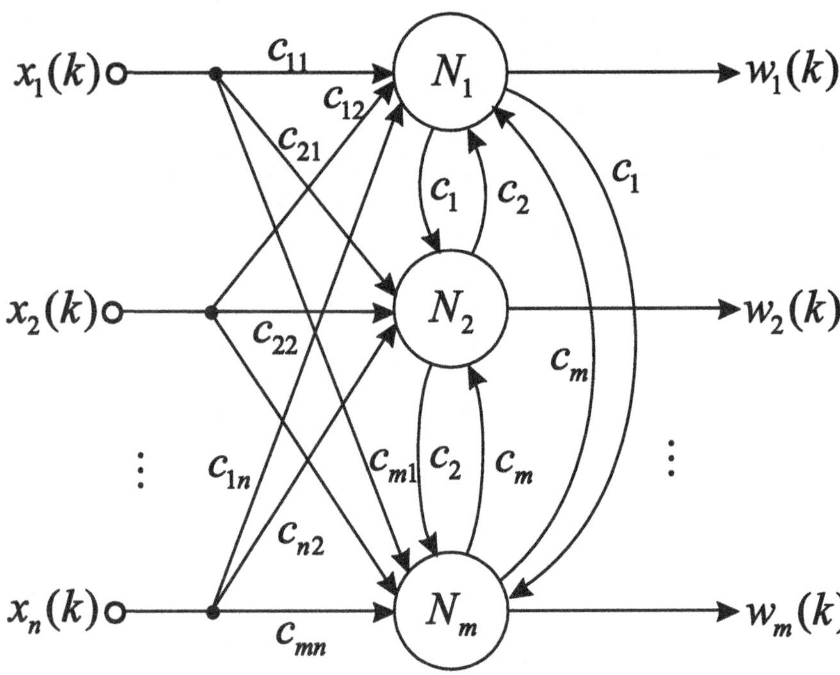

Fig. 5.6 Adaptive fuzzy-self-organizing neural network

observation $x(k)$ to each m formed clusters and is calculated using neurons N_j. Neurons exchange coordinates of $c_j(k)$ using lateral connections. These coordinates are needed to calculate membership levels $w_j(k)$.

The algorithm responsible for forming the membership levels of the observations to the formed clusters begins with the synaptic weights initialization formed by the generator of pseudorandom numbers. After the initialization of the network, in order to form the membership levels, the following main processes are launched:

1. Competition. For each input observation the relative values of the discriminant function are calculated. This function is the basis of competition among neurons.
2. Cooperation. The neuron-winner defines the spatial position of the topological neighborhood of neurons, thereby providing a basis for cooperation between these neurons.
3. Synaptic adaptation. The last mechanism allows excited neurons to increase the eigenvalues of discriminant functions in relation to input observations by means of appropriate corrections of synaptic weights. Adjustment occurs in such a way that the response of the neuron-winner to further application of similar patterns intensifies.

The basis of self-learning of adaptive fuzzy-self-organizing neural network is the probabilistic clustering method, based on the optimization of the goal function of the form [59]

$$E(w_j, c_j) = \sum_{k=1}^{N}\sum_{j=1}^{m} w_j^{\beta}(k)\left\|x(k) - c_j\right\|^2 \tag{5.88}$$

with constraints

$$\sum_{j=1}^{m} w_j(k) = 1, k = 1, 2, \ldots, N, \tag{5.89}$$

$$0 \le \sum_{k=1}^{N} w_j(k) \le N, j = 1, 2, \ldots, m \tag{5.90}$$

where $w_j(k) \in [0, 1]$, N is the number of processed observations.

The application of the standard apparatus of undetermined Lagrange multipliers and the non-linear programming Arrow-Hurwitz procedure, which is the procedure for locating the saddle point of the Lagrange function, leads to the probabilistic recurrent fuzzy clustering method

$$\begin{cases} c_j(k+1) = c_j(k) + \eta(k) w_j^{\beta}(k)\big(x(k+1) - c_j(k)\big), \ j = 1, 2, \ldots, m, \\ w_j(k+1) = \dfrac{\left(\left\|x(k+1) - c_j(k+1)\right\|^2\right)^{\frac{1}{1-\beta}}}{\sum_{l=1}^{m}\left(\|x(k+1) - c_l(k) + 1\|^2\right)^{\frac{1}{1-\beta}}} \end{cases} \tag{5.91}$$

where $\eta(k)$ is the learning rate parameter.

If in expression (4.38) the multiplier $w_j^\beta(k)$ is considered as a neighborhood function $\varphi(j, l, k)$, we can get the expression (4.39), which is the Kohonen self-learning rule based on the WTM-principle ("The Winner Takes More"):

$$c_j^{WTM}(k+1) = c_j^{WTM}(k) + \varphi(j, l, k)\big(x(k+1) - c_j^{WTM}(k)\big). \tag{5.92}$$

With $\beta = 0$ in (2.8), expression (2.8) will correspond to the standard WTA-Kohonen rule for the neuron-winner:

$$c_j^{WTA}(k+1) = c_j^{WTA}(k) + \eta(k)\big(x(k+1) - c_j^{WTA}(k)\big). \tag{5.93}$$

It is also easy to see that the recurrent procedure (5.93) minimizes the function of the form

$$E(c_j) = \sum_{x(k)\in Cl_j} \|x(k) - c_j^{WTA}\|^2 \tag{5.94}$$

(here Cl_j is the jth cluster, $j = 1, \ldots, m$).

And its direct optimization leads to a simple estimation of the arithmetic mean. Its recurrent form leads to

$$c_j^{WTA}(k+1) = c_j^{WTA}(k) + \frac{1}{k+1}(x(k+1) - c_j^{WTA}(k)) \tag{5.95}$$

(here $\eta(k) = \frac{1}{k+1}$).

Thus, in the final form, the adaptive self-learning algorithm of the fuzzy self-organizing network can be written as:

$$c_j(k+1) = c_j(k) + \frac{w_j^\beta(k)}{k+1}\big(x(k+1) - c_j(k)\big), \quad j = 1, 2, \ldots, m, \tag{5.96}$$

$$w_j(k+1) = \frac{\left(\|x(k+1) - c_j(k+1)\|^2\right)^{\frac{1}{1-\beta}}}{\sum_{l=1}^m \left(\|x(k+1) - c_l(k+1)\|^2\right)^{\frac{1}{1-\beta}}}, \tag{5.97}$$

which combines the computational simplicity and sequential processing of the Kohonen self-learning with the possibilities of fuzzy clustering.

If the goal function of fuzzy possibilistic clustering

$$E_k\big(w_j(k), c_j(k)\big) = \sum_{j=1}^m w_j^\beta(k) D^2\big(x(k), c_j\big) + \sum_{j=1}^m \mu_j\big(1 - w_j(k)\big)^\beta \tag{5.98}$$

is used as a self-learning quality criterion, then we get the possibilistic adaptive self-learning rule [45] described above:

$$\begin{cases} c_j^{POS}(k+1) = c_j^{POS}(k) + \eta(k) \\ \qquad \times \left(w_j^{POS}(k)\right)^{\beta}\left(x(k+1) - c_j^{POS}(k)\right), \; j = 1, 2, \ldots, m, \\ w_j^{POS}(k+1) = \left(1 + \left(\dfrac{D^2\left(x(k), c_j^{POS}(k+1)\right)}{\mu_j(k)}\right)^{\frac{1}{\beta-1}}\right)^{-1}, \\ \mu_j(k+1) = \dfrac{\sum_{\tau=1}^{k+1}\left(u_i^{POS}(\tau)\right)^{\beta} D^2\left(x(\tau), c_j^{POS}(k+1)\right)}{\sum_{\tau=1}^{k+1}\left(w_j^{POS}(\tau)\right)^{\beta}}. \end{cases} \tag{5.99}$$

Analyzing the first expression of the introduced self-learning rule, we can see that when considering a multiplier $\left(w_j^{POS}(k)\right)^{\beta}$ as a neighborhood function $\varphi(j, l, k)$, we get the Kohonen self-learning rule based on the "Winner Takes More" principle (WTM)

$$c_j^{WTM}(k+1) = c_j^{WTM}(k) + \eta(k)\varphi(j, l, k)\left(x(k+1) - c_j^{WTM}(k)\right), \tag{5.100}$$

where $\varphi(j, l, k)$ is bell-shaped neighborhood function, the argument of which is the distance in the accepted metric space between the centroids of the neuron-winner and the neuron N_i .

With fuzzifier $\beta = 1$ in (5.88) we get the k-means algorithm (the hard c-means, HCM), and with $\beta = 0$ we get the usual Kohonen rule based on the principle of "Winner Takes All" (WTA) [60, 61]

$$c_j^{WTA}(k+1) = c_j^{WTA}(k) + \eta(k)\left(x(k+1) - c_j^{WTA}(k)\right). \tag{5.101}$$

It should also be noted that the important property of the self-learning rules on the basis of the possibilistic approach is associated with the condition $\sum_{j=1}^{m} u_j^{POS}(k) \neq 1$, that, unlike the procedures of probabilistic fuzzy clusterization, allows to detect the emergence of new clusters in the process of neural network learning.

The parallel application of adaptive probabilistic and possibilistic algorithms leads to a combined procedure (for $\beta = 2$):

$$\begin{cases} c_j^{PR}(k+1) = c_j^{POS}(k) + \eta(k) \\ \qquad\qquad \times \left(w_j^{POS}(k)\right)^2 \left(x(k+1) - c_j^{POS}(k)\right), \\ w_j^{PR}(k+1) = \dfrac{\left\|x(k+1) - c_j^{PR}(k+1)\right\|^{-2}}{\sum_{l=1}^{m} \left\|x(k+1) - c_l^{PR}(k+1)\right\|^{-2}}, \\ c_j^{POS}(k+1) = c_j^{PR}(k+1) + \eta(k)\left(w_j^{POS}(k)\right)^2 \left(x(k+1) - c_j^{PR}(k+1)\right), \\ w_j^{POS}(k+1) = \dfrac{\mu_j(k)}{\mu_j(k) + \left\|x(k+1) - c_j^{POS}(k+1)\right\|^2}, \\ \mu_j(k+1) = \dfrac{\sum_{\tau=1}^{k+1} \left(w_j^{POS}(\tau)\right)^2 \left\|x(\tau) - c_j^{POS}(k+1)\right\|^2}{\sum_{\tau=1}^{k+1} \left(w_j^{POS}(\tau)\right)^2}. \end{cases} \tag{5.102}$$

The procedure (5.102) is the self-learning method of neuro-fuzzy system, formed by two parallel-acting adaptive fuzzy self-organizing neural networks that exchange information with each other as shown in Fig. 5.7.

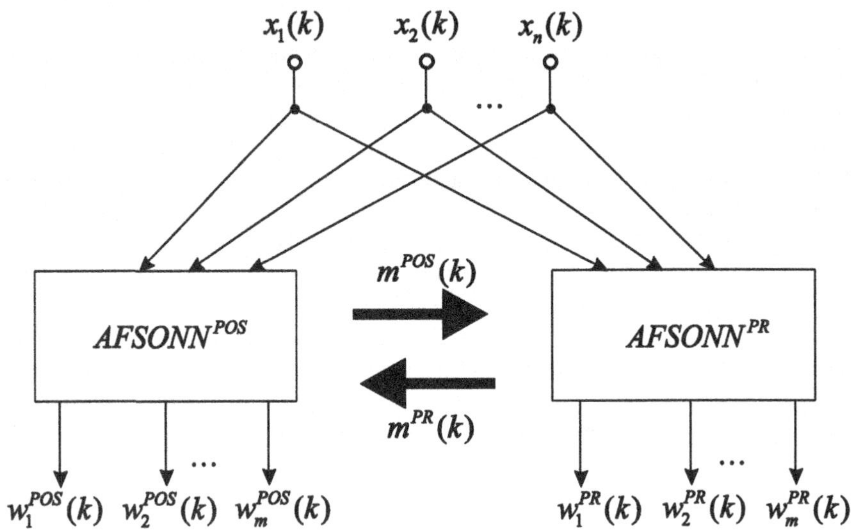

Fig. 5.7 Self-learning neuro-fuzzy clustering system

5.9 Conclusion

The problem of online clustering-segmentation of data streams sequentially received for processing under the conditions when the formed classes mutually overlap, is considered. The traditional probabilistic and possibilistic approaches to fuzzy clustering are considered and their recurrent adaptive modifications are introduced, which are gradient optimization procedures of the clustering goal functions. In situations where data under processing are corrupted by anomalous outliers, adaptive procedures of fuzzy robust clustering based on goal functions that suppress these outliers are introduced.

For tasks with a high level of a priori uncertainty in the input data, it is proposed to use ensembles of adaptive neuro-fuzzy Kohonen networks based on objective functions with polynomial fuzzifier. The use of neuro-fuzzy self-organizing maps allows to improve the quality of fuzzy clustering under the conditions of a priori uncertainty about the nature of the input data.

The considered methods of fuzzy clustering are simple in numerical implementation, are characterized by high speed and are designed to solve a number of problems arising in the Data Stream Mining.

References

1. Han J, Kamber M (2006) Data mining: concepts and techniques. Morgan Kaufmann, San Francisco, 800 p
2. Gan G, Ma C, Wu J (2007) Data clustering: theory, algorithms, and application. SIAM, Philadelphia, 466 p
3. Abonyi J, Feil B (2007) Cluster analysis for data mining and system Identification. Brikhauser, Basel, 303 p
4. Olson D, Dursun D (2008) Advanced data mining technologies. Springer, Berlin, 180 p
5. Xu R. Clustering, Wunsch DC (2008) Wiley & Sons, Hoboken, 358 p
6. Bow ST (2002) Pattern recognition and image preprocessing. Marcel Dekker, Inc., S. T.Bow, N.Y., Basel, 699 p
7. Forsyth D, Ponce J (2011) Computer vision: a modern approach. Pearson, 800 p
8. Szeliski R (2001) Computer vision. algorithms and applications. Springer, London. 813 p
9. Aggarwal CC, Reddy CK (2014) Data clustering: algorithms and applications. Chapman and Hall, CRC, 652 p
10. King RS (2014) Cluster analysis and data mining: an introduction. Mercury Learning & Information, 300 p
11. Bezdek JC (1981) Pattern recognition with fuzzy objective function algorithms. Plenum Press, N.Y., 272 p
12. Hoeppner F, Klawonn F, Kruse R (1996) Fuzzy-clusteranalyse: Verfahren fuer die Bilderkennung, Klassification und Datenanalyse. Vieweg, 280 S
13. Hoeppner F, Hoeppner F, Klawonn F, Kruse R, Runkler T (1999) Fuzzy cluster analysis and image recognition. Wiley & Sons, Chichester, 289 p
14. Bezdek JC, Keller J, Krisknapuram R, Pal NR (2005) Fuzzy models and algorithms for pattern recognition and image processing. Springer Science+Business Media, Inc., N.Y., 776 p
15. MacQueen J (1965) On convergence of k-means and partitions with minimum average variance. Ann Math Statist 36:1084

16. Cover TM (1968) Estimates by the nearest-neighbor rule. IEEE Trans Inf Theory 14:50–55
17. Bodyanskiy YV, Gorshkov YV, Kokshenyev IV, Kolodyazhniy VV (2002) On adaptive algorithm for fuzzy data clustering [in Russian]. Adaptive systems of automatic control: inter-branch scientific and technological digest. Syst Technol, Dnipropetrovsk 5(25):108–117
18. Gorshkov YV, Kokshenev IV, Rudnyeva OO (2003) Fast neuro-fuzzy classifier. In: Proceedings of the 1st international IEEE EMBS conference, neural engineering 2003, IEEE Press, Capri, Italy, 20–22 Mar 2003, pp 549–552
19. Hoeppner F, Klawonn F, Kruse R, Runkler T (1999) Fuzzy cluster analysis: methods for classification, data analysis and image recognition. Wiley & Sons, Chichester, 300 p
20. Gustafson EE (1979) Fuzzy clustering with a fuzzy covariance matrix. In: Proceedings IEEE CDC, San Diego, California, pp 761–766
21. Klawonn F, Kruse R, Timm H (1997) Fuzzy shell cluster analysis. In: Della Riccia G, Lenz HJ, Kruse R (eds) Learning, networks and statistics. Springer-Verlag, Wien, pp 105–120
22. Yager RR, Filev DP (1994) Approximate clustering via the mountain method. IEEE Trans Syst, Man Cybern 24–37:1279–1284
23. Gorshkov Y, Kolodyazhniy V, Bodyanskiy Y (2009) New recursive learning algorithms for fuzzy Kohonen clustering network. In: Proceedings of the 17th international workshop on nonlinear dynamics of electronic systems (NDES-2009), Rapperswil, 21–24 June 2009, pp 58–61
24. Bodyanskiy Y, Gorshkov Y, Kokshenev I, Kolodyazhniy V (2008) Evolving fuzzy classification of non-stationary time series. In: Angelov P, Filev DP, Kasabov N (eds) Evolving intelligent systems methodology and applications. Wiley & Sons, New York, pp 446–464
25. Bodyanskiy Y, Gorshkov Y, Otto P, Pliss I (2009) Medical image analysis using neuro-fuzzy network. In: Proceedings of the Internationales Wissenschaftliches Kolloquium (IWK-2009). Information technology and electrical engineering – devices and systems, materials and technology for the future, vol 54, 07–10 Sept 2009, TU Ilmenau, Ilmenau, p 7
26. Angelov P (2002) Evolving rule-based models: a tool for design of flexible adaptive systems. Physica-Verlag, Heidelberg-New York, 213 p
27. Dvoretzky A (1956) On stochastic approximation. In: Proceedings of the 3rd Berkeley symposium mathematics statistics and probability. University of California Press, Berkeley, California, vol 1, pp 39–55
28. Pau LF (1981) Failure diagnosis and performance monitoring. Marcel Dekker, N.Y, 448 p
29. Kohonen T (1995) Self-organizing maps. Springer-Verlag, Berlin, 362 p
30. Rojas R (1996) Neural networks: a systematic introduction. Springer-Verlag, Berlin, 502 p
31. Abe S (1997) Neural network and fuzzy systems: theory and applications. Kluwer Academic Publishers, Boston, 258 p
32. Tsoukalas LH, Uhrig RE (1997) Fuzzy and neural approaches in engineering. Wiley & Sons, Inc., N.Y, 587 p
33. Harris CJ (1994) Advances in intelligent control. Taylor and Francis, London, 373 p
34. Schalkoff RJ (1997) Artificial neural networks. The McGraw–Hill Comp., Inc., N.Y., 422 p
35. Ritter H, Schulten K (1986) On stationary state of the Kohonen self-organizing sensory mapping. Biol Cybern 54:234–249
36. Ritter H, Schulten K (1988) Convergence properties of Kohonen's topology conserving maps: fluctuation, stability, and dimension selection. Biol Cybern 60: 59–71
37. Grossberg S (1974) Classical and instrumental learning by neural networks. Proceedings of the progress in theoretical biology. Academic Press, N.Y, vol 3, pp 57–141
38. Ball GH, Hall DJ (1967) A clustering technique for summarizing multivariate data. Behav Sci 12(2):153–155
39. Abonyi J, Feil B (2007) Cluster analysis for data mining and system identification. Birkhauser, Basel, 303 p
40. Hartigan JA, Wong MA (1979) A k-means clustering algorithm. Appl Stat 28:100–108
41. Vuorimaa P (1994) Fuzzy self-organizing maps. Fuzzy Sets Syst 66:223–231
42. Rutkowski L (2008) Computational intelligence: methods and techniques. Springer-Verlag, Berlin, Heidelberg, 514 p

43. Mumford C, Jain L (2009) Computational intelligence: collaboration, fuzzy and emergence. Springer-Vergal, Berlin, 726 p
44. Gath I, Geva AB (1989) Unsupervised optimal fuzzy clustering. IEEE Trans Pattern Anal Mach Intell 11:773–781
45. Krishnapuram RA, Keller J (1993) Possibilistic approach to clustering. IEEE Trans Fuzzy Syst 1:98–110
46. Krishnapuram R, Keller J (1994) Fuzzy and possibilistic clustering methods for computer vision. Neural Fuzzy Syst 12:133–159
47. Bodyanskiy Y, Kolodyazhniy V, Stephan A (2002) Recursive fuzzy clustering algorithms. In: Proceedings of the 10-th East-West fuzzy colloquium 2002. HS, Zittau/Goerlitz, pp 276–283
48. Klawonn F, Kruse R (1997) Constructing a fuzzy controller from data. Fuzzy Sets Syst 85:117–193
49. Arrow KJ, Hurwitz L, Uzawa H (1958) Studies in linear and nonlinear programming. Stanford University Press, Stanford, CA, 242 p
50. Chung FL, Lee T (1994) Fuzzy competitive learning. Neural Netw 7(3):539–552
51. Park DC, Dagher I (1994) Gradient based fuzzy c-means (GBFCM) algorithm. In: Proceedings of the IEEE international conference on neural networks, IEEE Press, Orlando, FL, USA, pp 1626–1631
52. Bodyanskiy Y, Gorshkov Y, Kokshenev I, Kolodyazhniy V (2006) Outlier resistant recursive fuzzy clustering algorithms. In: Reusch B (ed) Computational intelligence, theory and applications. advances in soft computing. Springer-Verlag, Berlin; Heidelberg, vol 38, pp 647–652
53. Schwefel HP (1995) Evolution and optimum seeking. Wiley & Sons, New York, 350 p
54. Kureichik VM (1999) Genetic algorithms: state of the art, problems and perspectives. J Comput Syst Sci Int 38(1):137–152
55. Hlybovec NN, Medvid SA (2003) Genetic algorithms and their application for schedule making task (in Russian). Cybern Syst Anal (1):95–108
56. Kuo-Lung W, Miin-Shen Y (2005) A cluster validity index for fuzzy clustering. Pattern Recogn Lett 26:1275–1291
57. Minsky M, Papert S (1969) Perceptrons. The MIT Press, 258 p
58. Klawonn F, Kruse R (1992) Constructing a fuzzy controller from data. Fuzzy Sets Syst 85:117–119
59. Kohonen T, Kaski S, Lagus K, Salojärvi J, Honkela J, Paatero V, Saarela A (2000) Self organization of a massive document collection. IEEE Trans Neural Netw 11:574–585
60. Charikar M (2002) Similarity estimation techniques from rounding algorithms. In: ACM Symposium on theory of computing, pp 380–388
61. Haykin S (1999) Neural Networks: a Comprehensive Foundation. Prentice Hall, Inc., Upper Saddle River, N.J, 842 p

Chapter 6
Fuzzy Systems in Data Mining Tasks

Valentin Filatov and Andriy Kovalenko

Abstract Original method for implementing fuzzy models using relational systems is considered and theoretically justified. Unlike the known approaches, the proposed method enables solving the problems of data mining of integrated systems—databases and fuzzy systems. A method and technology are developed for organizing, storing and searching for fuzzy data by means of relational systems. An original information technology to search for fuzzy data by means of relational systems is proposed, which, unlike the well-known approaches, enables using a structured query language (SQL) to solve a complex data mining problem. Since standard self-organizing maps solve the problems of clear clustering, neuro-fuzzy self-organizing maps are introduced and it is shown that the neighbourhood function is actually a membership function in fuzzy inference systems. A robust modification of a neuro-fuzzy self-organizing card and an algorithm for its self-learning is introduced.

Keywords Information system · Database · Relational databases · Intelligent systems · Fuzzy system · Data mining

6.1 General Principles of Information Systems Building

Intensive theoretical and applied works in the development of information systems are dated back to the 70s of the twentieth century and were caused by the following reasons:

- technical: appearance of mass storage devices on magnetic carriers;

V. Filatov · A. Kovalenko (✉)
Kharkiv National University of Radio Electronics, 14 Nauky Ave, Kharkiv 61166, Ukraine
e-mail: andriy_kovalenko@yahoo.com

V. Filatov
e-mail: valentin.filatov@nure.ua

V. Mashtalir et al. (eds.), *Advances in Spatio-Temporal Segmentation of Visual Data*, Studies in Computational Intelligence 876,
https://doi.org/10.1007/978-3-030-35480-0_6

- social: introduction of computing technology in various sectors of the economy, which necessitated the creation of various information systems, as well as the growth of requirements for tools for the implementation of such systems;
- technological: means of the existing programming systems, file systems and operating systems did not meet these requirements and did not fully use the new properties of direct access memory devices.

There appeared a new class of information systems that is focused on creating, updating, searching, maintaining integrity, which has been called the database control system (DBCS). Since then, information systems have developed in two directions. The first of them can be described as a continuation of the line of conventional procedural programming languages—Algol, Fortran, Cobol, which make it possible to create various programs for processing relatively simply organized data files unrelated to each other. However, in applications, when the volume and structural complexity of data increased, and, most importantly, the need arose to process the same data with different programs, it was impossible to preserve the existing data processing technology. There was a need to move to a new technology-oriented database technology [1, 2]. As a result of this development, DBCS, which are based on programming languages, have appeared. This line of development has led to the emergence of such concepts as the data definition language (DDL) and the database scheme based on it, the unified data manipulation language (DML).

The second direction of development of the DBCS is related to the information systems oriented to the large arrays of stored data. They have passed a long way from the simple systems that specialized in a single application with a fixed data structure and a fixed set of functions to the universal systems that give the user a wide range of data structures and operations on them. The user communicated to such a system through an intermediary-specifier, which translated tasks from a natural language to a system-friendly representation of an information language. Naturally, such a narrow approach to the design of information systems could not satisfy either their users or developers.

Therefore, as the development of system programming in general, and an understanding of the place of information systems in computer software in particular, developed a tendency to generalize their functions and structures of processed documents, which eventually led to the creation of general information systems [3]. The following set of standard functions of a typical information system has been formed [4].

1. The possibility of describing and forming the required data structures, manipulating, storing and protecting them within the system.
2. Providing opportunities for data exchange, both with the user and with another information system.
3. Availability of a wide range of data processing and access methods in the interests of the user.
4. The presence of the analysis apparatus of the information system itself.
5. The ability to adapt the system to different operating conditions.
6. User interaction with the system in a dialog mode.

The basis for building systems has become a non-procedural input language in which the user can formulate what needs to be done without specifying how to do it. The constructions of such a language are aimed at providing the user with a specific set of operations without involving traditional programming. The task made in this language is clothed in a form close to the form of constructions of a natural language. On the other hand, it became possible to build an information system as a programming system based on a certain input language, i.e., based on translator design methods.

A number of papers [5, 6] are devoted to the theoretical foundations of building systems based on programming languages. By now, several formal methods have been proposed for the specification of software systems in general, models, and data types in particular. The algebraic approach [7] received the greatest popularity as a formal justification and a tool for the specification of data types.

In the general case, an algebraic system can be represented as:

$$U_a = \langle A, \Omega_f, \Omega_p \rangle, \tag{6.1}$$

where A—set of basics; $\Omega_f = \{F_1, \dots F_k\}$—set of operation names, given on the set A; $\Omega_p = \{\pi_1, \dots \pi_m\}$ —the set of predicates defined on the set A. System U_a can be written in short as $U_a = \langle A, \Omega \rangle$, in case of the union of sets $\Omega = \Omega_f \cup \Omega_p$.

The set A is called the carrier or the main set, operations F_k and predicates π_m unlike other operations and predicates are called primary or main.

Studies in the field of data models of information systems show that the concept of data type has now become the central. The problems of creating new programming languages and the introduction of modern data organization technologies are connected to that. Of the variety of approaches to determining the data type, the following looks most constructive: the data type defines a set of values through a set of operations [8].

Because of the special role of operations in the definition of data structures and the functioning of systems, consider this concept in more detail. For the formal definition of the data type, the concept of $\sum$ signature as a pair consisting of a set of operation names F_k and a set of operation descriptions Ω_o is introduced. Then the signature can be defined as $\Sigma = \langle F_f, \Omega_o \rangle$.

Definition. Data type T of the signature Σ is a pair: specification of the signature data type and implementation of the corresponding data type

To a certain extent, this approach gives a new view on data types, according to which a set of values of a type is characterized by a set of operations, allows designing operations on types not only to the creators of the language, but also to programmers creating their own types. Since a data type as an object consists of two main components: specification and implementation, then type operations deal with the components of these components. Such actions may include increasing or decreasing the number of operations, replacing implementations of all or some operations, changing the type of presentation and related operations.

Thus, the definition of the data type reflects both of its aspects: user (specification) and machine (implementation). The programmer, while composing his program, sees the type as a specific specification; the compiler, translating this program into object code, deals with both aspects of the data type. The processor, executing the object code, interacts only with implementations of the corresponding data types.

To create a type you need to build a specification and associate an appropriate implementation with it. In this case, of course, it is necessary to ensure that the proposed implementation satisfies this specification.

A special place is given to the data model as the basic component of any information system. According to L. A. Kalinichenko, at present, the data model is interpreted "as a set of methods and means of determining the logical structure of the database and the dynamic state of the subject area in the database." Date [9] identifies three most significant components in the data model:

- a set of means for determining acceptable data structures;
- a set of operations applicable to the valid state of the database for searching or modifying data;
- sets of integrity constraints, explicitly or implicitly determining the set of admissible database states.

Through a combination of these properties, the data model provides users with a means of describing, manipulating, and controlling integrity, expressed in one or more database languages. The task of developing a functionally complete set of operations on types requires careful research and cannot be considered solved now. Some ways to solve it are indicated in [7, 8].

According to the proposals of the American National Standards Institute, there are three levels of data description: conceptual, internal and external.

The conceptual level defines a conceptual model, based on which a specific conceptual scheme is created for each database containing a general description of the database, as presented by the developer, taking into account the interests of all users. The internal level is set by the internal model and determines the relationship with the operating system, the physical level of storage and access to data on external media. And finally, the external level is defined by the external model, on the basis of which one or several external schemes (subschemes) are created for each database, describing certain parts of the database that are interested for the relevant user groups.

A specific data model network, hierarchical or relational, guides modern universal DBCS, as a rule. This forces users either to artificially adjust their needs to the capabilities of a developed DBCS, or to master a new one with a more suitable data model, which, according to research, is a rather laborious task [10, 11]. The classification of structured data models is shown at Fig. 6.1.

The data models supported by the DBCS can be divided into several subclasses, distinguished by the structural characteristics of the data model—the way the objects are represented and their interrelationships in the database.

The defining role in the classification is played by the conceptual basis of the data model, its approach to the representation of real entities, to the category of objects or relationships between objects.

Significant differences between data models are manifested in the adopted methods for determining the semantics of data manipulation operators, in the means of

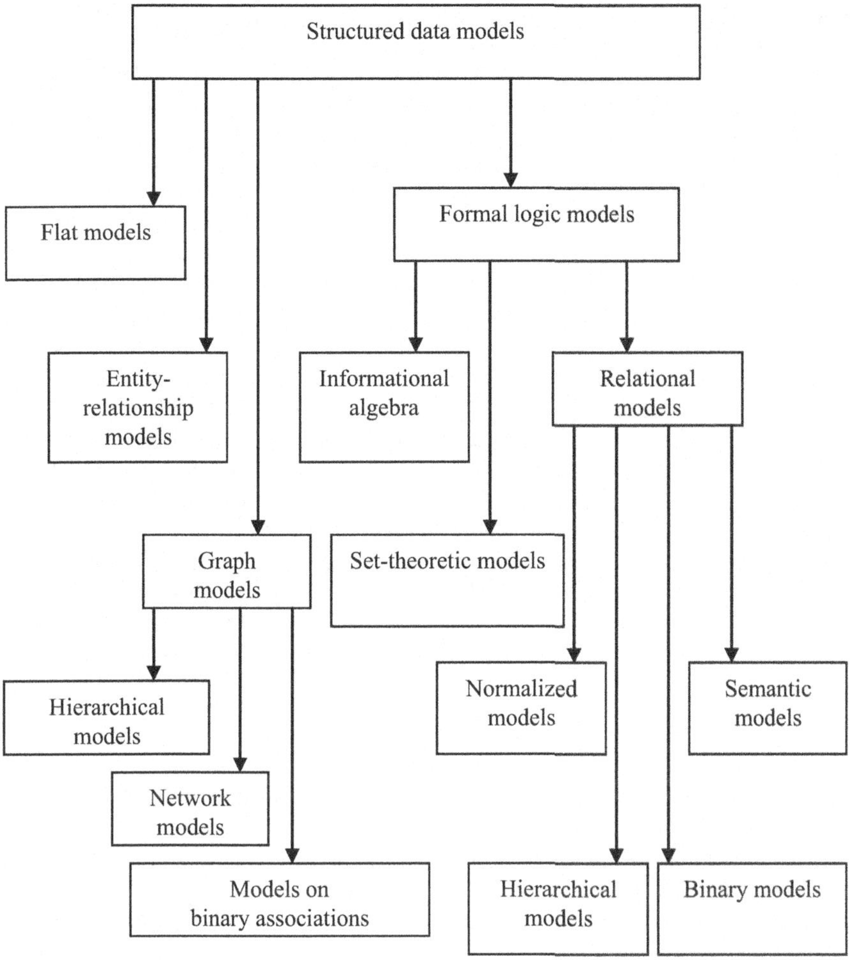

Fig. 6.1 Classification of structured models

ensuring the invariance of specified properties of object types and types of relationships between objects during the entire lifetime of the database. Consider the general classification of data models according to their structural characteristics, understanding that data models themselves are very complex objects and any classification is conditional. Examples of specific representatives of the entered classes of data models are given in Table 6.1.

The simplest is a class of flat data models, in which objects are represented by records that have a fixed set of fields containing elementary values of object properties. Links between objects in the database are not supported by DBCS in this case. This function is assigned to the application programmer, who can set them

Table 6.1 Data model classes

No	Data model class	Class representative
1	Flat data model	Data model on flat files
2	Entity-relationship data model	Chen data model
3	Hierarchical data model	Data model IMS
4	Network data model	The CODASIL data model, Bachmann's chains TOTAL data model
5	Data Model on Binary Associations	DIAM data model Falkenberg data model
6	Information algebra	Information algebra KODASIL
7	Data model on positional sets	Childs data model Hardgrave Data Model
8	Relational Normalized Data Model	Codd data model
9	Relational binary data model	Abrial data model Pelegatti Bracci data model
10	Semantic relational data model	Smith data model RM/T Codd Borkin data model

and change them arbitrarily, using the values of unique identifiers of objects in the database as properties.

In data models of the class "entity-relationship" [12] the subject area is represented by entities with attributes. Relationships between entities are interpreted by means of connections—usually binary associations between objects. In the most developed models of this class, communication can be endowed with attributes and to be treated like objects. For example, the relationship between an employee and a specific project is characterized by a fraction of the time the employee spends on the implementation of this project. Each entity can participate in an arbitrary number of links of a different type. The actions performed by database editing operators (placing new entities in the database, changing entities, deleting entities) depend on the links in which such entities are involved.

At the boundary of the class of flat data models and "entity-relationship" models, there are data models in which relationships between entities (objects) are established "by association" of the values of specified attributes for a pair of entities of various types. For example, the types of entities AIRPLANE and AIRPORT may have an associative relationship on the TYPE-AIRPLANE attribute, which determines the types of aircraft that the airport can take. Such relationships are supported for the pairs of objects of a given type automatically through associative associators formed in the dynamics of database changes. Relationships established in a similar manner are used to identify objects by their interconnection with other objects and do not have any influence on the actions performed when the database changes. An example of a model of this type is the data model DBCS ADABAS.

In graph models, the state of the database is interpreted by a directed graph, in which vertices—records or segments—depict objects, and marked arcs—connections between objects. Each link is a representative of a certain type of link defined on two types of objects. In a number of data models (hierarchical, network), this attitude is functional.

In binary graph models, objects are treated as follows: elementary objects (scalar values of a certain type—numeric or string) represent the vertices of the graph in the same way as complexly structured values. The latter are constructed from elementary objects by linking them into an appropriate structure through binary associations. The orientation of the graph reflects the fact that the connections between objects are usually asymmetric, and each object in the connection plays a certain role. In a number of—network models, data models on binary associations—such roles are explicitly indicated. Hierarchical and some network models of data limit the allowable type of graph: in a hierarchical model, for example, it is a collection of trees.

Various classes of formal logical data models are based on certain mathematical concepts: algebras, set theory, relation theory, mathematical logic. It should be noted that algebraic and set-theoretic data models are common abstractions and are still not supported by effective implementations of the DBCS. Therefore, such classes of models are of methodological rather than practical interest. According to informational algebra, which is the early work of KODASIL [2], the properties of objects are considered as maps $p_k : E \to V_k$, where E—a set of domain objects, a and V_k—a set of property values p_k.

The object is represented in the model by the ordered value of its properties. Properties $p_1, p_2 \ldots, p_n$ are interpreted as the coordinates of the universal information space $V = V_1 x V_2 x \ldots x V_n$; all types of objects can be represented there. Each point of such a space is identified with a specific object. Values of properties that do not make sense for a specific object have uncertain values. In the information space, segments consisting of points V are distinguished and areas $A \subset V$, representing the arbitrary subsets of points. Above the specified objects, the operations of combining, intersecting, difference, operations of associative sampling, operation of creating new areas by the means of functions defined at points of existing areas can be implemented.

The idea of introducing set-theoretic data models belongs to D. Childs [12]. In set-theoretic models, the state of the subject domain is represented by sets, each of which corresponds to a set of objects of a certain type. In such models, a canonical representation of ordered sets is introduced, allowing one to introduce elements of sets, which in turn are ordered sets. It is easy to imagine, for example, that the elements of a set are tuples of property values of objects. Algebraic operations are defined on sets: unions, intersections, differences, additions, Cartesian products, special operations on sets whose elements are the ordered pairs.

Relational data models are divided into subclasses according to the types of relationships they support:

- n-ary relations on simple domains (normalized relational models);
- n-ary relations on arbitrary—simple and compound domains (hierarchical relational models);
- binary relations (binary relational models).

In binary models, the concepts of two kinds of objects and named binary relations are sufficient. Interconnections in which more than two objects participate are in turn interpreted as objects. Since in binary models there is no explicitly defined concept "object property", the properties of objects are interpreted by means of a binary relation defined on a set of objects, by means of which the values of properties.

In formal-logical models, relations between objects are established by means of tuples of one kind or another, containing the values of the attributes—of the main or secondary keys-relations between which connections are established. For a considerable period, normalized revision data models were developed, in which such relationships were given implicitly, and the semantics of operations on relationships was determined without considering them.

As a result, the maintenance of arbitrary types of interrelationships of objects, reflecting fixed semantic constraints of the subject area, was considered the lot of the application program.

This deficiency was overcome in the so-called semantic relational models, which can rightly be considered the second-generation relational models. In them, relationships are no longer considered in isolation, but in combination. The resulting relationship systems are an effective means of depicting various connections of real-world objects, such as aggregates, sets of interrelated diverse objects, generalizations of generic objects with respect to elementary, primary objects. It is important that the semantics of the operations of changing relations in such models are significantly dependent on the relations between relations. Similar links are set by means of special declarative rules (axioms) that establish semantic restrictions for the system of relations that determine the state of databases.

The models considered above belong to the class of structured data models; however, in intellectual decision-making support systems, recently, much attention has also been paid to unstructured models.

Unstructured data models are conditionally divided into descriptor, descriptor with grammar, models on semantic networks, frame models. Descriptor models are quite common in a number of documentary information retrieval systems (IRS), where they are used in coding search images of documents. In the simplest cases, the content of the document is represented by a list of keywords (descriptors) selected from a dictionary (thesaurus) with a fixed or free composition of words. Each descriptor is a value of the type of a character string representing a word or a word combination of natural language. Descriptors in the list are not related to each other.

In more developed descriptor languages, it becomes possible to introduce various classes of descriptors in the dictionary, assign the role of a descriptor in the list of descriptors representing a document, and establish connections between descriptors in the list. Thus, the introduction of the classification of descriptors allows to specify the <aspect>: <descriptor> pair in the search image of the document, where the aspect is the name of the descriptor class (for example, NAME, YEAR OF BIRTH). The role of the descriptor in the context of the document is usually given by an additional descriptor specifying its value. For example, in the following list of descriptors, the role pointers are specified in parentheses: KEY (WATER), KEY (LOCK), KEY (CIPHER), BOW (GUN), and BOW (PLANT).

Relationships can be established between descriptors and the type of communication in this context is indicated, which results in the text content being represented as a sequence of triads including a communication pointer, a role pointer and a descriptor. This is how the presentation of unstructured data in DBCS BASIS is organized, which makes it possible to establish hierarchical and network structures of relationships between descriptors that characterize a certain object (situation) or represent the content of a document text.

Famous representatives of descriptor languages with grammar are the language of RX-codes and the language SINTOL. In the language of RX codes, between the terms (descriptors) X, binary relations are established through the relation R. As a result, stylized sentences are constructed

$$X_i = R_1 X_1 R_2 X_2 \dots R_n X_n, \tag{6.2}$$

or the triad form of presentation of statements in the form of triplets $X_i R_k X_j$ is used. A careful analysis reveals that the most developed descriptor data models are interconnected with semantic networks- one of the intensively developed ways of representing the states of a subject domain in databases

Semantic networks were developed by a number of authors in relation to knowledge representation systems, systems of visual perception and speech perception, natural language "understanding" systems, artificial intelligence, were also used by advanced information retrieval systems, as well as the basis for building data models in a DBMS. The difference in goals, approaches, and the search for new forms has led to the creation of numerous variants of languages for describing the subject area in the form of a semantic network.

The reason for such a wide application of such ideas is the universality of the conceptual basis used in them: the state of the subject area is depicted as a directed graph with marked vertices and arcs, with the vertices corresponding to elementary objects, and arcs are semantic relations between these objects. The minimum unit of information in a database is an elementary object represented by a scalar value of a particular type (for example, a character string, a numeric value, a logical value). The set of elementary objects is subdivided in the database into subsets categories. Each category has a unique name—aspect. Each elementary object belongs to a specific category of objects. It is assumed that within a certain category all elementary objects are different and are uniquely identified by their presentation. Elementary objects that do not belong to any category are allowed. For consistency, they assume that such objects belong to some special category of name objects, which is an empty sequence of characters. A pair <aspect, scalar value> is called an attribute.

So, the vertices of the semantic network—elementary objects are represented by attributes. The graph arcs are labeled with the names of binary relations that define the type of connection between elementary objects (or the type of association of elementary objects). Symbolically, each arc of a graph is expressed by a triplet, which is a sequence (the name of a relationship, an attribute, an attribute). Character values in the composition of attributes are mnemonic names of objects, or simply natural language words (descriptors). Such meanings are often interpreted as references to a dictionary (thesaurus), through which paradigmatic relations are established between natural language words (synonymy relations, generic relations, part-whole relations, etc.). Elementary objects play a certain role in each association, fixed for this type of binary relation.

An important feature of the representation of a state in the form of a semantic network is its proximity to the level of a natural language. Subgraphs of the semantic network can directly correspond to certain facts expressed in natural language. Semantic networks allow expressing arbitrary semantic relations, connecting individual objects, events, properties of objects into sentences, as well as sentences into texts. Taking into account the multi-level construction of texts in natural language, the semantic network is conveniently interpreted as an ordered set of attributes, triplets and semantic networks, where attributes correspond to isolated vertices of the network, and nested networks are subgraphs corresponding to nested texts or components of objects that have a hierarchical nature.

In various implementations of semantic networks, various types of semantic relations are used:

- linguistic (corresponding roles and interrelations of words in sentences);
- set-theoretic ("to be a subset" relation, "to be an element of a set" relation, equivalence relation);
- logical (implies, identically, equivalent).

When expressing the meaning of texts in natural language, the case grammar of Fillmore is often used [13]. In accordance with this approach, the "deep meaning" of the sentence is revealed when it is divided into a verbal (predicative) phrase and a series of nominal phrases—one for each case. To describe the meaning of sentences, a limited set of cases is considered sufficient, such as an agent, tool, addressee, object, source and purpose of the action, expressed by arcs of the semantic network with the necessary marks.

The development of this approach is the semantic memory of Schenk-Rieger in the MARGIE system [14], in which, along with the binary ones, ternary semantic relations are used, which are used in the analysis and construction of natural language sentences.

Frame models are used to represent data on the subject area in accordance with the theory of frames by Minsky [15]. According to this theory, in the case when a person is confronted with a certain situation, the person extracts from memory a certain structure, called a frame. Several types of information are associated with each frame. This is, for example, information on how to use this frame, what to expect in the next moment, what to do if these expectations are not confirmed.

The stereotypical situation represented as a frame has the form of a semantic network, the "upper levels" of which (non-terminal vertices) are fixed and contain facts that are always true in the intended situation. Terminal vertices are filled with specific data. Each terminal vertex may be associated with the conditions that the data must satisfy in order for such a filling to take place. Typically, such conditions are subframes of a certain type. Frame systems, in turn, are connected by an information retrieval network, which allows you to define a different frame in cases when the proposed frame cannot be brought into line with the existing situation.

To support the frame models a number of experimental languages was created. It can be expected that in its development frame models will have to integrate into themselves the best features of models of both structured and unstructured data.

Samples of the specific implementations of an IMS DBCS with unstructured data are SEARCH-1 and ASPID (descriptor data model), BASIS (descriptor model with grammar), SIZIF (data model on semantic networks).

In practice, the concept of an object-oriented approach [16], based on the following concepts, has recently become widespread in building information systems:

1. An object. Objects have many operations and a state that remembers the action of operations. Object-oriented programs consist of sets of interacting objects that communicate with each other by sending messages to perform operations, presented in the form of methods.

   If, until now, objects were considered as passive elements created by operations of the corresponding data type, and operations were linked directly to the data type, then the object-oriented approach considers an object as an active element having the entire set of operations (methods) of its class. It does not matter that, in the implementation, all operations of the same class will be cut off from the objects and concentrated in the definition of the class.

   Thus, we are talking about changing the programming paradigm, the need to consider a program not as a set of processes that process data, but as a set of active "machines" interacting with each other. The concept of presenting programs in the considered context was not introduced by the authors of object-oriented languages, but was proposed by P. Parnass as a new means of breaking programs into modules.

   Actually, the innovation of the object-oriented approach should be considered a syntactic rule, according to which the object itself is put in the first place, followed by the operation that it launches.
2. The class. Classes are the samples from which the objects with the properties specified in the sample can be created with the Create or New operations. Objects of the same class have the same operations and, consequently, the same behavior. Classes have several interfaces that list operations available to clients through this interface. The class body contains implementations of all operations listed in the interface.

   This class definition gives every reason to consider a class as a synonym for data type. Indeed, a class, like a type, defines a set of objects and contains operations for creating them. A class interface is a data type specification; a class body is a data type implementation.

3. An inheritance. A class can inherit superclass operations, and its operations can be inherited by subclasses.

Thus, from the analysis we can draw the following conclusion: as a methodological basis for the development of data models of information systems that implement a distributed technology for managing information resources, it is advisable to use the technology of object-oriented programming.

6.2 Basic Concepts of the Theory of Fuzzy Systems

The mathematical theory of fuzzy sets and fuzzy logic continue to attract the attention of researchers in the field of intelligent, expert systems, as well as decision support systems. These concepts were first proposed by the American scientist Lotfi Zadeh. The main reason for the emergence of a new theory was the presence of fuzzy and approximate reasoning when a person describes processes, systems, objects [17, 18].

The fuzzy approach to modeling complex systems has received worldwide recognition; more than a decade has passed since the inception of the theory of fuzzy sets. In addition, on this path of development of fuzzy systems, it is customary to allocate several periods. The first period is characterized by the development of the theoretical apparatus of fuzzy sets (L. Zadeh, E. Mamdani). In the second period, the first practical results appear in the field of fuzzy control of complex technical systems. At the same time, attention was paid to the construction of expert systems based on fuzzy logic, the development of fuzzy controllers. Fuzzy expert systems for decision-making support are widely used in medicine and economics. Finally, in the third period, which lasts from the end of the 80s and continues to the present, software packages for building fuzzy expert systems appear, and the fields of application of fuzzy logic are noticeably expanding. It is used in the automotive, aerospace and transport industries, in finance, analysis and management decision-making, and many others [19–21].

The traditional way of representing an element of the set A is to use a characteristic function $\mu_A(x)$ that is equal to 1 if this element belongs to the set A, or equal to 0 otherwise. In fuzzy systems, an element may partially belong to any set. The degree of belonging to the set A, which is a generalization of the characteristic function, is called the membership function $\mu_A(x)$, and $\mu_A(x) \in [0, 1]$. The values of the membership function are rational numbers from the interval [0, 1], where 0 means no membership in the set, and 1 means full membership. The specific value of the membership function is called the degree or coefficient of membership. This degree can be defined explicitly as a functional dependence or discretely—by specifying a finite sequence of $X \in \{X_n\}$ values. In the theory of fuzzy sets, in addition to variables of the digital type, there are linguistic variables with values assigned to them.

6.2.1 Fuzzy Inference Rules

The basic "if—then" derivation rule is called fuzzy implication, taking the form if x is A, then y is B, where A and B are linguistic values identified in a fuzzy way through the corresponding membership functions for variables x and y. The part "x—is A" is called the condition (prerequisite), and "y—is B"—the consequence (conclusion). The implication can be written in abbreviated form $A \rightarrow B$. Fuzzy reasoning is a procedure that allows you to define a conclusion that follows from the set of "if-then" rules. Such a set with N variables can take the form:

$$\text{if}\, x_1 - \text{is}\, A_1\, \text{and}\, x_2 - \text{if}\, A_2\, \text{and} \ldots \text{and}\, x_N - \text{is}\, A_N,\ \text{then}\, y - \text{is}\, B. \tag{6.3}$$

The variables x_1, x_2,..., x_N form the N-dimensional input vector X, constituting the argument of the condition in which A_1, A_2,..., A_N and B denote the values of the corresponding membership coefficient $\mu_A(x_i)$ and $\mu_B(y)$. It should be noted that there are individual membership functions for each variable x_i and separately for y.

The random value of the membership function $\mu_A(x)$, where x is the vector $x = [x_1, x_2,..., x_N]$, related to the condition of implication (the level of activation of the rule) should be subsequently interpreted using the fuzzy operations introduced earlier. Possible interpretation in the form of a logical product of sets or in the form of an algebraic product: interpretation in the form of a logical product:

$$\mu_A(x) = \min_{i=1,\ldots,N}(\mu_A(x_i)), \tag{6.4}$$

algebraic interpretation:

$$\mu_A(x) = \prod_{i=1}^{N} \mu_A(x_i). \tag{6.5}$$

The assignment of a single value of the membership function describing a multi-dimensional condition will be called a prerequisite aggregation. Each implication $A \rightarrow B$, defined by expression (6.3), can also be assigned a single value of the membership function $\mu_{A\rightarrow B}(x, y)$. The most popular interpretations of this function also take the form of a logical or algebraic product:

$$\mu_{A\rightarrow B} = \min\{\mu_A(x), \mu_B(y)\}, \tag{6.6}$$

$$\mu_A(x) = \prod_{i=1}^{N} \mu_A(x_i). \tag{6.7}$$

Elements of the theory of fuzzy sets, the rules of implication and fuzzy reasoning form a system of fuzzy inference. It is possible to distinguish many fuzzy rules used in the system, a database containing descriptions of the membership functions, as well as an output and aggregation mechanism, which is formed by the applicable rules of implication. It should be noted that in the case of the technical implementation, the measured values are the input and output signals, which unambiguously compare the corresponding output values to the input values.

To ensure the interaction of sets of these two types, a fuzzy system with a fuzzifier (a converter of a set of input data into a fuzzy set) is entered at the input and a defuzzifier (a converter of fuzzy sets into a specific value of the output variable) at the output.

The fuzzifier converts the exact set of input data into a fuzzy set, determined using the values of the membership functions, while the defuzzifier solves the inverse problem—it forms a unique decision regarding the value of the output variable based on many fuzzy inferences generated by the fuzzy system executive module. The output of this module can be in the form of M fuzzy sets, which determine the range of variation of the output variable. The defuzzifier converts this range into one specific value, taken as the output of the entire system.

It should be noted that there are also fuzzy inference systems in which the actuator directly generates clear values that are no longer required to be defused (for example, the Takagi-Sugeno-Kang system).

The generalized functional structure of the system can be represented in an extended form, which explicitly demonstrates fuzzy inference rules. Since the application of a set of fuzzy rules is allowed, it also provides for an aggregation block, most often implemented as a logical adder (Mach operator). The described output system is called the Mamdani-Zadeh system. It is very popular in conventional (non-adaptive) fuzzy systems. As a rule, the following operators are present in the Mamdani-Zadeh model:

- logical or arithmetic product operator to determine the resulting activation level, which takes into account all components of the vector x condition;
- logical or arithmetic product operator for determining the value of the membership function for the entire implication $A \rightarrow B$;
- logical sum operator as an aggregator of equivalent results of the implication of many rules.

In the last decade, hybridization of intellectual information processing methods has been observed. Soft computing combines areas such as fuzzy logic, artificial neural networks, knowledge extraction, databases, probabilistic reasoning, evolutionary algorithms, and a number of others. They complement each other and are used in various combinations to create hybrid intelligent systems.

The researchers in the field of databases did not stay aside from the interesting and modern trend. Fuzzy relational algebra and special extensions of a structured language for fuzzy queries are developed. In this area, the European scientists D. Dubois and G. Prade carry out intensive research. Forming a promising direction in modern information processing systems—fuzzy queries to databases (fuzzy queries) [22].

In this context, we can consider two main questions that are most relevant now: how to design, where and in what structures to store fuzzy data of systems of this class.

Solving these problems will open up ways for integrating accumulated colossal amounts of information from relational databases and systems based on fuzzy logic. Thus, the development of methods for storing and processing fuzzy data by means of relational systems is relevant. Special attention should be paid to the justification of the choice of the relational data model scheme for representing the membership functions of linguistic variables.

6.3 Formal Approach to the Choice of Data Model

The choice of formal tools for modeling the subject area (SA) due to the requirements that applies to databases (DB). Since the main purpose of the DB is to reflect the facts of SA, its formalization is determined by the relationship of information objects in both SA and DB. At the same time, it is necessary to ensure uniformity of such views. In addition, the chosen uniformity should provide the possibility of implementing a data description language and a data manipulation language, while an important requirement for languages is their isomorphism when moving from one model to another.

The formal tools being developed are based on the ideas of the connection of calculi of information objects with abstraction and implication operators.

The main step in building a model is to fix categories for SA analysis. First of all, the uniformity of reasoning is determined by specifying data objects and relations between them, then the mathematical spectrum is presented for analysis of various SA, presented in the form of various categories.

Information presented below examines formalized models and methods for describing SA, specifications of user information requirements, and analyzes user information structures.

The considered models and methods provide:

- formalized description of the SA of users in the form of basic components and relations between them;
- formalized description of the specifications of user information requirements in the form of sets of relationships between software information objects;
- analysis of information structures of local and distributed databases.

The analysis of the semantic capabilities of data models, conducted in the works of Schmidt and Swenson [23], Kent [24], Langefors [25], McLeod [26] and other authors initiated the creation of semantic DB models, that allows to define new requirements for the functioning of the IS and software modeling in general.

As for the objects so there are two problems: identification and adequate description of objects [27]. For a detailed study of methods for identifying objects of their syntactic and functional analysis, the reader can refer to [28, 29]. Further research will be aimed at solving the second problem—an adequate representation of objects and determining their semantic properties.

For the description of properties, predicates of the first order are used, which allows one to abstract from the data model of the information system (IS) and to represent the DB as a set of literals. For a complete understanding of the information content of IS it is proposed to explore an extended set of information objects, including both basic elements and calculated.

6.3.1 Data Presentation Model

The set of stored data in the IS will be considered as a randomly structured DB. The DB structure is determined by a finite set of attributes expressing the software properties that associate a value from the set of valid values of this attribute with each object. The sequence of attributes will be referred as DB schema. In this case, the values in each attribute may change, which entails a change in the state of the DB, while the scheme remains unchanged.

In any case, the attribute type is not important for the interpretation of information. In other words, if the type of attributes for each software object is changed, then the new state should represent the same information. For this reason, it is convenient to assume that there is some fixed set of abstract attributes, from which attributes are chosen, that describe a piece of software. We will use symbols of some fixed alphabet as the notation for attribute names, and we will consider the selected notation as a software universe.

To describe the DB, we will use the methods of predicate logic, and when describing software, it is necessary to analyze possible statements that are active in the specified area and the logical relationships that exist between these statements.

By fixing an alphabet, it is possible to describe the relationship between attributes by using formulas. The set of generated formulas will be referred as axioms or rules. The rules express the fact that from a certain combination of data in a certain database, ones can output a set of data included in another database, or that the query will be executed. DB semantics will be called such an interpretation of the set of rules in which each rule is true.

Imagine the information component of the IS as a family of sets:

$$D = \{D_1, D_2, \ldots, D_n\}, \tag{6.8}$$

where $D_i = \{d_1, d_2, \ldots, d_m\}$—a set of valid values. And the relationship scheme is

$$R(A_1, A_2, \ldots, A_n), \tag{6.9}$$

where $A_1, A_2, \ldots, A_n$—a set of attributes.

The correspondence between the attributes (6.9) and the values (6.8) is defined as the mapping of the following form:

$$\Psi : R \to D. \tag{6.10}$$

The mapping (6.10) determines which value from D corresponds to an attribute from a relation R.

Thus, in a simplified form, the structural component of an information system representing a database can be represented as SDB $= \{R, D_2\Psi\}$ [30]. As a means of defining a structural component, declarative specifications, formulas for calculating statements or calculus of first-order predicates can be used. Data objects that satisfy specified conditions constitute a valid DB state.

We will consider the DB as a set of predicates. Unlike arithmetic and logical functions, where the range of values and the range of changes of the arguments by type are the same, that is, homogeneous, the predicates of the range of values of the function are logical, and the range of changes of the arguments is objective. Thus, the predicate is a non-uniform function and can be used to simulate a database.

In predicate logic, an elementary object with a truth value is an atomic formula. An atomic formula consists of the symbolic designation of a predicate and terms acting as this predicate. In general, the predicate can be represented as:

$$p(t_1, t_2, \ldots, t_n), \tag{6.11}$$

where p—predicate, $t_1, t_2, \ldots, t_n$—terms.

The number of terms determines the dimension of the predicate, that is, in this case, the predicate p is n-local. Essentially, a predicate is a function that returns a boolean value of "true" or "false" depending on the value of the term.

Similarly to (6.10), we represent the single-place predicate $p(t)$ as the mapping:

$$\varphi : p \to t. \tag{6.12}$$

The mapping (6.12) establishes what value of t must correspond to the predicate p in order for the formula $p(t)$ to take the value "true". Then expression (6.11) will correspond to a single predicate, and the database can be described as a set of single predicates:

$$R(p_1(t_1), p_2(t_2), \ldots, p_n(t_n)), \tag{6.13}$$

where the predicate $p_i(t_i)(1 \le i \le n)$ takes the value "true" if t_i is a DB value and "false"—otherwise.

Fix an alphabet $\Re$ containing constants, variables and predicates. For a single predicate p, the formula $p(t)$ is called a positive literal l, and the formula $\neg p(t)$ is called a negative literal -l. A base literal is a positive or negative literal that does not contain variables. Thus, expression (6.13) can be written as:

$$R(l_1, l_2, \ldots, l_n). \tag{6.14}$$

Integrity constraints will be expressed by a set of rules

$$L = \{l \leftarrow l_1, l_2, \ldots\ldots, l_m\}, \tag{6.15}$$

where $l, l_1, l_2, \ldots, l_m$—literals ($n \geq 1$).

The two-place logical connection "←" is an implication and can be read as the expression "if $l_1, l_2, \ldots, l_m$ holds, then l is executed". The condition of satisfiability of the integrity constraints is that if all literals $l_1, l_2, \ldots, l_m$ are included in R, then l should also be in R. If this condition is not met, then integrity may be violated.

At the content level, the set R includes software objects, and L are properties that these objects must satisfy. The main condition for the correct functioning of the database is that the DB and the integrity constraints are joint. Consistency is the absence of the same positive and negative literal in R. Rules that define valid values define the semantics of the database.

In the following, the initial state of the DB will be denoted by R, and the state reflecting the DB semantics by S. For example, if $R = \{a, b\}$, and $L = \{\neg b \leftarrow a\}$, then R and L are consistent, and semantics $S = \{a, b, \neg b\}$ incompatible.

By modifying the DB, we mean the operation of adding or deleting a literal, during which the DB remains joint. Adding a literal means that l must be present in the semantics of the modified DB, and deletion means that l must not be present in the semantics of the modified database.

6.3.2 Logical Consequences of the Rules

Some rules are executed in all states in which the rules from L are fulfilled; we will call such rules consequences. Let L^* denote all the consequences of the rules L or the closure of the set L.

The theory of L—rules is based on the fact that in some set R between L—rules there are semantic patterns with which one can derive some rules from others, that is, to draw conclusions about the fulfillment of some rules based on the knowledge that for the set R other rules are fulfilled.

Denote by l_i the set of literals obtained by applying the rules (6.15), and $\{l_j\}i$ are the literals defining l_i. We will present the rules (6.15) in the following form:

$$L = \left\{l_i \leftarrow \left\{l_j\right\}i\right\}, (1 \leq i \leq n, 1 \leq j \leq m). \tag{6.16}$$

Let's say that the set S satisfies the rules (6.16) if all the elements l_i are in S, that is, $l_i \subseteq S$. Consider two extremal types of rules:

$$\emptyset \leftarrow \{l_j\}I, \tag{6.17}$$

$$l_i \leftarrow \emptyset. \tag{6.18}$$

Rule (6.17) trivially satisfies any L. Rule (6.18) satisfies such S, in which all elements $l_i \in S$. In the future, such rules will not be considered.

For the set S at any moment there is a certain set of rules L with which this set satisfies. Let two sets S_1 and S_2 be given and let L satisfy S_1 and not satisfy S_2. It is necessary to identify all valid rules from L (in the notation L') that satisfy S_1 and S_2 (or show the absence of such a set of rules).

To find L', semantic knowledge of S_1 is needed. This knowledge is determined by the set L, since the rules are primary in relation to the DB and essentially set restrictions on the DB objects. The closure of the L^* rules applicable to the set R is finite, since there are a finite number of subsets of the set R. Thus, it is always possible to find all the rules of L that satisfy S_1 by going through all possible rules.

However, this approach is time consuming. If some rules $L \in L'$ are known, then other rules can be derived. The set of rules L implies the rule $l_i \leftarrow \{l_j\}i$ if all objects from S that satisfy all the rules from L also satisfy the rules $l_i \leftarrow \{l_j\}i$. The rule definition is a procedure that states that if S satisfies certain rules, then it must satisfy some other rules that are not included in L.

Define inference rules.

1. Reflexivity. If $l \in R$, then $l \leftarrow l$. A set R always has at least one element, therefore $l \leftarrow l$ has its place in R.
2. Additivity. If $l_1 \leftarrow l$ and $l_2 \leftarrow l$, so $l_1, l_2 \leftarrow l$. Since the set of rules determines the semantics of S, that is, the properties of the l_1 literals, then from $l_1 \leftarrow l$ and $l_2 \leftarrow l$, then l_1 implies that l_1 and l_2 have the same properties defined by the literal l. This implies that $l_1, l_2 \leftarrow l$ satisfies the semantics S.
3. Transitivity. If $l \leftarrow l_1$ and $l_1 \leftarrow l_2$, then $l \leftarrow l_2$. The generated literals are the result of applying the rules to defining literals and the subsequent application of these rules to generated literals. Therefore, the literal 1 is the result of applying the rules that determine the sequential dependence of the literals. Therefore, $l \leftarrow l_2$ satisfies S.
4. Replenishment. If $l \leftarrow l_1$, then $l \leftarrow l_1, l_2$. Since to generate a literal l, it suffices to apply the specified operations to the base literal l_1, the rule of the form $l \leftarrow l_1, l_2$, including the additional base literal l_2, also holds for generating l, although it is redundant.
5. Pseudo-transitivity. If $l_1 \leftarrow l_2, l \leftarrow l_1, l_3$, then $l \leftarrow l_2, l_3$. The proof of this axiom follows from the axioms of reflexivity, completion, additivity, and transitivity. If $l \leftarrow l_1, l_3$ and $l_1 \leftarrow l_2$, then due to the axiom of reflexivity we have $l_3 \leftarrow l_3$. According to the axiom of completion, $l_1 \leftarrow l_2, l_3$ and $l_3 \leftarrow l_2, l_3$. By the additivity axiom, we have $l_1, l_3 \leftarrow l_2, l_3$. Applying the transitivity axiom, we get $l \leftarrow l_2, l_3$.
6. Projectivity. If $l_1, l_2 \leftarrow l$, then $l_1 \leftarrow l$ and $l_2 \leftarrow l$. Axiom 6 is somewhat inverse to Axiom 2. Since the literals l_1 and l_2 have the same properties defined by the literal l and this rule satisfies S, in the rules $l_1 \leftarrow l$ and $l_2 \leftarrow l$ literals l_1 and l_2 will also have the same properties that satisfy S.

It is obvious that $L \subseteq L^*$ and that $L^{**} = L^*$. Two sets S_1 and S_2 are logically equivalent (in the notation $S_1 \equiv S_2$) if $L_1^* = L_2^*$. But the construction of L^* corresponds to the enumeration of all subsets of the set L, which takes exponential time [31].

One of the ways to reduce the time to check whether a literal is entered in S is to construct the closure of a set R in concern to the rules of L. By the closure of the set R, there is a set of literals R^* for which $l^* \leftarrow l \in L^*$ and there is not a single literal from R that depends from l and did not belong to l^*. It is known that obtaining a closure is a time-consuming task for analytical calculations. Therefore, we will consider an algorithm that allows automating this process for rules that do not have interdependent and cyclic properties.

Closure calculation R: Entrance. $R, L = \{l_i \leftarrow \{l_j\}i\}$. Exit. R^*.

In general, the algorithm for calculating the closure R can be represented as follows. We will use the additional variable M to save the set of literals. Let $M := R$. By consistently revising the right sides of the rules of l_i, we check the condition $l_i \subseteq R$. If the condition $\{l_j\}i \subseteq R$ performed, then we modify $M := R \cup \{l_j\}i$, exclude $l_i \leftarrow \{l_j\}i$ from L and continue to revise the right parts of the rules, starting with the first rule in the modified set L. If no rules from the right side $\{l_j\}i$, for which $\{l_j\}i \subseteq R$ are found, algorithm is finished.

In order to make sure that the semantics of the two sets R_1 and R_2 are equivalent, it suffices to construct a closure for one of them and check the occurrence of each element of the second set in the closure. Thus, the condition $S_1 \equiv S_2$ holds if $R_1 \subseteq R_2^*$ and $R_2 \subseteq R_1^*$.

Findings. The time complexity of the considered algorithms for calculating closures depends on the size of the input set L. Thus, a smaller set of rules ensures faster execution of these algorithms.

The analysis of requirements for information systems shows that, as a rule, L contains a fairly large number of rules that significantly slow down the operation of the algorithm. This situation gives rise to the development of more efficient methods for determining equivalent databases. In particular, select a set of basic rules, eliminating the trivial and redundant rules and thus reducing the size L.

The obtained results confirm the adequacy of the proposed method and algorithm and can be used to improve access and data processing technologies, as well as the implementation of database representation methods for outputting query results by means of relational data management systems. Further research in this area should focus on the organization of queries in database systems with a heterogeneous structure, and the development of methods for constructing and maintaining queries for heterogeneous information systems and distributed databases.

6.4 Relational Data Model

Consider the classic approach to building a relational relationship and highlight the basic properties of the relationship when expanding the set of domains.

The main structural component of the data in the relational data model (RMD) is the n-ary relation, which is a subset of the tuples of the Cartesian product of domains, that is, the set of values of the data elements. For given finite sets $D_1, \ldots, D_n$ (not necessarily different in type), the Cartesian product $D_1 \times \ldots \times D_n$ is the set of products of the form $d_1, \ldots, d_n$, where $d_1 \in D_1, \ldots, d_n \in D_n$. A relation R defined on sets $D_1, \ldots, D_n$ is a subset of a product (Cartesian product) $D_1 \times \ldots \times D_n$, that is, $R \subseteq D_1 \times \ldots \times D_n$.

The set $D = \{D_1, \ldots, D_n\}$ is called domain. Domains are single-type semantically single-valued (identical in meaning) values of data elements. The elements of a Cartesian product $d_1, \ldots, d_n$ are called tuples, the number n determines the degree of the relation, the number of tuples determines the power of the relation.

A relation R scheme is an expression $S(A_1, \ldots, A_n)$ in which all attributes are different. In this case, the relationship instance $R(S)$ is defined as a subset of the Cartesian product of domains $r_i \subseteq \rho(a_i) \times \cdots \times \rho(a_n)$.

The instance of the relationship with the scheme R_i will be denoted as $R_i(r_i)$. Note that the permutation of attributes in the scheme does not generate a new state of the DB. Thus, the set of attributes $\{A_1, \ldots, A_n\}$ defines the type of relationship and determines its properties. The DB schema will be denoted as a set of relationship schemes $U = \{R_1, \ldots, R_n\}$, where all $R_i \in R$ and all R_i are different. Accordingly, the database instance will be denoted as a set of instances of relations $U = \{r_1, \ldots, r_n\}$.

Conceptually, a relational database is an information domain model (DM), such that each instance corresponds to a certain DM state at a certain point in time. Each state is modeled by an ordered set of data element values corresponding to the values of properties of DM objects. An object of a certain type corresponds to a relation tuple. Objects have a specific set of properties that are defined by a relational schema, and properties have certain sets of possible values that are given by the mapping ρ.

Access to relational data is carried out using relational algebra or equivalent relational calculus. Relational algebra is a set of operators that use relations as arguments and return relations as a result. Each relationship must have a unique name within the database. Traditionally, eight relational operators are defined that are combined into two groups. Set-theoretic operators: union, intersection, subtraction, Cartesian product. Special relational operators: selection, projection, connection, division. Not all of them are independent, i.e. some of these operators can be expressed through other relational operators.

6.4.1 The Main Properties of the Extension of Relational Algebra

Consider one of the approaches to the expansion of relational algebra (ERA) operations on fuzzification relations [32]. Through $\widetilde{R}$ we denote the set of all fuzzification relations and through the R set of ordinary relational (natural) relations, schemas of which are selected from some fixed database schema. For each relationship $\tilde{r} \in \widetilde{R}$, we define a function $N(\tilde{r}) = \{r|r - relationalattitude\}$. This feature will be used to explore various extensions of the scope of relational operations on $\widetilde{R}$. If, for example, it is necessary to define a join operation on a relation with fuzzy data, then for an extended join operation $\tilde{\bowtie}$, equality (6.1) for $\tilde{r}, \tilde{s} \in \widetilde{R}$ should be hold.

$$N(\tilde{r} \tilde{\bowtie} \tilde{s}) = N(\tilde{r}) \bowtie N(\tilde{s}). \tag{6.19}$$

Formulate a number of conditions necessary for the strict performance of the expanded operation.

Statement 6.1. Let an operation ξ defined on R and an operation $\tilde{\xi}$ on $\widetilde{R}$ is given. The operation $\tilde{\xi}$ is a natural extension ξ when the following conditions are met:

- if ξ and $\tilde{\xi}$ are unary operations, then $\xi(r) = \xi(r)$ for each $r \in R$ for which $\tilde{\xi}(r)$ is defined;
- if ξ and $\tilde{\xi}$ are binary operations, then $r \xi s = r\tilde{\xi}s$ for each $r, s \in R$ for which $r \xi s$ is defined.

Statement 6.2. Let an operation ξ on R and an operation $\tilde{\xi}$ on $\widetilde{R}$ is given. The operation $\tilde{\xi}$ is an exact extension ξ about the function N under the following conditions:

- if ξ and $\tilde{\xi}$ are unary operations, then $Nl(\tilde{\xi}(\tilde{r})) = \xi(N(\tilde{r}))$ for each $\tilde{r} \in \widetilde{R}$;
- if ξ and $\tilde{\xi}$ are binary operations, then $N(\tilde{r}\tilde{\xi}\tilde{s}) = N(\tilde{r})\xi\, N(s)$ for each $\tilde{r},\tilde{s} \in \tilde{R}$.

Statement 6.3. Let an operation ξ on R and an operation $\tilde{\xi}$ on $\widetilde{R}$ is given. The operation $\tilde{\xi}$ is adequate for ξ about the function N under the following conditions:

- if ξ and $\tilde{\xi}$ are unary operations, then $N(\tilde{\xi}(\tilde{r})) \supseteq \xi(\mathrm{N}(\tilde{r}))$ for each $\tilde{r} \in \widetilde{R}$;
- if ξ and $\tilde{\xi}$ are binary operations, then $N(\tilde{r}\tilde{\xi}\tilde{s}) \supseteq N(\tilde{r})\xi\mathrm{N}(\tilde{s})$ for all $\tilde{r},\tilde{s} \in \widetilde{R}$.

Statement 6.4. Let an operation ξ on R and an operation $\tilde{\xi}$ on $\widetilde{R}$ is given. The operation $\tilde{\xi}$ is limited for ξ about the function N under the following conditions:

- if ξ and $\tilde{\xi}$ are unary operations, then for each $\tilde{r} \in \widetilde{R}$ there is no such $\tilde{s} \in \widetilde{R}$, that $N(\tilde{\xi}(\tilde{r})) \supset N(\tilde{s}) \supseteq \xi(N(\tilde{r}))$;
- if ξ and $\tilde{\xi}$ are binary operations, then for each $\tilde{r},\tilde{q} \in \widetilde{R}$ there is no such $\tilde{s} \in \tilde{R}$, that $N(\tilde{r}\tilde{\xi}\tilde{q}) \supset N(\tilde{s}) \supseteq N(\tilde{r})\xi\, N(\tilde{q})$.

Obviously, if the operation $\tilde{\xi}$ is an exact extension ξ, then the operation $\tilde{\xi}$ is adequate and limited to ξ. In the cases where exact expansion is not possible, we will use adequate and limited extensions. It is also necessary to take into account that extended operations retain the properties of ordinary relational operations, such as commutativity and associativity.

6.4.2 *Operational Specification of Extensions of Relational Algebra for Fuzzification Relations*

Consider extensions of some relational operations for fuzzification relations.

The set-theoretic operations considered here are an extension of operations on fuzzy sets that are applied to the original relational schemes $\widetilde{R}$, as a result of which new relations are obtained, defined on the same data. Thus, set-theoretic operations will be considered as applied to the schemes (6.20).

$$\widetilde{R} = \{\mu_{\widetilde{R}}(x)/x\} = \widetilde{R}[A(a_i)], \tag{6.20}$$

where $\mu_{\widetilde{R}}(x)$ is a membership function and $A(a_i) = \{a_i | i = \overline{1, n}\}, x = \text{dom}(A(a_i))$.

The main operations of ERA $C_{\widetilde{R}}(\widetilde{R}[\text{A}(\text{a}_\text{i})])$, as a rule, interact not with one relationship scheme $\widetilde{R}$, but with two, that is, they are binary. If $A(a_i)$ and $B(b_i)$ bearers of relationship schemes, it can be defined by an expression (6.21):

$$C_{\widetilde{R}}\left(\widetilde{R}A[A(a_i)]\right) = \widetilde{R}X[A(a_i \cup b_i)], \tag{6.21}$$

where $a_i \in A, b_i \in B, \widetilde{R}\ X$—relation defined on a fuzzy domain dom $(A(a_i) \cup B(b_i))$.

The membership function of the set $\tilde{R}\ X$ is defined as an expression (6.22).

$$\mu_{\widetilde{R}_X}(u, v) = \mu_{\widetilde{R}_A}(u)|\forall(u \in \text{dom}(A(a_i), (u, v) \in \text{dom}(a_i \cup b_i))). \tag{6.22}$$

The operation of projection relationship of fuzzification is $\widetilde{R}\ A[A(a_i) \cup B(b_i)]$, where $\widetilde{R}\ A$ is a ratio defined on fuzzy domains $(u, v) \in \text{dom}(A(a_i)) \times \text{dom}(B(b_i))$, on a fuzzy set $\text{dom}(A(a_i))$, is defined by an expression (6.23).

$$\pi_{B_i}\left(\widetilde{R}A[A(a_i \cup b_i)] = \widetilde{R}X[A(a_i)]\right), \tag{6.23}$$

where $\widetilde{R}\ X$ is a fuzzy domain relation $\text{dom}(A(a_i))$ and a membership function (6.24):

$$\begin{aligned}\mu_{\widetilde{R}_X}(u) = \max\{&\mu_{\widetilde{R}_A}(u, v)|u \in \text{dom}(A(a_i)), v \in \text{dom}(B(b_i)),\\ &(u, v) \in (\text{dom}(A(a_i)) \times \text{dom}(B(b_i)))\}.\end{aligned} \tag{6.24}$$

Operation of fuzzing relations product $\widetilde{R}\ A[A(a_i)]$ and $\widetilde{R}\ B[B(b_i)]$, where $A(a_i) \cap B(b_i) = \emptyset$ is represented by expression (6.25):

$$\widetilde{R}A[A(a_i)] \times \widetilde{R}B[B(b_i)] = C_{b_i}\left(\widetilde{R}A[A(a_i)]\right) \cap C_{a_i}\left(\widetilde{R}B[B(b_i)]\right). \tag{6.25}$$

The operation of the sum of the same fuzzification relations is determined by the combination in the form of (6.26):

$$\widetilde{R}A[A(a_i)] + \widetilde{R}B[B(b_i)] = C_{b_i}\left(\widetilde{R}A[A(a_i)]\right) \cup C_{a_i}\left(\widetilde{R}B[B(b_i)]\right). \tag{6.26}$$

The selection operation allows to form a new fuzzification relation $\widetilde{R}\ X[B(b_i)]$ in regards to $\widetilde{R}\ A[A(a_i)]$ based on checking the condition of θ, where θ is a formula constructed using logical comparison operations ($<,>$, $=$, $\leq,\geq$, $\neq$) and logical connectives $\neg$ (no), $\wedge$ (and), $\vee$ (or).

As a result of the selection operation from a variety of fuzzy domains dom($B(b_i)$) that satisfy the constraint (6.27) are selected for which condition θ is true.

$$\left\{d_1^1, d_2^1, \ldots, d_n^1, d_1^2, d_2^2, \ldots, d_m^2, \ldots, d_1^p, d_2^p, \ldots, d_k^p\right\} \subset \mathrm{dom}(A(a_i), \tag{6.27}$$

At the same time, the membership functions of these domains do not change. The general expression of the selection operation can be represented as a formula (6.28):

$$\sigma\,\theta\left(\widetilde{R}A[A(a_i)]\right) = \widetilde{R}X[B(b_i)]. \tag{6.28}$$

Operation of comparison of two fuzzification relations $\widetilde{R}\ A[A(a_i)]$ and $\widetilde{R}\ B[B(b_i)]$ on the operation θ is true if condition is met $A(a_i)\,\theta\,B(b_i)$ and the following expression is correct (6.29):

$$\forall u\ \in\ \mathrm{dom}\big(A(a_i)|\mu_{\widetilde{R}_A}(u) = \mu_{\widetilde{R}_B}(u). \tag{6.29}$$

In general, the comparison operation is represented by the following expression (6.30):

$$\widetilde{R}A[A(a_i)]\theta\,\widetilde{R}B[B(b_i)], \tag{6.30}$$

where $\theta = \{<,>, =, \leq,\geq, \neq\}$.

The membership operation consists in checking for the inclusion (content) of the set $\widetilde{R}\ A[A(a_i)]$ on the set $\widetilde{R}\ B[B(b_i)]$. The operation determines the true result if $A(a_i) = B(b_i)$ and the expression (6.31) is true:

$$\forall u\ \in\ \mathrm{dom}\big(A(a_i)|\mu_{\widetilde{R}_A}(u) \leq \mu_{\widetilde{R}_B}(u). \tag{6.31}$$

In general, the operation of belonging can be represented by the following expression (6.32):

$$\widetilde{R}A[A(a_i)] \subseteq \widetilde{R}B[B(b_i)]. \tag{6.32}$$

In addition to operations that advise relational algebra, traditional set-theoretic operations can be applied to fuzzification relations: addition, union and intersection.

The addition operation on the fuzz relation $\widetilde{R}$ for a fuzzy domain allows to obtain a new relation when performing constraints of the form (6.33):

$$\neg\widetilde{R} = \bigcup_{u\in\widetilde{R}} (1 - \mu_{\widetilde{R}}(u)/u). \tag{6.33}$$

The operation of combining two fuzzification relations $\widetilde{R}_A$ and $\widetilde{R}_B$ over fuzzy domains allows us to form a relation obtained in accordance with the expression (6.34):

$$\widetilde{R}_A \cup \widetilde{R}_B = \bigcup_{u \in \widetilde{R}} \max[(\mu_{\widetilde{R}_A}(u), \mu_{\widetilde{R}_B}(u)]/u. \tag{6.34}$$

The operation of intersection of two fuzzification relations $\widetilde{R}_A$ and $\widetilde{R}_B$ across fuzzy domains is mapped to the relation obtained in accordance with the expression (6.35):

$$\widetilde{R}_A \cap \widetilde{R}_B = \bigcup_{u \in \widetilde{R}} \min[(\mu_{\widetilde{R}_A}(u), \mu_{\widetilde{R}_B}(u)]/u. \tag{6.35}$$

Thus, the conditions for expanding the basic operations of relational algebra for a class of systems designed on the basis of fuzzy logic were obtained. Using the results obtained, it is possible to effectively apply the advantages of database management systems based on the classical relational model for intelligent systems with elements of fuzzy logic.

6.4.3 *Fuzzification Relationship Structure*

For data analysis tasks, we define an additional type of relationship that determines the belonging of existing data to some given numerical segment, which determines the characteristic of the information object under consideration.

Any line on the coordinate plane can be represented as a binary relation, where Dom *R* is described by abscissa axis values, and Im *R* is described by ordinate axis values. In the task of fuzzification, the diagram contains three indicators that must be considered when forming the relationship.

By a fuzzy variable, we mean the set (N, X, Y), where N is the name of the variable, X is the area of reasoning, Y is the fuzzy set on X. Using this definition, we define three domains corresponding to the elements of the variable.

Let $N = \{n_1, \ldots, n_m\}$, $Y = \{0, 0.1, \ldots, 1\}$, $X = \{x_0, \ldots, x_k\}$. The values of X and Y correspond to the chosen scale of discretization of the coordinate axes and describe the area of belonging to the parameter N. For each parameter N, we will sample the data on the values of X and build on the basis of real data on the number of dissertations defended (Fig. 6.2).

For the considered case, we define the corresponding domains to represent the fuzzy variable:

- $D_1 = \{n_1, n_2, n_3, \ldots, n_m\}$;
- $D_2 = \{x_0, \ldots, x_8, x_9, \ldots, x_{14}, \ldots, x_i, \ldots, x_{20}, \ldots, x_{30}, \ldots, x_m\}$;
- $D_3 = \{0, 0.1, \ldots, 1\}$.

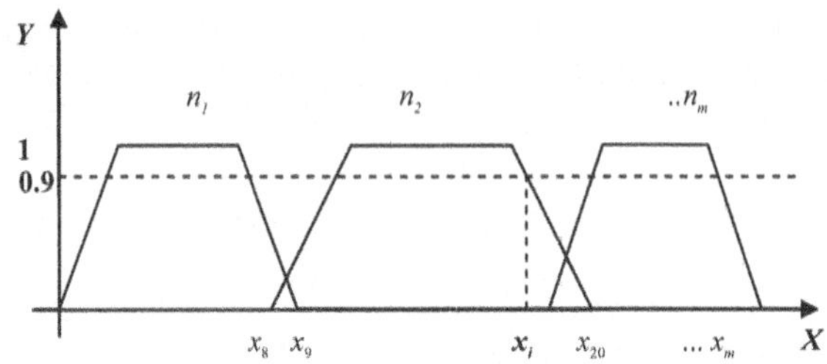

Fig. 6.2 Example of linguistic variable representation

Fig. 6.3 Data fragment for the selected indicators

A_1	A_2	A_3
n_2	x_{14}	0
n_3	x_{14}	1
n_2	x_i	0.9
n_3	x_i	0.9
n_2	x_{20}	1
n_3	x_{20}	0
...	...	...

We assign a set of domain names and construct a mapping. For the set of names $A = \{A_1, A_2, A_3\}$ the map ρ: ($A_1 \to D_1$; $A_2 \to D_2$; $A_3 \to D_3$) defines the set of attributes $A = \{A_1, A_2, A_3\}$ and the corresponding relational scheme $S(A_1, A_2, A_3)$. The structure of the ratio *Rf* in a tabular form will be called the fuzzification relation.

Possible intersections of fuzzification graphs determine the fact that all values can be repeated relative to each other. For example, consider the relation fragment for the indicators $\{x_{14}, x_i, x_{20}\}$ is presented in Fig. 6.3.

Thus, we can speak of a relation that includes the full set of tuples of the Cartesian product of domains $D_1 \times D_2 \times D_3$ (the complete Cartesian product, in this case, is determined by the finiteness of the coordinate system values). Based on this fact, we can conclude that the key of the relationship will be the set of all attributes $K = \{A_1, A_2, A_3\}$.

It is obvious that the informativeness of tuples is determined by the values of the linguistic variable on the graph of fuzzification functions. On the basis of the problem under consideration, it is necessary to take into account one more indicator—the set of sample values from the database for which the graph is plotted. That is, it is necessary to establish a connection between the domain of the *Rf* ratio, which reflects the values of the abscissa axis and the domain of the database, containing fuzzification parameter values.

Thus, the next task that needs to be addressed is related to the integration of the *Rf* relationship with the database storing the main indicators.

6.4.4 Operational Specification on Relationships for Mamdani-Zade Output Systems

The General functional structure of the system output Mamdani-Zadeh (Fig. 6.4) can be represented in the extended form which permits the use of a set of fuzzy rules and provides for the aggregation of data in a logical adder.

In problems of aggregation of inference rules when there are several fuzzification tables, logical summation can be used to aggregate fuzzy rules with regards to specific variables x_i, which form the vector X, and at the level of $A_2 \rightarrow A_3$ implication for single rules [33–35]. Aggregation is carried out using the joining of tables, in which a clear value of y is matched with vector values X.

Using the notation introduced above, we represent the set of fuzzification relations for each membership function as $\Re f = \left\{R_1^f, \ldots, R_p^f\right\}$. Logical summation of fuzzy values is reduced to combining fuzzification relations with the DB U, which can be implemented in two ways.

It follows from the symmetry property that the connection operator is commutative and associative. So the identity will be fair (6.36):

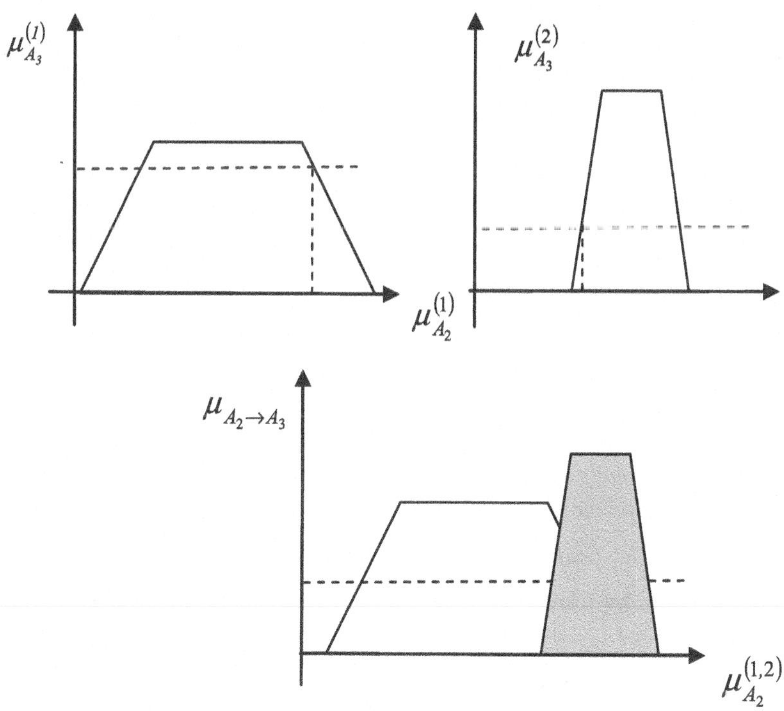

Fig. 6.4 Mamdani-Zade output system

$$\left(R_1^f \rhd\lhd \ldots \rhd\lhd R_p^f\right) \rhd\lhd U = \left(R_1^f \rhd\lhd U\right) \rhd\lhd \ldots \rhd\lhd \left(R_p^f \rhd\lhd U\right). \qquad (6.36)$$

We introduce some notation for multiple connections. Let $\left\{R_1^f, \ldots, R_p^f\right\}$ be the set of fuzzification relations and let $S = R_1^f \cup \ldots \cup R_p^f$ be the union of the schemes of fuzzification relations. Further, let $t_1, \ldots, t_p$ be a sequence of tuples in which $t_i \in R_i^f$, $1 \leq i \leq p$. We say that tuples $t_1, \ldots, t_p$ are connectable on S if there is a $t \in S$ tuple such that $t_i = t(R_i^f)$, $1 \leq i \leq p$. The tuple t is the result of connecting $t_1, \ldots, t_p$ tuples on S.

It follows from the definition of a connection that the tuple t must belong $R_1^f \rhd\lhd \ldots \rhd\lhd R_p^f$. Conversely, if it is a tuple $R_1^f \rhd\lhd \ldots \rhd\lhd R_p^f$, then there must exist tuples $t_1, \ldots, t_p$ in $R_1^f, \ldots, R_p^f$, respectively, that $t_1, \ldots, t_p$ are connectable on S with the result t. Consequently, $R_1^f \rhd\lhd \ldots \rhd\lhd R_p^f$ consists of those tuples t that are the result of connecting two tuples $t_1, \ldots, t_p$ connected on S.

Using the property of associativity of a join operation, one can prove the following statement.

Statement. A join relation $R_1^f \rhd\lhd \ldots \rhd\lhd R_p^f$ consists of all tuples that are the result of connecting the $t_1, \ldots, t_p$ tuples connected in S, correspondingly, belonging to $R_1^f, \ldots, R_p^f$ [32].

In the considered Mamdani-Zadeh output problem, not every tuple of each relation can enter a connection. Fuzzification relations $R_1^f, \ldots, R_p^f$ will be called fully connected if each tuple in each relation is a member of some list of connected tuples on S.

Join and projection operations in general form are not mutually inverse, although they form additional functions. Let the join operation $R_1^f \rhd\lhd R_2^f$ with the result circuit $S_1\left(R_1^f\right) \cup S_2\left(R_2^f\right)$ is performed. Take $r_1 = \pi_{S_1}\left(R_1^f \rhd\lhd R_2^f\right)$. For any tuple $t \in R_1^f \rhd\lhd R_2^f$ is turned on and the projection will be a tuple of R_1^f, and $r_1 = \left\{\mathrm{t}\left(R_1^f\right) | \mathrm{t} \in R_2^f\right\}$. Inclusion becomes an equality when there is a tuple for each tuple $t_{R_2^f} \in R_2^f$ with $t_{R_1^f}\left(S_1\left(R_1^f\right) \cap S_2\left(R_2^f\right)\right) = t_{R_2^f}\left(S_1\left(R_1^f\right) \cap S_2\left(R_2^f\right)\right)$.

If we say that $r_2 = \pi_{S_2}\left(R_1^f \rhd\lhd R_2^f\right)$, then the conditions $r_1 = R_1^f$ and $r_2 = R_2^f$ mean the same as the condition R_1^f and R_2^f and are completely connected. Obviously, this result can be interpreted into more than two fuzzification relations.

Let $r_{11} = \pi_{S_1}\left(\left(R_1^f \rhd\lhd R_2^f\right)'\right)$ and $r_{22} = \pi_{S_2}\left(\left(R_1^f \rhd\lhd R_2^f\right)'\right)$, and there is $r' = r_{11} \rhd\lhd t_{22}$. In order to receive r', it is necessary to apply the projection-join procedure twice to $(R_1^f \rhd\lhd R_2^f)$. Let $T = S_1\left(R_1^f\right) \cap S_2\left(R_2^f\right)$, then $\pi_T\left(R_1^f\right) = \pi_T\left(\pi_{R_1^f}\left(R_2^f\right)\right) = \pi_T\left(R_2^f\right) = \pi_T\left(\pi_{R_2^f}\left(R_2^f\right)\right)$. Accordingly, R_1^f and R_2^f fully interconnected, as for any tuple $t_{R_1^f}$ at R_1^f there must be such a tuple $t_{R_2^f}$ at R_2^f, that $t_{R_1^f}(T) = t_{R_2^f}(T)$.

Thus, from the relation $t_{R_1^f}\left(S_1\left(R_1^f\right) \cap S_2\left(R_2^f\right)\right) = t_{R_2^f}\left(S_1\left(R_1^f\right) \cap S_2\left(R_2^f\right)\right)$ follows that r' and $(R_1^f \rhd \lhd R_2^f)$ must match.

From this, we can conclude that the projection-connection procedure is idempotent. This, in turn, determines the possibility of combining fuzzification relations for more than one membership function for given parameters in the Mamdani-Zadeh output system.

6.4.5 *The Integration Model of the Relationship of Fuzziification and Relational Data Structures*

Consider the problem in general. Let DB $U(R_1,\ldots, R_n)$ is the database storing the main data, $Rf(A_1, A_2, A_3)$ is the fuzzification relation. The task will make sense if in the DB U there is a parameter relative to the one for which fuzzification is performed.

In order to organize collaboration with the U and R databases, we formalize the integration procedure, based on stepwise normalization. The database structure U is obtained on the basis of the functional dependences $F = \{M_i \rightarrow N_i\}$, where $M_i, N_i \in U$. We single out one relationship, which includes an attribute with fuzzification parameters, and denote it as $W \rightarrow V$, and W and V can be sets. The ratio Rf contains one dependence of the form $F' = \{A_1, A_2, A_3 \rightarrow A_1, A_2, A_3\}$. Based on the axioms of the conclusion, we can get an equivalent set $F' = \{A_1, A_2, A_3 \rightarrow A_1; A_1, A_2, A_3 \rightarrow A_2; A_1, A_2, A_3 \rightarrow A_3\}$.

Let the fuzzification parameter correspond to the attribute A_2, then in order to determine the type of relationship, it is necessary to obtain the set $F = F \cup F'$ and consider two cases that affect the normalization rules.

$A_2 \in W$—search for incomplete dependencies: if the functional dependences $\xi \rightarrow \zeta$ and $\omega \rightarrow \zeta$, are fulfilled, moreover, $\omega \subseteq \xi$, then the dependence $\omega \rightarrow \zeta$ is incomplete.

$A_2 \in V$—search for transitively dependent elements: if the functional dependencies $\xi \rightarrow \omega$ and $\omega \rightarrow \zeta$, are fulfilled, then the element ζ is transitively dependent.

The existence of such dependencies will allow you to perform the correct decomposition and establish a connection between the database U and Rf.

If $A_2 = W$ or $A_2 = V$, then the decomposition process leads to a second or third normal form. If the equalities are not fulfilled, then it is impossible to organize the support of uniqueness of connected data, since the association between the relations will correspond to the type of "many-to-many".

As a rule, in practice, the conditions of equality are not fulfilled and for normalization it is necessary to single out the basis F and repeat the decomposition procedure. Taking into account the fact that the structure of the DB should not change, it is necessary to connect the fuzzification relation Rf and the DB U without restructuring the data schema. Using the "entity-relationship" model diagram, we represent Rf and U as entities (Fig. 6.5)

Fig. 6.5 Entity-relationship diagram between DB U and $R f$

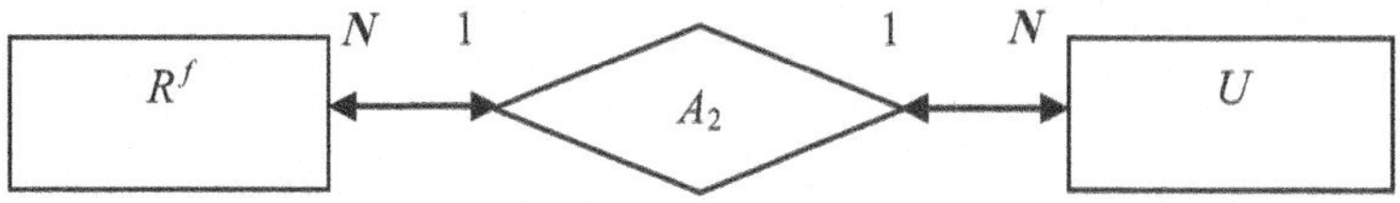

Fig. 6.6 Removing the "many-to-many" relationship

To eliminate the «N:N» relationship, we introduce an additional binding entity that solves the problem of data integrity by defining new types of connections [36]. Entity connection will contain one attribute—a binding for $R f$ and U, and for objective reasons, it will be key. Figure 6.6 shows a diagram of DB integration and fuzzification relations with the support for uniqueness of communication.

On the basis of the software conceptual scheme description, it is clear that for the correct connection of $R f$ and U it is necessary to build an intermediate table. This approach ensures data consistency for any type of fuzzification parameters.

We show that for the problem in question the results are quite correct when performing the connection of relations with an association of the type «N:N». Possible values of the attribute $A_1 \in U$ can be repeated as many times as this value crosses the boundaries of the fuzzification diagram along the ordinate axis. That is, each value of the A_1 attribute corresponds to a string of unique data. If A_1 is not a key, and the values are repeated, then, by the definition of a set, there must be at least one distinct value in the string. In terms of the problem to be solved, it is necessary to analyze all such strings. Attribute A_1 of the relation $R f$ also repeats the values that need to be analyzed, and in all possible combinations.

Thus, in general, to analyze the data accumulated in relational databases, it is enough to build a fuzzified relation and establish a connection with the attribute (attributes), the values of which require the corresponding analysis.

6.5 Conclusion

In the chapter, the original method for implementing fuzzy models using relational systems is considered and theoretically justified. In contrast to the known approaches, the proposed one allows solving the problems of data mining of integrated systems—databases and fuzzy systems.

The novelty of the materials reviewed in the chapter is as follows:

1. An effective method of designing a relational data model scheme for representing the membership functions of linguistic variables and a technology for representing these functions by means of the relational model is proposed; it allows analyzing data using fuzzy queries.
2. A method and technology have been developed for organizing, storing and searching for fuzzy data by means of relational systems. An original information technology for finding fuzzy data using means of relational systems is proposed, which, in contrast to the known approaches, allows using SQL to solve a complex data mining task.
3. An information technology and design methodology for information-analytical systems based on the relational data model have been developed to solve practical problems of intellectual analysis [37, 38].

References

1. Martin J (1980) Database organization in computing systems (trans from English). Mir, Moscow, 662p
2. Ulman J (1983) The basics of the database system. Finance and Statistics, Moscow, 335 p
3. Setlak G (2004) Intelligent decision support systems. Logos, K, 252p
4. Korneev VV, Gareev AF, Vasyutin SV, Reich VV (2001) Database. Intellectual information processing, 2nd ed. Nolidge, Moscow, 496p
5. Sichkarenko VA (2002) SQL 99 database developer guide. DiaSoftUP, Moscow, 816p
6. Rob P, Coronel K (2004) Database systems: design, implementation, and management (Trans from English). BHV-Petersburg, SPb. 1023p
7. Maltsev AI (1970) Algebraic systems. Nauka, Moscow, 392p
8. Filatov V., Rudenko D, Grinyova E (2014) Means of integration of heterogeneous data corporate information and telecommunication systems. In: Proceedings of the 24th international crimean conference microwave and telecommunication technology (CriMiCo-2014), 7–13 sept. 2014. Sevastopol, Ukraine, pp 399–400
9. Date K (2001) Introduction to database systems (Trans from English) Publishing House "Williams", Moscow, 1072p
10. Kharchenko VS, Illiashenko OA, Kovalenko AA, Sklyar VV, Boyarchuk AV (2014) Security informed safety assessment of NPP I&C systems: gap-Imeca technique. In: International conference on nuclear engineering, proceedings, ICONE, vol 3. https://doi.org/10.1115/ICONE22-31175
11. Kuchuk G, Nechausov S, Kharchenko V (2015) Two-stage optimization of resource allocation for hybrid cloud data store. In: International conference on information and digital technologies, IDT-2015, pp 266–271. https://doi.org/10.1109/DT.2015.7222982
12. Cycritis D, Lokhovsky F (1985) Data models (Trans from English). Finance and Statistics, Moscow, 344p
13. Fillmore CJ (1978) The case for case In: Universals in linguistic theory. Holt, Rinehart and Winston Inc., NY, 234p
14. Schenk R (1980) Processing conceptual information. Energy, Moscow, 268p
15. Gavrilova TA, Khoroshevsky VF (2000) Knowledge base of intellectual systems. Peter, SPb, 384p
16. Rumbaugh J, Blaha M (1991) Object-oriented modeling and design. Prentice Hall, NJ 348p
17. Filatov VA, Kasatkina NV (2009) On one approach to the intellectual analysis of relational data. Vestnik of kherson national tehnical university, ed 1(34), pp 157–161

18. Zadeh LA (1974) Basics of a new approach to the analysis of complex systems and decision-making processes. Math Today, Znanie, Moscow, pp. 5–49
19. Zadeh LA (1976) The concept of a linguistic variable and its application to making approximate decisions. Mir, Mocow, 165p
20. Asai K, Vatada D, Iwai S et al (1993) Applied fuzzy systems. In: Terano T, Asai C, Sugeno M (Eds) Mir, Moscow, 368p
21. Osugi S, Saeki Y (Eds) (1990) The acquisition of knowledge. Mir, Moscow, 304p
22. Dubois D, Prades A (1990) Theory of opportunities. Applications to the representation of knowledge in computer science. Radio and communication, Moscow, 288p
23. Schmid, HA, Swenson JR (1975) On the semantics of the relation model. In: Proceeding of ACM SIGMOD international conference management of data, pp. 211–223
24. Kent W (1981) Consequences of assuming a universal relation. ACM trans on database systems, vol3, pp 3–17
25. Langefors B (1974) Information systems. Information processing 74. North-Holland, Amsterdam, pp 937–945
26. McLeod D (1979) The semantic data model. MIT Press
27. Tsalenko MS (1989) Modeling semantics in databases. Nauka, Mocow, 288p (Main ed. ph.-mat.lit.)
28. Filatov V, Semenets V. Methods for synthesis of relational data model in information systems reengineering problems . In: Proceedings of the international scientific-practical conference "Problems of infocommunications. science and technology" (PIC S&T-2018), 9–12 oct. 2018. Kharkiv, Ukraine, pp 247–251
29. Langefors B (1980) Infological models and information user views. Inform Syst 5:17–32
30. Buslik MM (1993) Optimal image of a real database. Monograph, ISDO, K, 84p
31. Schreiver AA (1991) Theory of linear and integer numerical programming. vol 1 (Trans from English). Mir, M, 360p
32. Kasatkina NV, Tanyansky SS, Filatov VA (2009) Methods of storing and processing fuzzy data in relational systems. Automatics Automation Electrotechnical complex and system. Edition 2(24), 80–86
33. Borisov AN, Alekseev AV, Merkur'eva GV (1989) Processing of fuzzy information in decision-making systems. Radio and commun, 304p
34. Filatov V, Radchenko V (2015) Reengineering relational database on analysis functional dependent attribute. In: Proceedings of the X International Scientific and Technology Conference "Computer Science & Information Technologies" (CSIT'2015), 14–17 sept. 2015. Lviv, Ukraine, pp 85–88
35. Motvani R, Ulman J (2005) Introduction to the theory of automats, languages and calculations. Williams, Moscow, 528p
36. Chen P (1995) The "entity-relationship" model as a step towards a unified view of data. SUBD. 3:137–158
37. Zaitseva E, Levashenko V, Kvassay M, Barach P (2017) Healthcare system reliability analysis addressing uncertain and ambiguous data. In: Proceeding of the International Conference on Information and Digital Technologies (IDT 2017), 5–7 July, 2017, Zilina, Slovakia, pp 442–451
38. Levashenko V, Zaitseva E, Kvassay M, Deserno T (2016) Reliability estimation of healthcare systems using fuzzy decision trees. In: Proceeding of federated conference on computer science and information systems, pp 331–340

The manufacturer's authorised representative in the EU is Springer Nature Customer Service Centre GmbH, Europaplatz 3, 69115 Heidelberg, Germany. If you have any concerns regarding our products, please contact ProductSafety@springernature.com

Printed and bound by CPI Group (UK) Ltd, Croydon, CR0 4YY
30/07/2026
02181693-0009